Praise for

CHRISTINA, QUEEN OF SWEDEN

"Vividly persuasive . . . richly evocative. . . . You find yourself thinking of Christina as a prematurely modern spirit who couldn't manage to fit into a not-yet-modern world—one of those historical figures who make the past feel less distant and, in the hands of a sensitive writer like Veronica Buckley, fully alive."
—*New York Times Book Review*

"A diverting biography. . . . Both instructive and hugely entertaining."
—*Wall Street Journal*

"A stunning debut and an absorbing page-turner. Veronica Buckley writes with immense style, vitality, and great humanity. The fascinating tale she weaves is as compelling as the most riveting of novels."
—ALISON WEIR, author of *Henry VIII: The King and His Court* and *Eleanor of Aquitaine: A Life*

"An impressive first biography, *Christina, Queen of Sweden: The Restless Life of a European Eccentric* not only justifies its subtitle in fascinating biographical detail but also provides an interesting commentary on gender and rule."
—*The Times* (London)

"A highly entertaining, well-researched account of the proud, impetuous, and frequently frustrating queen's life and peregrinations, as she lurched from place to place in various stages of crisis, near bankruptcy, and (often surreal) scandal." —*Evening Standard*

"Sparkling. . . . Buckley's witty and highly coloured prose is eminently suited to such a bizarre and self-dramatised life, and she describes with great sympathy Christina's long, heroic, and unsuccessful attempt to find something in place of marriage, motherhood, and her queenly destiny. Her book is much less a debut than the highly polished work of a writer who has been thinking about and loving her subject for years, and her enjoyment in the writing of Queen Christina's life is wonderfully translated into our pleasure in reading it." —*The Sunday Times* (London)

"Buckley's skill, in her wonderfully rich and poignant book, lies in exploring how this wild eccentric, armed with an overriding sense of her own uniqueness and supported by all the privileges of her rank, managed to live, in the end, a life so ordinary. . . . As a biographical study of the fear of freedom, Queen Christina is peerless." —*The Guardian*

"Buckley comes close—perhaps as close as it is possible to come—to explaining Christina's enigmatic nature and the extraordinary potency of her personality. This is a splendidly robust and colourful account of a remarkable woman and the turbulent age in which she lived. Astonishingly, this is Veronica Buckley's first book. May she write many more." —*Daily Telegraph*

"Christina's was a grandiose and reckless life, and Veronica Buckley narrates it with great authority and skill. It is a remarkable debut. This biography is filled with tragedy, farce, and absurdity as popes, regents, mavericks, losers, philosophers, and soldiers all involve themselves in Christina's wayward and eccentric progress, not many enrich themselves in the process. As lives go, it certainly ain't dull." —*Literary Review*

"Buckley, obviously captivated by her subject . . . presents her in all her eccentric glory, shedding light on an historical character who will be completely new to most readers." —*Daily Mail*, Critic's Choice

"An enjoyably belles-lettriste biography. . . . [Buckley] has a flair for description and relates this extraordinary life with sympathy and engaging panache."
—*Sunday Telegraph*

"Written with unwavering intelligence and flashes of wit; it looks handsome and the pictures are excellently chosen. If this is Veronica Buckley's first book, it is a good argument for not rushing into print at the earliest opportunity."
The Spectator

"Newcomer Buckley catches in all its peculiarity the life of a woman who abdicated Sweden's throne. . . . With considerable polish—and an occasional tilt to the baroque that fits the subject—Buckley tells the story of Christina. . . . As good a case as any against the existence of royalty.   —*Kirkus Reviews*

"Buckley presents a wide-ranging, entertaining exploration of the dynamics of the queen's unusual life. . . . Christina emerges as a complex and difficult character who transcends the attempts of others to mold her to their uses and expectations."
—*Publishers Weekly*

CHRISTINA, QUEEN OF SWEDEN

The Restless Life of a European Eccentric

. . . . .

VERONICA BUCKLEY

HARPER PERENNIAL

NEW YORK • LONDON • TORONTO • SYDNEY

For CRB, my father,
who's always known how to tell a good story

HARPER ⬤ PERENNIAL

FIRST PUBLISHED IN GREAT BRITAIN IN 2004 BY FOURTH ESTATE.

The first U.S. edition was published in 2004 by Fourth Estate,
an imprint of HarperCollins Publishers.

P.S.™ is a trademark of HarperCollins Publishers.

HarperCollins books may be purchased for educational,
business, or sales promotional use. For information please write:
Special Markets Department, HarperCollins Publishers,
10 East 53rd Street, New York, NY 10022.

FIRST HARPER PERENNIAL EDITION PUBLISHED 2005.

Designed by Barbara Bachman

The Library of Congress has catalogued the hardcover edition as follows:

Buckley, Veronica.
Christina, Queen of Sweden: the restless life of a European eccentric / Veronica Buckley.
p.   cm.
Includes bibliographical references and index.
ISBN 0-06-073617-8
1. Kristina, Queen of Sweden, 1626–1689.   2. Kings and rulers—Sweden—Biography.   I. Title.
DL719.B83   2004
948.5'034'092—dc22
[B]   2004050619

ISBN-10: 0-06-073618-6 (pbk.)
ISBN-13: 978-0-06-073618-7 (pbk.)

05 06 07 08 09 ❖/RRD 10 9 8 7 6 5 4 3 2 1

contents

PART TWO | 161

. . .

ACKNOWLEDGMENTS

FIRST ENCOUNTERED Queen Christina more than twenty-five years ago, when I was a young student preparing an essay on the moral philosophy of Descartes. My tutor, the late Dr. Alec Baird, introduced me to the correspondence between the two—I believe it took me years to return the books—and since then I have had a hundred reasons to maintain my interest in the queen and the philosopher, and a hundred more to be grateful to Alec Baird and to his wonderful wife, Katherine—I owe them both many thanks for many kindnesses.

There are other debts that I would like to acknowledge: first, to my agent, Victoria Hobbs; to my editor at Fourth Estate, Courtney Hodell; and to my copyeditor, Ed Cohen. I must also thank the staff of the following libraries and archives: the Bibliothèque Sainte-Geneviève (La Nordique) in Paris; the Biblioteca Apostolica Vaticana in Rome; the manuscript collection of the Bibliothèque Interuniversitaire de Montpellier; the Österreichische National-bibliothek in Vienna; the Nationalmuseum in Stockholm; and Svenska Porträt-tarkivet, with particular thanks to Elisabeth Höier, Eva Karlsson, Emilia Ström, Sussi Wesstrom, and Kerstin Wiking. I am most grateful to them all for their help with so many of the illustrations.

Thanks are also due to the following people for their kind assistance along the way: Catherine Blyth, my editor during the early stages of the book; Dr. Jan Gerard Boecker, Director of the Internationale Komponistinnen Bibliothek Unna, for kindly sending me information about women and music in seventeenth-century Rome; Michel Brisson and David Carey, for very kindly permitting me the use of their house in Montpellier during my stay in that lovely city; my dear sister, Anne Buckley, for her analysis of Queen Maria

Eleonora's use of language; Jean-David Cahn, for drawing my attention to a late 1680s bust of Queen Christina; Dr. Görel Cavalli-Björkman and Louise Hadorph-Holmberg of the Nationalmuseum in Stockholm; Mercedes Ceron and Paul Gardner of the British Museum Department of Prints and Drawings; Thierry Demarquest, for kindly providing a photograph of Queen Christina's adult handwriting; Dr. Mary Frandsen of the University of Notre Dame, Indiana, for information concerning Queen Christina's music patronage, and for bringing to my attention a letter in the Sächsische Hauptstaatsarchiv concerning Queen Christina in the Spanish Netherlands; Kristina Hagberg and Britt-Marie Toussaint of the Centre Culturel Suédois in Paris; Lars Holmblad of Stockholm's Historika Museet; Anniina Jokinen, for drawing my attention to Marvell's poems of praise to Queen Christina; Steve Lum, for precious encouragement along the road; Klas Lundkvist of Stockholm's Stadsmuseet; Malin Lindquist of the Gotlands Fornsal; Anastasia Mikliaeva of the State Hermitage Museum, St. Petersburg; Dr. Françoise Monnoyeur of the University of Linköping and the Kungliga Tekniska Högskola in Stockholm; Dr. Stephen Paterson for kindly reading parts of the manuscript; Signore Giovanni Pratesi, for his information pertaining to the Roman bust of Queen Christina sculpted in the late 1680s; Antoinette Ramsay of Stockholm's Kungliga Biblioteket; Mariella Romagnoli of the Biblioteca Comunale Planettiana in Jesi, for kindly providing me with the catalogues of the Archivio Azzolino; Kristiina Sepänmaa of the Swedish Institute; Dr. Ulrich Sieg of Marburg University; Christina Sievert for information about the musicians of seventeenth-century Rome; Dr. Marja Smolenaars of the Koninklijke Bibliotheek at The Hague; Elizabeth Westin Berg of the Skoklosters Castle Library. My particular thanks are due to Karsten Thurfjell for his warm Scandinavian hospitality during my visit to the beautiful city of Stockholm.

I owe a special debt to my husband, Philipp Blom, without whom, I am sure, the manuscript would never have been finished—nor indeed started.

Finally, I would like to thank my dear friend and unwitting benefactor, Gerard Richardson, the "ideal reader" whom I kept in mind as I made my way through the book. I hope he will enjoy it.

Paris
August 2003

AUTHOR'S NOTE

URING QUEEN CHRISTINA'S LIFETIME, the Julian calendar was still in use in Sweden, as in other Protestant lands. By the modern Gregorian calendar, already in use in Catholic countries and gradually adopted throughout Europe, the date was advanced by ten days. Hence, for example, the Battle of Lützen, in which Gustav Adolf the Great was killed, was recorded in Sweden as November 6, 1632, but elsewhere in Europe as November 16, 1632. The locally recorded dates have been used throughout the text.

Unless otherwise stated, all translations are the author's own.

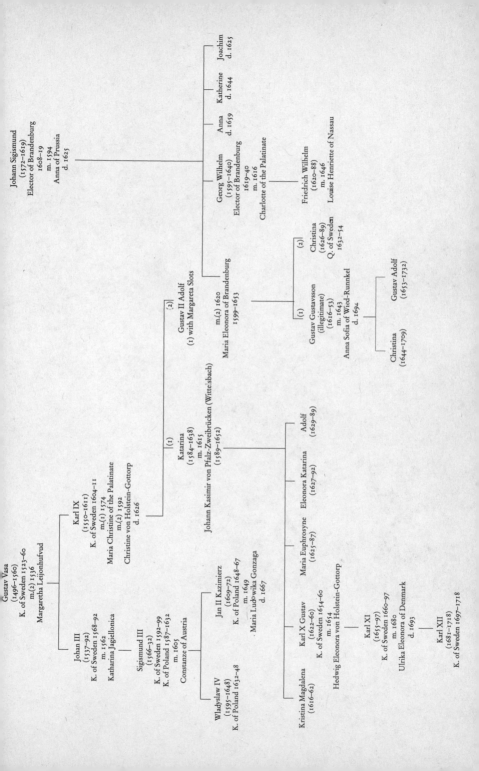

Gustav Vasa
(1496–1560)
K. of Sweden 1523–60
m.(2) 1536
Margaretha Leijonhufvud

Johan III
(1537–92)
K. of Sweden 1568–92
m. 1562
Katharina Jagiellonica

Karl IX
(1550–1611)
K. of Sweden 1604–11
m.(1) 1574
Maria Christine of the Palatinate
m.(2) 1592
Christine von Holstein-Gottorp
d. 1626

Sigismund III
(1566–32)
K. of Sweden 1592–99
K. of Poland 1587–1612
m. 1605
Constanze of Austria

Katarina
(1584–1638)
m. 1615
Johann Kasimir von Pfalz-Zweibrücken (Wittelsbach)
(1589–1652)

Gustav II Adolf
(1) with Margareta Slots

Wladyslaw IV
(1595–1648)
K. of Poland 1632–48

Jan II Kazimierz
(1609–72)
K. of Poland 1648–67
m. 1649
Maria Ludowika Gonzaga
d. 1667

Kristina Magdalena
(1616–62)

Maria Euphrosyne
(1625–87)

Eleonora Katarina
(1627–92)

Adolf
(1629–89)

Karl X Gustav
(1622–60)
K. of Sweden 1654–60
m. 1654
Hedwig Eleonora von Holstein-Gottorp

Karl XI
(1655–97)
K. of Sweden 1660–97
m. 1680
Ulrika Eleonora of Denmark
d. 1693

Karl XII
(1681–1718)
K. of Sweden 1697–1718

m.(2) 1620
Maria Eleonora of Brandenburg
1599–1655

(1)

Gustav Gustavsson
(illegitimate)
(1616–53)
m. 1643
Anna Sofia of Wied-Runnkel
d. 1694

Christina
(1644–1709)

Gustav Adolf
(1653–1732)

(2)

Christina
(1626–89)
Q. of Sweden
1632–54

Johann Sigismund
(1572–1619)
Elector of Brandenburg
1608–19
m. 1594
Anna of Prussia
d. 1625

Georg Wilhelm
(1595–1640)
Elector of Brandenburg
1619–40
m. 1616
Charlotte of the Palatinate

Anna
d. 1659

Katherine
d. 1644

Joachim
d. 1625

Friedrich Wilhelm
(1620–88)
m. 1646
Louise Henriette of Nassau

CHRISTINA, QUEEN OF SWEDEN

part One

prologue

OWADAYS, IF YOU HAVE a few dollars to spare, you can buy a copy of an old newspaper printed on the day you were born. Leafing through the pages, you glean something of the world as it was at the time of your own arrival. You see recorded the lives of those who made your world, their interests and values, what motivated them, and what they feared. You see the world that has shaped and bordered your life and, in significant measure, made you what you are.

What, then, was Christina's world? What forces shaped her? What ideas framed her singular mind? She was born in 1626 into a world overwhelmingly European, though the bounty and burdens of the great era of exploration had opened its eyes to other lands beyond. American silver framed the holy icons of the pious, and the soft white hand of many a countess sparkled with jewels from the East, while the first African "indentured servants" had begun their woeful voyage aboard a Dutch cargo ship. Knowledge had come, too, with the diamonds and the spices, but Europe's "gentleman-travelers" seldom ventured to very distant shores in search of it. It was left to the sailors and traders and priests to make the longest journeys, and to bring the tales back home.

Christina's was a cold world, the coldest time Europe had known for thousands of years—the "Little Ice Age" that balked the harvests and froze the seas. Fires blazed on ice-thickened rivers, and birds were seen to drop from the skies in midflight, frozen to sudden death. Christina's world was a dirty world of sudden illness and doubtful water and scanty, tainted food, where peasants and beggars faced hunger as routinely as the sunrise. It was a man's world, where women had little public power; high rank might soften the outlines, but too frequent childbirth was most women's lot. And it was a familiar world, a world of small towns where great families ruled, where faces were known and strangers few, and secrets hard to keep.

Above all, Christina's world was a world at war, the great Thirty Years' War, which raged across Europe from 1618 to 1648, claiming countless lives, including that of her own great father. Christina would grow quickly accustomed to it; during the whole of her life, Europe would know barely a single year of peace. Warfare in her world was a normal aspect of government. States and empires grew from it in a savage symbiosis, filling its maw with their choicest fruits, and drawing new wealth from its wake. Christina's contemporaries accepted it as a fact of life and reserved their greatest praise for those who waged it successfully.

The finest laurels were still worn by the Habsburg Empire of Spain, whose brilliant armies had dominated Europe for more than a hundred years. But, fearful of new ideas and disdainful of trade, Spain had now begun its long decline. The new road was being paved by its vibrant little brother along the western shores of the Continent; in the energetic, enterprising provinces of Holland, the ships and banks and warehouses of a new commercial prosperity were busily being built. Within a few decades, Spain's political supremacy would pass to France, whose brilliant star had yet to rise, but its military honors were even now being captured by Sweden itself, whose innovative armies, seemingly invincible, had pressed deep into Europe, captained by their own splendid king.

The Swedes' great enemy was the Austrian Habsburg Empire, a vast Catholic power that stretched from Poland to the Czech lands and from Bavaria to Croatia. Since the infamous defenestration of Catholic officials by Protestant reformers in Prague in 1618, the empire had been at war, alternately desultory and ferocious, with various Protestant powers. The many German lands that were not within its borders stood as independent states, either Catholic or Protestant, numbering in their confusing hundreds, each with its own loyalties to dynasty or faith.

No single land of Italy existed, but the marvelous Italian cities, Europe's most fabulous jewels, still dazzled eye and mind after centuries of cultural preeminence. Their most gifted sons had made their way to every corner of the Continent, leaving the fruits of their artistry in marble and on canvas, changing perspectives, opening minds, firing the imagination. The papal city of Rome itself had recently enjoyed a great artistic renaissance, encouraged and funded by successive popes intent on reestablishing the primacy of Catholicism after the Protestant Reformation.

England, though not isolated from European life, remained as yet peripheral. Its new king, Charles I, was beginning to set out his claim to absolute rule by divine right, an idea that would spark revolution and in time engender the downfall of Christina's world. The king's nemesis, Oliver Cromwell, was a young country gentleman, still unknown. Shakespeare had lain just ten years in his grave.

To the east, the first Romanov tsar sat upon the ancient throne of Muscovy. After its long "Time of Troubles," Russia now looked forward with hope, but for decades to come it would be outshone by its dual neighbor, Poland-Lithuania, the largest state in Europe and Sweden's longstanding threat from the east. And southward, linking Sweden with the mainland over the much-disputed Baltic Sea, lay the ancient enemy and former ruling power of Denmark.

Despite its military prowess, Sweden itself was undeveloped. In economic and social terms, it was essentially still a medieval land, overwhelmingly rural, exporting its ablest youth to more promising environments, and relying on foreigners for capital and enterprise at home. Throughout the country a series of cold fortress castles, grim stone on the outside and bare-walled inside, contained what little the kingdom possessed of scientific endeavor or cultivated living. But the war booty of recent years had at last allowed Sweden to begin its ascent into the light of culture and learning, and in the twenty years of his youthful reign, Christina's brilliant father would succeed in dragging and thrusting his backward homeland into the very heart of European life.

Christina's world was a world of vibrant learning, of philosophy and poetry, of religious scholarship and scientific experiment. It had begun its long, deep love affair with the world of classical antiquity, now resurgent after the exotic lures of the great age of exploration; on the foundations of this ancient world new temples were being built to Greek thinking and the soldierly virtues of Rome. Renaissance ideas persisted, too, not least in the widespread practice of alchemy, consuming fortunes and lifetimes in a misbegotten search for truth.

And, while Europe's princes fought among themselves, their Christian world faced two mightier enemies, from without and from within. The external threat was the great Ottoman Empire of the Turks, which at Christina's birth stretched from Algiers to Baghdad and as far west as Budapest. Late in her life she would hear of a vast Turkish army pitching its tents at the very gates of Vienna. But it was the internal enemy that in the long run would prove the more decisive. With their stumbling, excited experiments, Europe's "natural philosophers" had begun their challenge to religious orthodoxy. With increasing success, they now strove to provide materialist explanations of the natural world. Though most were repaid with hostility and persecution, and some even with death, no Church, and no state, could stop them. The great march of empirical science had begun, and all ears, willing and unwilling, heard the beat of its tremendous drum.

Christina's world was a crossroads world, where God still ruled but men had begun to doubt. She herself would stand at many crossroads, of religion and power, of science and society and sex. And she would prove a dazzling exemplar of her own quixotic era, an exemplar of great, flawed beauty, like the misshapen baroque pearl that would give its name to her vibrant, violent age.

BIRTH OF A PRINCE

IN THE SPRING OF 1620, a delegation of German nobles made their way along the river Spree toward the town of Berlin. The town was not what it had been; years of plague had depleted its people, and its once thriving trade had dwindled to the narrow service of luxury goods to its resident court. Now, among the low wooden buildings, only the vast old castle impressed upon the visitor that Berlin was still a place of power, the residence of the Hohenzollern family of Brandenburg, electors of the Holy Roman Empire. To them, together with six other princes, fell the privilege and the duty of electing the empire's ruler. In Berlin, a new elector, the young Georg Wilhelm, had held his stately office for just a year.

Now, toward the castle, the nobles rode, down the bridle path under the linden trees that would one day give their name to the town's most lovely thoroughfare. The delegation was led by Johann Kasimir, the Count of Pfalz-Zweibrücken, and in his train were two young gentlemen who had joined him from the homeland of his wife, the Princess Katarina of Sweden. One of these was "Adolf Karlsson," a strongly built and handsome man with the blond hair and keen blue eyes of the north. The other, his friend, was Johan Hand, an eager observer of all that passed and who kept a lively record of the journey in the pages of his personal diary.

The count was related to the elector's wife, Elisabeth, and it was ostensibly to see this princess that he had made his present journey. The visit had been timed strategically, for the Elector Georg Wilhelm himself was not at home, nor did the count regret his absence. A matter of importance was now at hand, in which the elector's mother, the Electress Dowager Anna, would cast the deciding vote. The count had hopes of persuading her to his own views, and he knew that Anna would hear him more readily if her son was not there to speak against him. The matter at hand was no less than the marriage of Anna's daughter, Maria Eleonora, and the proposed bridegroom was the count's own brother-in-law, Gustav Adolf, King of Sweden. He had made the journey himself, just to have a look at the lady, for "Adolf Karlsson" was in fact the king.

A marriage between Maria Eleonora, now age twenty, and Gustav Adolf, five years her senior, had been under consideration for some years already. Offers for the hand of the young countess were not wanting: among her suitors she could boast Gustav Adolf's cousin, the Crown Prince Wladyslaw Vasa of Poland, and Prince Charles Stuart, heir to the English throne. Her father had been ambivalent toward a possible Swedish match, but his son, the new elector, had taken a clear stand against it. Sweden was a fiercely Protestant land, and he had no wish to antagonize the Catholic emperor, or the king of neighboring Catholic Poland, whose vast country lay only two days' march from Berlin. The Swedes were already at war there, and Georg Wilhelm thought little of their chance of victory. Though a Calvinist himself, and ruler of a Lutheran state, he felt his sister would do better to marry the Crown Prince of Poland. In the Habsburg lands, not so far to the south, the emperor had recently reasserted his power over the luckless Protestants of Bohemia, whose ill-starred "Winter King" was the brother of Georg Wilhelm's own wife. Religious neutrality seemed the wisest course as the match set in Prague began to kindle. But by family custom it was the privilege of the electress to decide her daughter's marriage, and on this the Swedes had pinned their hopes. An alliance with Brandenburg could strengthen their hand against Poland, and might hasten the formation of a new bloc of Protestant states against the Catholic Habsburgs. The elector's fear was Gustav Adolf's hope.

For his journey now, however, the young king had paid a great personal price. A spirited and warmhearted man, he had been passionately in love with the daughter of one of Sweden's noblest families, the beautiful Ebba Brahe. Ebba had returned his love, but the king's strongminded mother had felt that a match between them would not serve Sweden's diplomatic interests. Intriguing and determined, she had set to with a will to break off the romance, at one point even laying her own violent hand on the lovers' go-between. In due course, she had succeeded. Ebba was married off to the scion of another noble family. The sad and disappointed king dispatched a beautiful letter of farewell, wishing his love "a thousand nights of gladness" in her husband's arms, and at length he turned his thoughts toward Brandenburg, where his mother's gaze had long been fixed.

Happily, the object of his present attentions was well formed and incited new passion in the young man's heart. Maria Eleonora was a genuine beauty, her figure rounded, her face soft and full, with a sweet bow mouth, a strong nose, and large, beautiful eyes. She was blond, and her manner was lively, giving an impression of girlish gaiety to all those who saw her.

At first, though, it seemed that her young suitor might not succeed in seeing her at all. Her father had died in the previous December, and the court was still in mourning. Dark hangings draped the rooms, and the few permitted candles flickered on his doleful, black-garbed retainers. Five months after his death, the old elector's body lay, embalmed but still unburied, in the castle chapel. The usual bustling life of the court was suspended, and visitors received only the simplest civilities. But the pulse of youth was strong in the burgeoning spring, and besides, Gustav Adolf could not afford to wait; there was too much to do at home. For a bribe of three hundred ducats, he acquired a portrait of the young countess, and, duly encouraged, arranged a secret rendezvous. It was a Sunday, and all the court was at church, all except Maria Eleonora, who had found some pretext for absenting herself. The Swedes, being Lutheran rather than Calvinist, could not, of course, attend, and soon the meeting was effected in the shade of the trees in the castle park. The countess, at least, was not disappointed, as the king's friend would later remind him. "Where the girl's thoughts were, I couldn't say," he wrote, "but she didn't take her eyes off Your Majesty."[1]

There was not much else, it seems, in Maria Eleonora's head. She had chafed at her school lessons, and she had no interest now in learning or literature. But she was lighthearted, prettier than most girls, and had at least a genuine love of music and art. No doubt these things were spoken of in the further meetings that were soon arranged between the two, for the king himself enjoyed them both; he was interested in painting, and he played the lute well. Johan Hand records that the couple met privately several times, that they dined together and conversed at length, and that Gustav Adolf did not depart unkissed. On the whole, he was pleased to have made the journey. The girl's grandfather and great-uncle had been insane, it was true, but this could hardly count against her, for had not his own uncle and aunt been the same?[2]

For her part, Maria Eleonora was delighted. She soon discovered the true identity of the handsome "Adolf Karlsson," and, turning her heart where duty lay, promptly fell in love with him. In this, at least, she showed good judgment, for the young king was among the very finest men of his age: able and cultivated, brave, strong, and generous, courteous, farsighted, conscientious, and just, an inspired military leader, and a man of profound religious humility, amply deserving the epithet that his dazzled contemporaries would one day accord him—Gustav Adolf the Great. Had he lived in a time of peace, his many gifts might have borne yet finer fruits, but in 1611, when he had come to the throne, at the age of only sixteen, his tiny country was already at war with Denmark, and by 1618, at the outbreak of the Thirty Years' War, Sweden had embarked on warfare to last a generation, in which the greathearted king, the "Lion of the North," was to lose his own life.

But for now, Gustav Adolf's fine soldier's reputation can only have added to his attraction. For Maria Eleonora, he seemed the fulfillment of a dream, indeed, the fulfillment of a prophecy, for her father's own astrologer had once predicted that she would grow up to marry a king. The king himself was not so sure. Though he wanted to marry quickly, a Brandenburg connection was not the only possibility. In the ripening spring, he made his way southward, pausing in the vibrant town of Frankfurt am Main, where books and silks and jewelery were traded in the busy streets beneath the great cathedral. While there, he took the time to purchase a magnificent diamond necklace, at a value of almost nine thousand riksdaler—the price of three thousand cows, no less—borrowing the money from his brother-in-law to do so. As yet, however, he had not decided whose neck the lovely item would adorn.

From Frankfurt, he made his way to Heidelberg, there to cast his eye upon an alternative marriage candidate, the Princess Katarina, sister of Friedrich V, Elector and Prince Palatine of the Rhine, the unhappy "Winter King" of Bohemia. The Swedish king had maintained his incognito, but he was now dressed as an army captain, and disguised by the simple acronym of Gars— *Gustavus Adolphus Rex Sueciae* (Gustav Adolf, King of Sweden). The princess, a young lady of generous circumference but, it seems, no great perspicacity, failed to recognize her prospective suitor. She mistook his interested approach for impertinence, declaring to her sister, in imperious French, "What intrusive people these Swedes are!" Alas, among the eleven languages understood by the clever king, French was not the least.[3] Gustav Adolf decided that the pretty little countess of Brandenburg would suit him better, and in due course he made his way back to Stockholm, dispatching his friend and chancellor, Axel Oxenstierna, to complete the arrangements in Berlin, while the countess herself sat down to pen an excited letter to her brother, who found himself angrily obliged to accept the match his mother had made. "The whole journey was like a play," wrote Johan Hand in his diary.[4]

But if the journey was a romantic comedy, it was not without its dramatic aspects. His search for a wife in the German lands had allowed Gustav Adolf to assess the strengths and weaknesses of the various princes who served as a Protestant bulwark against the Catholic Habsburgs. He was not impressed, and he returned contemptuous of their "feebleness, cross-purposes, selfishness, and military incompetence," an ill omen for the Protestant alliance that he would later attempt to forge.[5]

In the autumn, Maria Eleonora and Chancellor Oxenstierna set out upon the northward road, accompanied by the Electress Dowager and her youngest daughter, Katharina, together with the bride's personal secretary and many ladies-in-waiting. They traveled in some comfort, their journey assisted—and the bride's dowry increased—by the pawning of valuables that the Electress Dowager had raided from the Brandenburg state treasury. At Kalmar, not far from the Danish border, they stopped, for here the king himself had come to meet them, pausing en route to purify the land for his bride by torching a number of plague-stricken houses in the surrounding countryside.

At Kalmar, the party passed several days of alternate rest and celebration in the beautiful castle beside its placid harbor. It was a historic place, for here, more than two hundred years before, the triple-queen Margareta had united Sweden with the neighboring lands of Norway and a dominant Denmark, a union against which Gustav Adolf's own grandfather had led his people to rebel.[6] Kalmar was Sweden's architectural jewel, a castle of fairy-tale beauty and among the finest in Europe, built with a sure artistic sense by the king's Renaissance forebears. Many of its rooms were beautifully decorated, with painted moldings and inlaid wood, and finely made furniture from the lands to the south. No doubt it was all displayed with pride to the newcomers, and perhaps, too, the young bride was teased with horror stories of the murders the same rooms had witnessed, half a century before.[7] If so, they did not deter her. The bridegroom set out for the Tre Kronor Castle, thoughtfully going ahead to give his personal attention to the heating of Maria Eleonora's rooms, and soon she set out after him with her own entourage on the long, hard journey to Stockholm, three hundred miles northward, the winter closing in around them.

If the sophisticated ambience of Kalmar had reassured the young bride, her composure was soon to be tested as she made her way through her new-found country, for as yet Sweden had little to impress a German countess. Its climate harsh and its people few, it was overwhelmingly rural, with small clusters of farmsteads thinly spread over the less inhospitable southern areas. Lakes and forests dwarfed and isolated all but the largest settlements. Almost all the Swedes, about a million souls in all, were peasants, with a few tens of thousands living in small and undeveloped towns, and even the nobles mostly choosing to live in the countryside, putting their modest incomes back into the land. The very crown revenues, including taxes, were still paid in kind; grain and fish and butter, hides and furs, and iron and copper from Dutch-owned mines, all poured into the royal warehouses, and out of them, too, for the crown's own servants and even foreign creditors were paid in kind, as well. In the early days of Gustav Adolf's reign, meetings of parliament had taken place in the open air, while at the Tre Kronor Castle, the monarch's own residence, the doors remained open to all comers.

As the weary train arrived in Stockholm, the young bride's deepening disappointment turned to dismay. Not yet the country's formal capital, the grand northern city where she had thought to make her home was in fact scarcely more than a backward country town, its muddy streets lined with basic wooden houses, unwarmed as yet by the ubiquitous red paint that would one day turn their roughness to charm. Goats wandered on the brown turf rooves, nibbling at the roots and grass, sending a plaintive bleating into the chilly air. Inside, the dwellings of rich and poor alike were largely bare, with little covering on the floors and less upon the walls, and now, in the gathering winter, reliably cold. Though the king himself, like his forefathers, was genuinely interested in architecture and the fine arts, there had been little excess wealth for great public buildings or lavish artistic patronage; native literature and music remained rudimentary, theater almost unknown, paintings and sculpture rare, and Sweden's nobles, in their bare-walled, bare-floored houses, largely unconvinced of the need for any of them. To the citizens of the superbly cultured towns of Italy, or to those of Holland with its advanced financial system and its plethora of cheap goods, Sweden seemed a desperate outpost at the ends of the earth. To Maria Eleonora, accustomed to the rich heritage of Brandenburg, and with cultural pretensions of her own, disdain was now added to disappointment. She conceived a contempt for the land and its people, her husband only excepted, and in so doing garnered much ill will for herself.

In December of 1620, the marriage took place, and three days later, before the silver altar of Stockholm's Storkyrka, Sweden's new queen was crowned. Though her title was ancient, her accoutrements were new, for the former queen, Gustav Adolf's mother, Christine, had refused to hand over her regal insignia. Though she had sought the marriage energetically, she was reluctant now to accept its implications. In some haste, a new crown had been beaten out of gold, a new scepter and orb provided, studded with rubies and diamonds, the red and white of the queen's native Brandenburg. The king was dressed in the colors of his own land, in a blue robe embroidered with gold. Liveried pages and knights in pearled helmets paid homage, as the resentful Queen Mother looked on.

LATE IN THE SUMMER of the following year, Maria Eleonora gave birth to a stillborn daughter. The king was away, campaigning in Livonia,[8] taking advantage of a Turkish attack on southeast Poland to harangue his old enemy from the north, when the news arrived that his wife had been "too soon and untimely" delivered of the child. From his camp outside Riga, he sent a grieving letter to his brother-in-law, lamenting the "misery" that had befallen the queen and stricken his royal house. "May God be kind to her," he wrote, "and help her quickly back to health."[9]

Health of a kind did return to the queen, but not quickly, and it was more than two years before she was brought to bed again, of a second daughter, who was named Kristina Augusta. "The little girl is doing well," she wrote, and in the summer of 1624, after almost three years in Sweden, Maria Eleonora's mother decided that she and her youngest daughter could safely return home to Brandenburg. But this hopeful time was not to last. In the autumn, the child fell suddenly ill, and before she had reached her first birthday, she died, an unhappy reminder of Maria Eleonora's own three youngest siblings, all of whom she had seen die within their first year of life.

Maria Eleonora passed a sad winter, bereft of her mother and sister, her little daughter dead, and her husband, to whom she had begun to cling with a desperate fondness, too often preoccupied and too often away. In February came a further blow, the death of her younger brother, at the age of just twenty-one. As the spring approached, happier times seemed promised; the days lengthened and a mild sun shone down, and another baby quickened in her womb. But in April, news arrived from Berlin of the Electress Dowager's death. The queen was deeply affected, and for some weeks she lay sorrowing and ill, mourning her mother, wearied by her pregnancy. Toward the end of

May, she rallied. The king was again in Stockholm, and in the fine spring weather an inspection of the Swedish fleet was to take place in the surrounding harbor. The royal couple would attend together, reviewing the ships from aboard their own small yacht. The fleet lay at anchor off the little island of Skeppsholmen, and as the king and queen sailed past, a sudden squall blew up around them, rocking their yacht from side to side until it almost capsized. Though the mooring was soon reached, the queen was carried back, frightened and ailing, to her rooms in the castle, and there she endured the bitter conclusion of the day. For within a few hours, her labor had begun, too early; the morning light would break upon her weeping women, and her little stillborn son.

The king recorded the tragedy with pious resignation. "Disaster has befallen me," he wrote. "My wife has brought a dead child into the world. It is because of our sins that it has pleased God to do this."[10]

For his Vasa dynasty, at least, it was indeed a disaster. In this fifth year of his marriage, and despite the queen's three confinements, Gustav Adolf had as yet no living heir. Three years before, his younger brother had been killed in battle in Poland, and the king of that same country, Gustav Adolf's cousin, Sigismund III, now stood to inherit the Swedish throne. Moreover, Sweden's enemy heir had two adult sons of his own, through whom a Catholic dynasty might be foisted upon the unwilling Swedes, raising once again the specter of civil war.

But the lack of an heir was not the only disaster to have befallen the king. His wife's behavior was becoming increasingly eccentric. During his many absences on campaign, she would be ill and depressed, then would bound out of her dismal moods with cravings for sweet foods and lavish spending on gifts for her favorites that the treasury could not afford. She had always been passionately fond of her husband, but now her attachment became obsessive, and she pleaded repeatedly with Axel Oxenstierna to persuade him to return. "Please help me, if you can help me," she wrote to the exasperated chancellor. One courtier, describing her as "unimaginably" hysterical, attributed her behavior, sympathetically, to simple loneliness. Maria Eleonora herself felt sure of the source of her malaise. "When I know that my most beloved lord is coming," she wrote, "then all my sickness and panic fall away."[11]

The queen's extreme behavior was not the only sign that she was now far from well. Her very odd use of language was becoming the subject of comment by many at court. Far from having mastered the language of her adopted country, since coming to live in Sweden she had become incapable of using even her native German correctly. Whether speaking or writing, she confused syllables and made up strange concoctions of words that resembled but did not match those of any language she had learned. Although no one regarded the queen as intelligent, and many spoke of her extravagant flights of hysteria, her unusual difficulty with language suggests a possible neurological problem. It may be that, during one of her confinements, she had suffered some kind of stroke; certainly there was no mention of any language problem before her marriage, and her own father had suffered several strokes that had left him increasingly debilitated. Whatever the reason for the queen's muddled speech, it no doubt added to her growing sense of desperation—even her handwriting, once straight, in lines of even spacing, now showed a pronounced downward slope, the graphologist's telltale sign of depression.[12]

The queen's unhappiness can only have been increased by the knowledge that, only a few hours' journey from Stockholm, her husband's nine-year-old illegitimate son was living with his Dutch mother and stepfather, Margareta and Jakob Trello, at Benhammar, an estate in the king's gift. The king was evidently proud of the boy; he had named him, after all, Gustav Gustavsson.[13] His existence was no secret, and indeed, rumors abounded that the affair between the king and Margareta was still ongoing; Margareta herself had written to Gustav Adolf to reassure him that she herself was not the source of them. There does not seem to have been any truth in the rumors, but the boy's bright and sturdy presence in itself must have been a constant reminder to the queen of the son she herself still lacked.

The king, though courteous and considerate, had by now abandoned any hope of a genuine companionship with his wife. In public, he spoke of her affectionately, but in private he referred to her as his *malum domesticum*, a "domestic cross" he was obliged to bear. To his friends, it seemed, he regarded her as "more or less a child," to be attended to and watched over, but from whom no mature, reciprocal feeling could ever be expected. Still in her twenties, Maria Eleonora had already begun to assume the sad mantle of old age, confused in her speech, prey to every illness, trying to those about her.

Further troubles now beset Gustav Adolf, for this was 1625, a plague year, and his own troops in the east had not been spared. In December came news of his mother's death. It was late in the spring before he could return to bury her; through the long months of winter her body lay in state in Nyköping. But on his arrival, the king brought joyful news: the queen was expecting another child. Pitying her pleading, and no doubt only too aware that an heir had yet to be produced, the king had agreed to her joining him after a Swedish victory had provided a pause in the fighting. As the year progressed, every precaution was taken to ensure Maria Eleonora's safety, and in November, a few weeks before the expected birth of the new baby, Gustav Adolf's illegitimate son was tactfully dispatched to the university at Uppsala, in the care of the king's own boyhood tutor. It was not in any sense a dismissal; the young Gustav would retain his place in his father's affections, but for now, it seems, he was best out of the way.

DECEMBER IN STOCKHOLM, the cold, dark winter of the north, and a new moon glimmered on the frozen river. Around the castle, the plain wooden dwellings stood huddled and low, as if to shelter themselves from the bitter weather. Above, in a black sky, the stars were aligned just as they had been more than thirty years before at the birth of Gustav Adolf; now, once again, the Lion ascendant cast its faint reflection on the old stone tower's three golden crowns. Within the castle, torches flamed and fires blazed, striving against the darkness and the damp. Courtiers paced and servants dozed, while the queen consulted her astrologers, and the king dreamed of a son.

It had been an anxious time. Gustav Adolf and Maria Eleonora had been six years married, and they had as yet no living child. The birth of a boy was now predicted, but as the queen drew near her confinement, the astrologers foresaw death as well. The child would die, or if he lived, he would cost the life of his mother, or even his father, who lay ill, feverish and troubled as the hour of the birth approached. If the boy lived, he would be great, they said, and the queen took comfort, remembering the signs of her pregnancy, the omens in the stars, and her husband's dreams.

It was the eighth of December,[14] a Sunday, and as night fell, a night of bitter cold, the queen began her labor. She was not strong, and the birth proved difficult, but as the clocks neared eleven, the baby emerged, alive, into the eager hands of the midwives. That the child was strong and likely to survive was clear—a lusty roar announced a determined entry into the world—but it was covered from head to knee in a birth caul, concealing the crucial evidence of its sex. The caul was removed at once, and the queen's attendants, delighted to meet the expectations of the court, declared the child a boy; its siblings were dead, and it was, after all, sole heir to a valiant warrior king. The mother and father were duly informed, and through the cold midnight air the castle rang "with mistaken shouts of joy."

The nurse came confidently forward, the exhausted queen lay back, but for the disconcerted midwives it would be no night of rest or sweet, familiar work. A closer look at the baby had revealed their error; it was in fact a girl. Through the dark night hours they waited, for no one dared tell the king. As the morning light dawned weakly over the castle, the baby's aunt decided to take the matter in hand. She took the child up in her arms, went to her brother's sickroom, and lay the child directly on the king's bed, sans swaddling clothes or, as the baby herself was to describe the event, "in such a state that he could see for himself what she dared not tell him."[15]

Legend has it that the king expressed no disappointment, indeed, not even surprise, at this extraordinary turn of events. He calmly took up the child and kissed it, then spoke to his sister in accents of tender stoicism. "Let us thank God," he said. "This girl will be worth as much to me as a boy. I pray God to keep her, since He has given her to me. I wish for nothing else. I am content." The princess reminded him that he was still young, as was the queen, that there would surely be other children, surely a son, but the king merely replied, "I am content. I pray God to keep her for me," and he blessed the baby and kissed her again, as if to emphasize his contentment. "She will be clever," he added, smiling, "for she has deceived us all."[16]

The legend has its source in the pen of Christina herself, though she claimed to have heard the story "a hundred times" from her aunt and also from her mother, who, at the time of this exchange, lay perilously weak in her own room. It is not likely to be true, though the princess may well have softened the tale for the lonely little girl whom she later took into her care. In fact, the birth of a daughter was a desperate disappointment for Gustav Adolf and his followers, and it threw into question the very survival of the shaky Vasa dynasty. The king's calm acceptance, if calm it really was, is more likely to have been the result of his fever, the lassitude or lethargy of a draining illness, or even of quiet relief to have at least a living child. As for the queen, it was some time before she was considered strong enough to withstand the sorry news. After four pregnancies and the deaths of three infants, and this latest, most difficult birth, she was "inconsolable" to find that she had not borne a son after all. She rejected the child out of hand, and began her own descent into a profound mental disarray.

Whatever his private feelings, and despite his fever, the king soon rallied.

A *Te Deum* was commanded in thanksgiving for the birth, and the baptism was quickly arranged. The child was christened Kristina Augusta,[17] the same names as had been given to the elder sister who had died three years before. "Christine" had been the king's mother's name, and his grandmother's, too, and it was also the name of a Finnish noblewoman with whom he had once been in love—the memory of that young beauty may now have brought a smile to his lips as he announced the name he had chosen for his little daughter.[18] The baby's second name, Augusta, perhaps a loose rendering of "Gustav," may have been the queen's choice. She is not likely at any rate to have liked the baby's first name; there had been no love lost between herself and the king's late mother.

Many years later, needing to emphasize her Catholic credentials, Christina was to claim that, during her baptismal ceremony, the pastor had inadvertently blessed her baby forehead with a sign of the cross, so enrolling her unwittingly in the "happy militia" of Rome. But in fact, this kind of blessing had remained fairly common in Sweden through the early decades of Lutheranism. The pastor's sign, far from a presaging, was a gesture made instinctive from the force of long habit. And Christina's claim, as so much of her life was to be, was no more than a ruse to persuade her audience, and perhaps even more, to persuade herself.

WHY HAD IT BEEN so difficult for Maria Eleonora's attendants to determine the sex of her newborn child? The large caul would surely have been removed at once to establish the answer to this most important of dynastic questions. The baby's loud voice, the "extraordinary, imperious roar" may have been a sign of strength, but not more. It is more likely that the experienced midwives were for once confronted with something unfamiliar in the squalling little person of a baby of ambiguous sex. Though the child had been born before midnight, they waited until the morning to make their final—altered—decision.

Was the little girl really a boy? Was she a hermaphrodite, or a pseudohermaphrodite, with female organs outside, and male inside? Diagnoses of this kind, at a distance of centuries, must always be conjectural. It is possible that Christina was born with some kind of genital malformation, and she may even have been what would now be called intersexual. Our own statistically minded age records that about one in every hundred babies is born with malformed genitals of varying degrees of ambiguity, making it often difficult, and sometimes impossible, to determine the baby's sex. There are various disorders that can cause such malformations;[19] in the case of a baby girl, the most common of them would produce a perfectly healthy infant with normal internal sex organs, but often with an enlarged clitoris and partially fused labia, easily confused at first glance with the small infant penis and scrotum of a longed-for male child.[20]

Whatever the case, Christina's sex, like her sexuality, was to remain ambiguous to others and ambivalent to herself throughout her tempestuous life. It would distort her relations with her mother and her father, poisoning the one and tainting the other. And in the first years of her life, it would precipitate a dynastic crisis from which she would emerge an acclaimed crown prince.

DEATH OF A KING

$\mathscr{I}$N HIS DIARY, looking back to the years of his childhood, John
Evelyn records:

> I do perfectly remember . . . the effects of that comet, 1618 . . . whose
> sad commotions sprang from the Bohemians' defection from the Em-
> peror Matthias: upon which quarrel the Swedes broke in, giving um-
> brage to the rest of the princes, and the whole Christian world cause
> to deplore it, as never since enjoying perfect tranquillity.[1]

The English diarist's "comet" of 1618, eight years before Christina's birth,
was no less than the beginning of the Thirty Years' War, set in slow motion by
the infamous "defenestration of Prague," when the city's two unhappy Habsburg governors were thrown from a window of the Hradčany Castle by angry
Protestant reformers.[2] The governors, ignobly landing on a dungheap, survived unhurt, disappointing many of the emperor's supporters of two early
martyrs to his cause. But in the following years there had been no lack of martyrs on either side, indeed, on all sides, for the war was proving less a struggle
for or against imperial power than a muddled conflict of shifting alliances, religious, territorial, political, and personal. No one, it seems, had wanted war;
fear had motivated most. But defensively, preemptively, unwittingly, dozens
and then scores of combatant armies were gradually dragged or preached or
bribed into the lists of the perverse, ancient battle for peace.

For generations, the Holy Roman Emperors of the German Nation had
been successively elected from the Catholic Austrian House of Habsburg.[3]
The empire, a loosely linked archipelago of hundreds of principalities and estates, cities, and bishoprics, both Catholic and Protestant, was by no means

exclusively German; territories as far afield as Lombardy had allowed it to claim its "Roman" title, and it had once encompassed even the papal states. But since the beginning of the Reformation, a hundred years before, its tenuous cohesion had been threatened by growing Protestant objections to the rule of a Catholic emperor.

Of the empire's seven electors, three were Catholic bishops, three Protestant princes, and the seventh was the elected king of Bohemia, in recent decades always Catholic and always a member of the Habsburg family. But in the early spring of 1619, as the aged and childless Emperor Matthias lay dying, the restive Protestants of Bohemia saw their chance at last. On the emperor's death, a new king of Bohemia would be elected, a new voice for the choosing of the next Holy Roman Emperor. They determined that the voice would not be Catholic, nor would it be the voice of a Habsburg, and they set their sights on Friedrich, the Calvinist elector of the Palatine.

On Matthias's death, his titles of Archduke of Austria and King of Bohemia were assumed by his Habsburg cousin, Ferdinand of Styria, in the full expectation that the title of Holy Roman Emperor would also soon be his. But the Protestant Bohemians countered by deposing Ferdinand, and elected Friedrich as their king in his place. Ferdinand's response was ferocious. In the autumn of 1620, at the great Battle of the White Mountain at Bíláhora, near Prague, the Bohemian army was destroyed. Ferdinand exacted a terrible revenge: the gates of Prague were closed, and for a week his troops were licensed to take whatever they could. The city was sacked, and the gates of the Hradčany Castle itself were more than once blocked with wagonloads of plunder. For the rebels themselves, there was no mercy; the native nobility was simply wiped out, most by execution, the rest by confiscation of their lands and subsequent exile—many found their way to Sweden. Bohemia was forcibly re-Catholicized, while Friedrich's expected allies, the Union of German Protestant princes,[4] stood anxiously by, shaking their heads.

Friedrich appealed to Gustav Adolf to adopt his cause and take up arms against the Habsburg forces, but the Danes had already answered the call, and the Swedes could not be persuaded to fight alongside their old enemies and former overlords. The hapless "Winter King" continued a disheartened and desultory search for help, while his own Palatinate lands were occupied by Spanish Habsburg troops, cousins to Ferdinand's Austrians. Thenceforth the greater part of Europe was gradually sucked into the vortex. The Dutch, seizing their chance to strike at the distracted Spaniards, fanned the flames with their plentiful banknotes. Catholic France, no friend to Catholic Austria or to Catholic Spain, joined the fray on the Protestant side, while every German field and town paid its pound of flesh.

IN THE MONTHS BEFORE Christina's birth, the Spanish Habsburgs had been making a last attempt to reassert their own imperial strength, forging closer links with their Austrian relatives and trying to construct a united bloc of powers friendly to both Habsburg dynasties. The jewel now loosening from the Spanish imperial crown was the Dutch United Provinces—broadly, the northern area of today's Netherlands. Since the end of their truce with Spain in 1621, the Dutch had been fighting once again for independence; their wealthy towns, with their enterprising immigrant populations, progressive administration, and advanced banking systems, had become a trading and financial nexus for Europe and far beyond. Such a prize the Spanish empire, long declining, could not afford to lose. The Spaniards hoped that combined Habsburg forces might seize the ports along the coast of northern Germany; from there, a strengthened Austrian-Spanish navy could control the Baltic Sea, cutting off the Dutch from the rich trade that was financing their military resistance.

The Austrian Habsburgs responded as their Spanish cousins had hoped. In April 1627, the Emperor Ferdinand II conferred on his general, Count Wallenstein, the title of Generalissimo of the Baltic and Open Seas. The new generalissimo was already in control of several territories in northern Germany, and by November he had installed himself in the Baltic port of Wismar, where he set to work to build up the imperial navy. In the same month, Gustav Adolf wrote anxiously to his chancellor, Axel Oxenstierna:

> The popish league comes closer and closer to us. They have by force subjugated a great part of Denmark, whence we must apprehend that they may press on to our borders, if they be not powerfully resisted in good time.[5]

The chancellor agreed. Imperial forces had by now captured the whole of mainland Denmark, and the Danish king had been forced to retreat to his nearby islands. From Denmark an attack might easily be launched against Sweden itself, on its own territory. The situation, Oxenstierna remarked, "makes my hair stand on end."

In January 1628, a Secret Committee of the Swedish Senate agreed to an invasion of the emperor's German lands if the king should deem it necessary. A preemptive attack, to draw the imperial forces away from their present too threatening position, had been Gustav Adolf's own suggestion. In the face of the Habsburg threat, Poland was demoted to a secondary enemy, and Oxenstierna was accordingly dispatched to conclude a peace in the east, so that Swedish forces might be deployed elsewhere. After almost two years of negotiating, and twelve years of war, the Poles agreed to a truce.[6] Since their king, Sigismund III, would not renounce his claim to the Swedish throne, a real peace remained elusive, but for Gustav Adolf a halt to the actual fighting was for now just as useful. It was an opportune moment for the Swedes to become involved at last in the great conflict that had been gathering pace in the Habsburg lands for more than a decade already. Protestant Germany had found no champion, and many exiled voices were calling for Swedish help. Now the armistice with Poland released thousands of battle-hardened men, ready for active service elsewhere.

Gustav Adolf's decision met with loud applause from the Dutch; they had their Baltic trade to protect. But they were not the only power to welcome the idea of a Swedish march against the empire. The French encouraged it, too, and promised to assist with subsidies; Catholic France was no friend to Catholic Austria, and Richelieu had hopes of using the Swedes as a pawn in his own ongoing game against the emperor. But his terms were unacceptable to Gustav Adolf, and toward the end of 1629, preferring to find other allies, the king sent his own emissaries to the various courts and free cities of Europe; all returned empty-handed. The German Protestant princes, who had most to gain by a Swedish invasion, also declined to help, for by the same invasion, or so they feared, they also had most to lose.

Sweden was a small country, with not many more than a million souls. Despite many recent reforms initiated by the king and his able chancellor, it was still poor, with commerce and industry struggling to develop, and the state coffers empty after years of war by land and sea. It could not afford to fight alone against the resourceful Habsburg Empire. Bereft of allies, Gustav Adolf hesitated. Then, paradoxically, the very lack of money that had stayed his hand now forced it. In Prussia, squadrons of German cavalry who had fought for him against the Poles stood waiting; they were mercenaries, and, though their Polish campaign was over, they could not be disbanded, for there was no money to pay them off. If they were kept in service, payment could be delayed, and so it was decided. The cavalry would be sent to Pomerania, now occupied by imperial troops, and there the rest of the Swedish army would join them.

The forces ranged against the Swedes were led by the Czech Count Wallenstein and General Count Tilly, the latter a Dutch nobleman and a professional soldier, a Jesuit *manqué* whose devotion to the Virgin Mary and strict personal morality had earned him the epithet of "the monk in armour."[7] Wallenstein, though he led his own armies, was neither by nature nor by training a military man. Modestly born, through an advantageous marriage and the cheap purchase of no fewer than sixty-six estates confiscated from the defeated Bohemian rebels a few years before, he had become one of the wealthiest men in Europe. He was consequently able to raise and pay large armies of his own, and, owing to his administrative brilliance, to keep them fully supplied, as well.[8]

The Swedes pressed inland, and on a hot and windy day in September of 1631, they drew up at Breitenfeld, near the Saxon city of Leipzig, where imperial forces commanded by Tilly were already waiting. At the eleventh hour, the wavering Elector of Saxony, Johann Georg, had thrown in his lot with the Swedes; his own land was now at stake, and he had arrived to do battle himself at the head of his ranks of young noblemen, with their new-polished armor and their gaily colored cloaks—"a cheerful and beautiful company to see," said Gustav Adolf, and so indeed they must have seemed by comparison with his own hardbitten men in their torn and dusty outfits.

Tilly's forces had begun to fire as soon as their opponents came into sight, but the imperial general, despite his great experience, was soon disconcerted by the novel "chessboard" maneuvers of the Swedes, whose agile little squares of alternating cavalry and infantry swiveled to fire in all directions, easily outmaneuvering Tilly's traditional forward-facing lines.[9] Despite a dazzling sun against them, and despite the hasty departure of the frightened Saxon elector and most of his novice troops, the Swedes achieved a resounding victory, in no small part due to the brilliant planning and indefatigable energy of their own remarkable king.[10]

And by morning, of the host of imperial soldiers who had survived the battle only to be taken prisoner by the Swedes, many thousands had enlisted in the service of their yesterday's foe. After the battle of Breitenfeld, mercenaries from all parts of Europe flocked to the Swedish standard. By 1632, as well as substantial forces in Prussia and the Baltic, on the seas and at home on

Swedish territory, Gustav Adolf had some 120,000 men fighting in the German lands. Of his great army, perhaps one-tenth were native Swedes. The remainder, mostly mercenaries, recruited on no more than a promise of pay, were drawn from east to west: Finns and Germans; Scotsmen, English, and Irish; Frenchmen; Dutchmen; Czechs and Poles and Russians; their motives for fighting as varied as their origins.

THE FORTUNES OF WAR of the Emperor Ferdinand were now at their lowest ebb. The Swedes' position seemed unassailable. At this point, Gustav Adolf could have offered a peace settlement, but he chose to fight on, expanding his territories and claiming hesitant allies among the German princes, both Protestant and Catholic. In the spring of 1632, his soldiers cut a triumphal path through southern Germany toward Bavaria. In early April they crossed the Danube River, leaving in their wake a devastated countryside from which no pursuing army might take sustenance. By the middle of May they stood at the gates of Munich, where they met with no resistance; a huge ransom had purchased the safety of the city and its people. From Munich, Gustav Adolf hoped to entice the emperor's forces into battle, and then to march on the imperial capital of Vienna.

On the Bohemian border, the generalissimo Wallenstein waited with his own army. He had himself raised it, equipped it, and paid it, and at length he moved it into Prague itself, barring the Swedish army's way to Vienna. Gustav Adolf's allies wavered, and in June the hesitant king withdrew to Nuremberg. There, over the next few days, he revealed his plan for the future of Germany. The lands of the Holy Roman Empire were to be completely reorganized. The power of the Habsburg dynasty would be broken, and a new, dominant body of Protestant princes, the *Corpus Evangelicorum*, would take its place under an elected president, Gustav Adolf himself. The ban on Protestant worship was to be withdrawn, and religious toleration practiced throughout the empire. Peace would be maintained by a strong standing army.

The Corpus Evangelicorum was an idea born of crisis, an interim plan to provide cohesion and leadership for the duration of the war. It implied no long-term political objectives and was not intended as a blueprint for a Swedish empire in the German lands. So at least said the Swedes, but few of the group's proposed members regarded it so innocently. As the German campaigns had progressed, it had seemed to them increasingly clear that Gustav

Adolf harbored major dynastic ambitions for himself, ambitions that had much to do with their own German territories.

The Corpus Evangelicorum itself may have been an interim plan, but it seems that something of its kind was, after all, intended to endure. For more than a year already, Gustav Adolf had been negotiating a betrothal between his five-year-old daughter and her cousin Friedrich Wilhelm, the eldest son of the Elector of Brandenburg. This, the king hoped, would achieve what his own marriage to the elector's sister had so far not achieved: unite Swedes and Germans in a new northern bloc, which would shift the whole balance of power in Europe away from the Catholic Habsburg south and toward a new Protestant Swedish-German dynasty.

HONORED AND BELOVED FATHER,

As I have not the happiness of being with Your Royal Highness, I am sending you my portrait. Please think of me when you look at it, and come back to me soon and send me something pretty in the meantime. I am in good health, thanks be to God, and learn my lessons well. I pray God will send us good news of Your Majesty, and I commend you to his protection.

I remain,
Your Royal Highness,
Your obedient daughter,
CHRISTINA.[11]

So read the king, seated on a campstool in his tent at Fürth, on the outskirts of Nuremberg. The summer was drawing to a close. For almost three months, his army had been encamped there, while on the ridge above them, the imperial force stood waiting. Wallenstein had followed the king to Nuremberg, and now held him trapped with his weakening army. Though the king's thoughts may have turned often enough to his little daughter at home, he cannot have had much leisure to think of sending "something pretty" to her, for his supply lines were poor, food and water were scarce, and his men were beset by disease and discouragement. An attempt to fight their way out had ended in disaster; of their cavalry alone, three-quarters had been lost. The camp was full of rumors that the king's allies were turning from him, and among the men, for the first time, his popularity began to fade.

The time had come to offer peace terms, and accordingly, Swedish envoys were sent out to Wallenstein, in their hands the plan for the Corpus Evangelicorum, with plenty to placate the besieger. The generalissimo chose not to accept it, as he could well afford to do, with Gustav Adolf and his once invincible men penned in beneath the ridge. Without fresh supplies, the Swedes could not survive the cold weather that would soon be upon them, and in mid-September the desperate king decided to attempt a retreat from the camp. If he succeeded, he would march toward Austria, where new rebellions were rumored to have started against the Habsburg powers.

The retreat began, the Swedes fearing every moment the onslaught of the imperial troops. But Wallenstein did not attack. Instead, he turned his army toward Saxony, to the lands of Gustav Adolf's halfhearted ally, the Elector Johann Georg. The Swedes themselves turned back to help the elector, and by mid-October they were once again in Nuremberg, the scene of their own grim defeat only weeks before. Now, passing through the abandoned imperial camp, they found, to their horror, the remnants of Wallenstein's army, the unburied dead and, worse, the starving wounded, still lying there. The king gave instructions for the occupation of the area before the winter should set in, and moved his army on toward Saxony. They marched via Leipzig, then west some fifteen miles to the little town of Lützen, where, so they had heard, Wallenstein was encamped with a reduced army. There the revitalized Swedes would engage them, sure of victory with their novel fighting tactics and their superior numbers.

But Gustav Adolf's information was only partially correct. Wallenstein had only just dismissed his 12,000 allied forces. Learning of the Swedish king's advance, he had sent for them to be recalled. His remaining army alone numbered some 14,000, and they spent the night setting up their cannon and their barricades. In the early hours of the next morning they were still to be seen, making their way by torchlight, digging their trenches and hoisting their defenses, while, outside the town, in the fields nearby, the 16,000 Swedish troops lay sleeping.

It was the sixth of November. By eight in the morning, in clear light, the first shots had been fired, while the king still stood before his army, offering prayers for a Protestant victory. Wallenstein had drawn up his army in traditional formation, with infantry in the center, protected by artillery, and cavalry on the wings, while the Swedes stood ready in the flexible squares which had served them so well at Breitenfeld. It was not until ten that the two armies engaged, and by then the battleground was covered in a thick mist, alternately providing cover and hampering visibility. The Swedes charged first, and the desperate struggle began.[12]

Later, those who had fought that day could not agree when the imperial reinforcements had arrived; some thought midday, others thought not until the evening. In fact the cavalry arrived first, led by the legendary Count Gottfried Pappenheim, hero of the imperial army and idol of his own soldiers. They attacked immediately, beating the Swedes back over the territory they had won. Pappenheim was shot through the lung and retreated from the battle to die, choked with blood, in his coach behind the lines. It was rumored that Gustav Adolf had also been hit. His horse had been seen, wounded in the neck, plunging wildly across the battlefield. The imperial general Piccolomini, himself grazed seven times by bullets, swore that he had seen him lying on the ground. Duke Bernard of Saxe-Weimar took command of the Swedish force, and by nightfall, the imperial troops had been driven back. Wallenstein retreated to the nearby town of Halle, leaving his men in disarray behind him. The battle had been inconclusive, but the Swedes now occupied the field, and the victory was held to be theirs.

In the darkness, the Swedish soldiers began the terrible search for the body of their king. Beneath a heap of the dead, naked but for his shirt, they found him. He had been killed by a shot through the temple, but his body showed other wounds: a dagger thrust and another shot in his side, two shots in the arm, and a shot in the back.

Rumors spread that the king had been betrayed, killed by his own men under cover of battle. Some recalled the Bloodbath of Linköping, saying that the sons of those beheaded by his father had succeeded in claiming a tardy revenge.[13] Others held that his murder had been ordered by Cardinal Richelieu, determined to be rid of the "impetuous Visigoth" who had bettered him at his own political games. It seemed impossible that the great Gustav Adolf could have died like any ordinary soldier, shot and stabbed as he fought his way through enemy lines. Bernard of Saxe-Weimar gave out that the king was not dead but only wounded, and for days afterwards merchants in London were placing bets that he was still alive. Waiting in Erfurt, Maria Eleonora learned the truth on the tenth of November; she collapsed with grief. The following day, in Frankfurt, Chancellor Oxenstierna heard the news, and passed the first sleepless night of his life.

Gradually it emerged that, leading a cavalry charge early in the battle, the king had been shot in the arm, and had lost control of his horse. In the thick mist covering the battlefield, he had been separated from his escort of cavalrymen. Wounded again, he had fallen from his saddle, but his boot had caught in a stirrup and he had been dragged along the ground. Falling free, he had been unable to rise, and had been shot in the head where he lay.

His body was carried, on a powder wagon, to the little village of Meuchen near the battlefield, and there it was washed clean of dirt and blood. The king's reverent soldiers stored the blood itself in the village church, marking the place with his coat of arms. Overnight, the body lay before the altar, and when morning broke, the village schoolmaster, who served as the local joiner as well, set to work to build a wooden coffin. In this the king's body was carried to the town of Weissenfels, some ten miles distant. There, in the bay-windowed room of a local guesthouse, it was laid out and embalmed by the king's own apothecary. Among those who saw the body there in its simple coffin was Gustav Gustavsson, the king's illegitimate son, now sixteen years old and serving in the Swedish army.

Back in Meuchen, one Swedish soldier, recovering from his own wounds, arranged a primitive memorial to his lost commander-in-chief. With the help of local peasants, he rolled a large stone—the "Swede's Stone"—to the place where his king had fallen.[14]

It is said that the emperor himself wept for his enemy, and ordinary people who had never set eyes on the great king wailed in the streets at the news of his death. "He alone was worth more," said Richelieu, "than both the armies together."

THE KING'S BODY HAD now to be transported back to Sweden, escorted, in death as in life, between footsoldiers and cavalrymen. From Weissenfels it was carried a hundred miles north toward Berlin, and in the middle of December, the cortege was met by the newly widowed queen. She was almost hysterical. For several weeks, fearing to aggravate her state, her attendants had kept her in Erfurt, preventing her from traveling to where her husband's body lay. Now, seeing his lifeless form, she gave way to an extravagant grief. The king's heart had been taken from his body, to be separately preserved; this Maria Eleonora now took to herself, wrapping it first in a linen kerchief and later placing it in a golden casket. She kept it with her constantly. At night, it hung above her bed, glowing in the light of vigil candles, while the queen wept desperate tears.

Northward the Swedes continued their sorrowful journey. At Wolgast, they paused; the Baltic Sea was frozen, and for many months there would be no passage across to Sweden. Maria Eleonora's behavior became increasingly bizarre; the eccentric traits she had shown for some years had been intensified, it seemed, by the shock of the king's death. Now, disregarding the entreaties of those around her, she began to make plans for an elaborate funeral, spending wildly on one scheme after the next. Her stranded little court began to disintegrate into chaos, while day and night Maria Eleonora clung, often literally, to her husband's mortal remains, until her attendants feared she had lost her reason. In February 1633, three months after the king's death, she wrote from Wolgast:

> Since We, God pity Us, were so rarely granted the pleasure of enjoying the living presence of His Majesty, Our adored, dearest master and spouse, of blessed memory, it should at least be granted to Us to stay near his royal corpse and so draw comfort in Our miserable existence.[15]

From Stockholm, the alarmed senators dispatched the chancellor's cousin, Gabriel Oxenstierna, to investigate the queen's entourage and to oversee the return of the king's body home to Sweden. Delayed by illness and the winter weather, Oxenstierna reached Wolgast only in the middle of May, and there he found the grieving queen "swimming in tears," and her little court in wretched disorder.

It was not until July that the royal flagship set off at last, and in early August the entourage arrived at the industrial town of Nyköping, on the eastern coast of Sweden. Here, furnaces blasted and foundries thundered, shipwrights and millworkers toiled and travailed. Once the country's capital, Nyköping now centered on a magnificent Renaissance castle, the queen's private residence. It was here, twenty years before, that Gustav Adolf had been proclaimed king, and it was here that his body now came to a temporary rest.[16]

It was in Nyköping, too, that Maria Eleonora at last saw again her six-year-old daughter. It had been some fifteen months since their last meeting; since the spring of the previous year, the queen had been in Germany, visiting her family and following her husband's campaigns. Christina had been left in Sweden in the care of her paternal aunt, the Princess Katarina, and she had now traveled to Nyköping "in person, with all the senators and all the noblemen and women," to meet the sad cortege. Dutifully, the little girl approached the unfamiliar, grieving woman who was her mother. "I kissed her," she was later to write, "and she drowned me with tears, and nearly suffocated me in her arms."[17]

The king's body was lain at first in the castle's Green Hall, but Maria Eleonora, now refusing any talk of burial, soon had it removed to her own bedroom. The coffin was covered by an elaborate pattern of oval pearls after her own design, but it remained unsealed, and it seems that it was not infrequently opened. More than a year after the king's death, the men of the Swedish parliament, shocked, embarrassed, and indignant, petitioned the Estate of the Clergy, asking "whether a Christian could in good conscience apply for and be granted the right to open the graves and the coffins of their dead and gaze at and fondle their bodies in the belief that through these acts they would receive some comfort and solace in their state of great heart-rending sorrow and distress."[18] But slow planning for the state funeral, and perhaps, too, some pity for their great king's widow, stayed any firm response.

After many delays, and constant opposition by Maria Eleonora, the king's body was at last interred on June 15, 1634, nineteen months after his death. Toward the Riddarholm Church in Stockholm, final resting place of Sweden's kings, the body was borne on a silver bier, encircled by military standards and captured enemy cannon and other symbols of the warrior-king's victories, including his bloodstained sword, just as it had been taken from the battlefield at Lützen. A vast crowd of people accompanied the procession, weeping, mourning, straining to see. And among the nobles and soldiers and court officials, some of them spied one very small figure—their new, seven-year-old queen.

Within a day of the king's interment, Maria Eleonora pleaded for the coffin to be opened again, asking that the king should not be buried while she lived.

Fifteen years before, when the handsome young "Adolf Karlsson" had come to court her, the body of Maria Eleonora's father had lain in state, months after his death, in the gloom of the castle chapel, while the drab accompaniments of a formal mourning oppressed his court. The king's own mother, too, had lain unburied through the long northern winter, awaiting her son's return from the conquests that would make his name famous and feared. But, even in an age of delayed burial and long months of mourning, Maria Eleonora's grief at her husband's death was felt to be excessive. Throughout the royal apartments, darkness reigned. Black fabrics draped the walls from ceiling to floor, and the windows were blocked with sable hangings; no daylight filtered through. Sermons and pious orations droned endlessly. The queen mourned day and night, relieved only by her troupe of dwarves and hunchbacks, dancing in the candlelight. Bereft of her husband, she now turned her attention for the first time to her little daughter, smothering her with newfound affection, and forcing her to live alongside her in the macabre atmosphere. She dismissed Christina's Aunt Katarina, who had looked after the child for the previous two years, and announced that from now on she herself would take care of her. The once rejected girl-child, ugly and "swarthy as a little Moor," was now found to be "the living image of the late King," and the queen scarcely let her out of her sight. By day the little girl struggled to escape to her books and her horses; by night she was obliged to share her mother's bed in the gloomy chamber, lying fearful and lonely beneath her father's encased heart. The king's death had set in train a melodrama of mourning in which Christina was to remain a virtual prisoner, until her rescue by the "five great old men" who were now to serve as Sweden's regents.

THE LITTLE QUEEN

*It was victory that
announced my name on the
fateful field of battle
—Victory, a herald at arms
proclaiming me King.*[1]

So CHRISTINA WAS TO write, many years later, at a time when she needed to call upon her every credential of greatness. She was, she continues, "the link, weak as it was, that united so many good men, so many diverse and opposing interests, all dedicated to sustaining the rights of the girl who began to reign at that fatal moment." All the generals, she says, all the men of the army, and "the great Chancellor," too, submitted to the name of Christina.

In rhetorical terms, there is some truth in this tale, but in reality the crown did not pass to the little girl quite so smoothly. Gustav Adolf's generals stood firm and announced their loyalty to his fragile Vasa dynasty from their battlefields in Germany, giving the chancellor, who now assumed power in the king's stead, the means to continue the war. But in fact there was no guarantee that Christina would inherit her father's throne at all. Only five years before, when she was just a year old and no male heirs seemed likely, Gustav Adolf had had to confirm her right, as a female, to succeed him.[2] His own royal line was not so ancient that he could be sure of its continuance against all odds; his cousin Sigismund, the Catholic king of Poland, had his own, arguably greater, claim to the Swedish throne. Moreover, heredity was not enough; for many centuries the Swedish monarchy had been elective, and the principle, established by Christina's great-grandfather, applied to males only. It was by no means certain that the Estates would accept a woman—indeed, a little girl—as their "king," as the Swedes always formally referred to their sovereign. There were even some who might have preferred to oust the monarchy entirely and install a republic in its place.

In the Senate, or so Christina was later to write, it was a different story. All the senators declared themselves in her favor. They all felt that her right to the throne was "incontestable." They were "only too happy" to have this child, who was "their only strength and Sweden's only hope of salvation at such a dangerous time."[3] Histrionic as the words may seem, they were probably true—indeed, the words "strength and salvation" comprised the Swedes' very definition of their monarchy. And it was certainly a dangerous time, with Swedish armies exposed in Germany and elsewhere, and the constant threat of the Catholic Poles taking power at home. It was no doubt this double peril that persuaded the senators now to support Christina's succession, for they had much to gain by opposing it. For the noble families from which every senator was drawn, the three generations of the Vasa dynasty had meant, above all, a steady waning of power. Their own grandfathers had only grudgingly accepted the first Vasa king, Gustav I Eriksson. Though he had driven out the Danes by his energy and bravery, they had regarded him as an upstart with no very ancient lineage. Resentment had rankled into the next generation: Gustav's son, Karl IX, had been determinedly opposed by the noble families. He had sought support instead from the common people, earning the nobles' disdainful epithet of "the rabble king." But the people's support had allowed Karl to govern on his own terms. Power had drained from the noble families and collected around the crown. In 1600, the king had finally secured his position in the infamous "Bloodbath of Linköping," where his five leading opponents, including four members of Sweden's highest nobility, were beheaded in the town's market square. It had required the extraordinary gifts and the no less extraordinary personality of Karl's son, Gustav Adolf, to quiet the outrage of the noble families and persuade them to support their malefactor's heir. But now the golden-haired king was gone, leaving no son to succeed him. His sole heir was female; the principle of heredity could at last be abandoned, and the nobles could reclaim their ancient right of electing their own grateful and malleable sovereign.

It says much for the senators' patriotic spirit, or perhaps for their fear of Poles and popery, that they decided to forgo this right and give their support to a continuing Vasa dynasty. But, although the Senate stood unanimously behind the little "king," she was not so quickly accepted by the men of the *riksdag*, a socially more diverse group with differing views of the perils facing their homeland. The *riksdag*, Sweden's parliament, comprised four Estates: the clergy, the nobles, the burghers, and the peasants. It was among these last, as Gustav Adolf had feared, that opposition to a female ruler now proved strongest. The story is told that, in March of 1633, four months after Gustav Adolf's death, when the *riksdag* was assembled to affirm Christina's succession to the throne, the marshal was interrupted in the middle of his address by a member of the peasants' Estate, a man bearing the almost symbolically Swedish name of Lars Larsson. The peasants, it seemed, were not convinced by the senators' arguments. "Who is this girl?" Larsson demanded. "We don't know her. We've never even seen her." Larsson was seconded by a growing number of the men, and the child was sent for. Happily for her, and for the senators, Christina's resemblance to her father was clear. Larsson recognized at once the great king's forehead, his blue eyes, and, starting out from the solemn little face, his long, distinctive nose. The succession was assured. Christina was unanimously acclaimed Elected Queen and Hereditary Princess of Sweden—"elected" as a warning to the Polish Vasas that their hereditary rights would not be enough to claim a Swedish throne.

The little blond-haired girl, just six years old, now bore the titles Queen of the Swedes, Goths, and Vandals; Great Princess of Finland; Duchess of Estonia and Karelia; and Lady of Ingria—the last owing to the Peace of Stolbova recently concluded with the Russians. If Christina's own story is to be believed, she bore them all, even at this early age, with appropriate aplomb. She did not really understand what was happening, she writes, but nonetheless she was delighted to see all the great men of the land—among them the Count Palatine, Johann Kasimir—on their knees at her feet, kissing her hand. Her delight is understandable, for Johann Kasimir was her uncle, and she had already spent a good deal of time in his castle at Stegeborg, in his care, no doubt kindly, but also under his no doubt authoritative eye. Here was a reversal indeed.

Christina has left a description, addressed to God, of the first convening of the *riksdag* in 1633, following her acclamation. Before all the men of the four Estates, she ascended the throne of her great father:

> The people were amazed by my grand manner, playing the role of a
> queen already. I was only little, but on the throne I had such an air,

such a grand appearance, that it inspired respect and fear in everyone. . . . You had planted on my forehead this mark of greatness. . . . I
never went to sleep during all the long ceremonies and all the
speeches I had to sit through. Other children have been seen going
to sleep or crying on occasions like this, but I received all the different
signs of homage like a grown-up person, who knows they are his
due. . . . [4]

Writing in her later years, Christina admitted that "it doesn't take much to admire a child, and even less a child of the great Gustav, and perhaps flattery
has exaggerated all this." But in fact "all this" reflects an idea that was to remain absolutely consistent throughout her life, the idea that sovereignty was
something she carried within herself. For Christina, kingship was a personal
attribute that had nothing to do with the rights and regalia of monarchy. Her
right to rule, she believed, was innate; she could not be divested of it. It was
not dependent on possession of the Swedish crown, or any other, since God
Himself had planted "this mark of greatness" on her forehead, and even in her
childhood it had inspired "respect and fear" in all who saw it.

A large delegation of diplomats from Muscovy supposedly observed this
inborn sovereignty at about the same time, and, we are told, it left them
quaking in their fur-lined boots. The Russians had arrived to offer their con-
dolences on the king's death and to extend a formal greeting to the new
monarch; they had also to ensure that the peace Gustav Adolf had made with
them at Stolbova, after eight years of fighting, would now be ratified.[5] According to Christina's "little story," the regents were anxious that their six-year-old
queen would not be able to endure the rigors of the formal reception with the
necessary *gravitas*. "I was such a child," she writes, "that they thought the Russians would frighten me with their strange clothes and their wild manners.
They were dressed very differently from us, they said. They had great big
beards, and they were terrible-looking. But I just laughed, and told them to
leave it all to me."[6] And when the Russians finally approached the little
queen, seated on her throne, looking "so assured and so majestic," they felt, or
so she believed, "what all men feel when they approach something that is
greater than they are." Whatever the Russians felt, they were rewarded with
the ratification they sought, and were "sent off with the usual tokens."

The ratification itself had been agreed by Christina's regents, the "five
great old men" who had accepted the charge of government until their little

queen should reach her eighteenth birthday. Though it had been a mighty blow, Gustav Adolf's death entailed no difficult transition for those who governed the country. During the king's frequent absences on campaign, the regular business of government had been left in the hands of ten nominated men of the Senate, and now, despite their loss, they adapted easily to the new situation. The king himself had chosen five of them to form a regency in the event of his death, five noblemen who were also to hold the five great offices of state: Grand Chancellor, Grand Treasurer, Grand Marshal, Grand Admiral, and High Steward. The government was now dominated by what amounted to Sweden's second royal family, the Oxenstiernas. The premier office of Grand Chancellor was held by Baron Axel Oxenstierna, the late king's close friend and undoubtedly one of the ablest administrators of the age. The Grand Treasurer was the chancellor's cousin, Baron Gabriel Bengtsson Oxenstierna, and the High Steward his younger brother, Baron Gabriel Gustavsson Oxenstierna.[7] The office of Grand Marshal was held by one of Sweden's finest generals, Count Jakob De la Gardie; to him Gustav Adolf had lost his former love, Ebba Brahe; their son Magnus was to prove a contentious figure during Christina's own reign. The Grand Admiral was Christina's uncle, Baron Karl Karlsson Gyllenhjelm, illegitimate half-brother of the late king. On his broad soldier's shoulders, and on those of his four fellow senators, the burden of government now lay.

Christina herself has left us a picture of her regents. Of Axel Oxenstierna, *primus inter pares*, she writes with respect and affection, indeed almost with awe: he was, she says, a man "of great capacity, who knew the strengths and weaknesses of every state in Europe, a wise and prudent man, immensely capable, and greathearted."[8] Tireless in the affairs of state, he nevertheless always found time to read, so continuing the studious habits of his youth. She notes that he was "as sober as a man can be, in a country and at a time when that virtue was unknown." Christina describes him as an ambitious but loyal man, and incorruptible, if a little too "slow and phlegmatic" for her taste; but she loved him, she says, "like a second father," and like a second father, too, he was to present a mighty challenge to her developing sense of self.

The chancellor's cousin, Gabriel Bengtsson, Sweden's Grand Treasurer, Christina regarded as "upstanding," and "capable enough" of his high office. Of the younger Oxenstierna brother, Gabriel Gustavsson, now High Steward, she writes that he was well liked and well spoken, but in the natural way of the Swedes, without the burden of much erudition, since he had "only a smattering of Latin." But he was, she adds consolatorily, "a very good man." The Grand Marshal, Jakob De la Gardie, is described as able and personally courageous; this preeminent soldier had distinguished himself in the Swedish campaigns against Poland and Russia. Christina notes that his personality was direct, even brusque, but that he liked to chat. He had been a favorite with her father, she says, and was always competing with Axel Oxenstierna for the king's favor. In Karl Karlsson Gyllenhjelm, the Grand Admiral, "bastard brother of the late king and my uncle," Christina recognized "a good, brave, old-fashioned man, a good Swede, bright enough," but worn down by the twelve years he had spent in irons in a Polish prison, refusing to abjure his Lutheran faith for the despised Catholicism of his captors.[9] He was "absolutely devoted to the house of Vasa," she writes, "and he loved me like his own child."

FOR THE NEXT TWELVE YEARS, the "five great old men" were to rule in their little queen's name, though in fact Christina may never have been intended to rule at all, or at least not to rule alone. The steps that her father had taken to ensure her succession to the throne had been, as it were, an emergency precaution, anxiously put in place as he himself prepared to return to the war from which he felt he would not return. The pious king, almost fearful of his extravagant successes in the sight of "a jealous God," had had premonitions of his own death. The succession must be assured if civil war, or worse, were not to overtake his homeland. A long period of regency was certain, but in time the girl would marry; her husband would rule alongside her, or even in her place. Besides, Sweden's name was now great in Europe; Gustav Adolf himself had made it so. A king's daughter was an opportunity incarnate to forge new alliances, and shift the balance of power.

Negotiations for the little girl's betrothal had consequently been in place for some time. The chosen prince was her own first cousin, Friedrich Wilhelm, her senior by seven years, the eldest son of the Elector of Brandenburg, and now, in the summer of 1633, thirteen years of age.[10] The boy was Protestant, and seemed promising, and, crucially, he stood to inherit the duchy of Pomerania in northeastern Germany, whose long coasts were strategically important for both trade and warfare. Pomerania was now, insecurely, in Swedish hands—Gustav Adolf had concluded a treaty with its Archduke Boguslav XIV—but Boguslav's heir was the Elector Georg Wilhelm, and in time the vital Pomeranian coasts would pass to his son, Friedrich Wilhelm. A marriage between Friedrich Wilhelm and Christina would thus ensure Sweden's continuing access to them. It would make Brandenburg a safe neighbor and, moreover, would serve as a mighty cornerstone for the new bloc of Protestant powers once envisaged by Gustav Adolf, and now promoted by Christina's regents. Above all, the marriage would give Sweden at last the almost mythical *dominium maris baltici*, the mastery of the Baltic Sea, that had lain at the heart of Swedish policy for generations.

The king had promoted the match with some energy, traveling to Berlin himself, when Christina was only four years old, to suggest the project personally to the elector.[11] Maria Eleonora, too, had been very much in favor of it. Her nephew, it was planned, would abjure his Calvinist religion and become a Lutheran; this had been agreed by the elector's own theologian. The boy would move to Sweden for the rest of his education and for his military training, learning the language and the ways of the Swedes while still in his impressionable years.

But the Berlin meeting had not borne much fruit. The elector distrusted Gustav Adolf; he had not wanted his sister to marry the king, and he did not want his son to marry the king's daughter. Unwilling to state the matter so plainly, he prevaricated: the religious clause was objectionable, he said; he had hoped instead for some kind of union between Calvinist and Lutheran believers. Besides, his son was too young to be sent away from home, and then Gustav Adolf might yet have a son of his own. Privately, Georg Wilhelm had sought the advice of other German princes, most of them still smarting from the Swedes' roughshod riding over their own territories in the recent years of fighting. Their advice was consistent: the elector should not pursue the plan; the pair were too young, and the political situation might be different by the time they had come of age. The Swedish climate was too harsh, and the Swedes themselves "not very nice people" who would not welcome a German king. Besides, Sweden's enmity with the Holy Roman Empire might drag Brandenburg into the same fearful morass. And the marriage would make Sweden much too powerful; the German princes, and many others even within Sweden itself, feared that Gustav Adolf would use it as a stepping-stone to the throne of the Holy Roman Empire. Despite encouragement from his own chancellor and renewed attempts on the part of the Swedes, the elector had decided to let the matter drift.

At the end of 1632, when the news arrived of Gustav Adolf's death at Lützen, the Danish king Kristian IV had decided to try his luck in arranging a marriage for his own son, the Archduke Ulrich, now in his early twenties, to the little queen of Sweden. It was a second attempt on Kristian's part; the previous year, his hopeful embassy had been rejected by the Swedish king himself. But reflection on the huge cost of recurrent war persuaded Kristian to try again, and now, it seemed, a window had opened in the house of his old enemy, through which he might insert some Danish influence. A measure of dissension among the land's new governors would serve his interests well; an official embassy of condolence would provide the perfect opportunity. Barely a week after the news had arrived, his envoy received instructions to seek a private audience in Wolgast with the late king's grieving widow.

Kristian hoped, at the very least, to create a rift between Maria Eleonora and Sweden's five regents, already in office for some time on account of the late king's long absences on campaign. Early in the new year, Chancellor Oxenstierna, still in Brandenburg, received a letter from the Danish king, relaying his renewed hopes of the match. Oxenstierna, unpleasantly surprised, replied that Christina was too young for any marriage plans to be made for her as yet, and added that there were "many other considerations" besides. But he took the precaution of writing at the same time to the regents in Stockholm to ensure that, if consulted, they would give the same reasons for declining Kristian's offer. The Danes were near neighbors, after all, and their alliance was very recent. It would be unwise to offend them, for they might also prove to be uncertain friends.

Meanwhile, amid the increasing chaos of the castle at Wolgast, Maria Eleonora was able to master her grief sufficiently to begin negotiations of her own with the Danish envoy. Though Friedrich Wilhelm was her nephew, that did not ensure her constancy now to his cause. As fervently as she had wished for the match while her husband was alive to promote it, so now, in the first months of her widowhood, she turned determinedly against it. She decided, or was persuaded, that it would never do; Christina was the daughter of a king: only a king's son could be a suitable husband for her.

An anxious Chancellor Oxenstierna wrote to Wolgast, urging the widowed queen to caution. Denmark was Sweden's oldest enemy, he reminded her. The two would never be brought together "without great bloodshed or the complete extinction of one or the other." The queen should speak to the envoy, or indeed anyone else, "only in the most general, noncommittal terms."[12] She replied, duplicitously, that she had "given no yes or no" to anyone. But throughout the winter and the spring she kept constant company with the

Danes, and the rumor spread that the young Archduke Ulrich himself was soon to visit Wolgast.

In April, at home in Stockholm, the Senate met to discuss the matter. There could be no better prince than Friedrich Wilhelm, said the Count Per Brahe. Sweden could find no better supporter, and the marriage would make Sweden formidable among all nations. On the contrary, said the chancellor's cousin, Gabriel Oxenstierna, it would be better to choose a poor Swedish nobleman, who would be more dependent on the Senate. Foreigners in the past had only tyrannized the country. He would rather have a local man. But, said Per Banér, if the foreigner were the husband of a Swedish princess, he would not tyrannize anyone. No foreign ruler had ever married a Swedish princess before. Quite true, said Jakob De la Gardie. A Swedish consort would only sow dissension, having his own support among the local people. However, said Gabriel Oxenstierna, a Swede would be more easily constrained by the law. On the other hand, a royal marriage was an excellent way for a nation to increase its power, and certainly a connection with Brandenburg would be politically advantageous, particularly with regard to Poland. It might be wise, then, said Per Brahe, to come to a decision soon. If the Brandenburgers thought they were being led around by the nose, they could turn their backs on Sweden and embrace the Poles instead.

A letter from the chancellor, favoring Friedrich Wilhelm, was then read once again to the assembled noblemen. They were duly impressed, and several senators now remembered that a Brandenburg marriage would keep Pomerania in Swedish hands *jure perpetuo*. That would be good security against the Dutch. They reassured one another that Friedrich Wilhelm would have a duty to appoint all his officials exclusively from Swedish families. All things considered, the elector's son was to be preferred to any other foreign prince.[13]

In short, the little queen was to serve as a chattel in the crudest old terms. One senator did remark that she might not actually want to marry Friedrich Wilhelm when she grew up; at only six years of age, she could hardly be consulted now. This was agreed, and a message sent to Axel Oxenstierna in Berlin, conceding him full powers of negotiation, but suggesting that he proceed slowly. He took the senators at their word, and kept the discussions going for a further fifteen years.

Nevertheless, by the beginning of 1634, six months after her regal reception of the Russian ambassadors, the betrothal of the now seven-year-old Christina

to her Brandenburg cousin was understood throughout Europe to be a fait accompli. Resigned shoulders shrugged in Copenhagen, and an anxious emperor paced the floors of his palace in Vienna. Only in Stockholm and Berlin
did doubt remain, for the two protagonists had in fact reached no agreement
at all.

Christina herself was never to mention her father's plan for the Brandenburg marriage, for it clearly indicated that he had seen no particular "mark of
greatness" planted on her childish forehead. He had not intended her to rule
alone, nor perhaps even to rule at all. She chose to dwell instead on the instructions that he had left for her upbringing, exaggerating them to accommodate her own profound need to be accepted, not as the little queen of
Sweden, but as its divinely appointed king.

IN THE TWO YEARS preceding the king's death, Christina had seen
equally little of her father and her mother. Whether following her husband on
campaign or visiting her family in Brandenburg, Maria Eleonora appears to
have given little thought to the child left behind. "My mother could not bear
the sight of me," Christina was to write, "because I was a girl, and she said I
was ugly."[14] Portraits of Christina in her early childhood depict nonetheless a
charming little girl, though most are conventional, and all are no doubt flattering. It is true, however, that she was slightly deformed. As a baby, she had
apparently been dropped, and her injuries had left her noticeably lopsided in
the upper body, with one shoulder higher than the other; the portraits show
her in tactful semi-profile. She herself was later to claim that this "dropping"
was no less than an attempt on her life commanded by Catholic sympathizers
among her cousins, and at times even suggested that it had been her mother's
own idea. Whatever the truth, the resulting deformity cannot have endeared
her to the beauty-loving queen.

In the absence of both mother and father, Christina had spent most of her
time with her family of Palatine cousins, who lived in unpretentious comfort
at Stegeborg Castle, to the south of Stockholm. Her aunt was the Princess
Katarina, the king's elder half-sister, and her uncle Count Johann Kasimir of
Pfalz-Zweibrücken, who had once accompanied Gustav Adolf to Berlin to
meet the young Maria Eleonora. The Princess Katarina, then in her late for-
ties, was the mother of five surviving children, the youngest still in his infancy;
Christina describes her aunt as a woman of "consummate virtue and wisdom."

She had settled in easily with the other children; among this lively half-dozen, she was fourth in age, with two little countesses, Maria Euphrosyne and Eleonora Katarina, a year or so either side of her. Her eldest cousin, some ten years older, was the Countess Kristina Magdalena, and there was a young boy, too, Karl Gustav, Christina's senior by four years, and the baby, Adolf.

Her father's untimely death had wrenched Christina from this comfortable environment and installed her, against her will, in her mother's bizarre and gloomy apartments at Nyköping Castle; here she had been closeted for a year or more. "It would have been a lovely court if it hadn't been spoiled by the Queen Mother's mourning," Christina was to complain. "There is no country in the world where they mourn the dead as long as they do in Sweden. They take three or four years to bury them, and then when they do, all the relatives, especially the women, weep all over again as if the person had only just died."[15] Maria Eleonora "played the role of grieving widow marvelously well," she writes, insisting at the same time that her mother's grief was sincere. "But I was even more desperate than she was, because of those long dreary ceremonies and all the sad and sorry people about me. I could hardly stand it. It was far worse for me than the King's death itself. I had been quite consoled about that for a long time, because I didn't realize what a misfortune it was. Children who expect to inherit a throne are easily consoled for the loss of their father."[16]

Consoled or not, in the midst of her mother's melodrama, Christina fell ill with the first of many maladies attributed by contemporaries, as by later scholars, to her distressed state of mind. She developed "a malignant abcess in my right breast, which brought on a fever with unbearable pain. At last it burst, releasing a great flow of matter. That did me good, and in a few days I was perfectly well again."[17]

After the king's burial, in June 1634, Maria Eleonora moved her court to the Tre Kronor Castle in Stockholm, near to the Riddarholm Church, where her husband's body was now entombed. Christina may have sensed a touch of theatricality in her mother's extravagant mourning, but to the four regents who remained in Sweden, it seemed real enough. Taking advantage of the move to Stockholm, they proposed to place the child in separate apartments within the castle, but the suggestion drew forth "pitiful tears and cries" from her mother. Axel Oxenstierna, writing from Germany, where he had remained to continue direction of Sweden's armies, urged the senators to insist: the child must be taken from her mother; the late king himself had warned that Maria Eleonora was not to be permitted any influence over her. The senators, it seems, were divided; some felt that the child should be left where she was; others wanted to send the Queen Mother back to Nyköping by herself. Every remonstrance with Maria Eleonora was met with fresh hysterics, so that the senators, torn between sympathy and exasperation, came to no conclusion at all; their wavering condemned the little queen to two further miserably cloistered years. In the meantime, with surprising initiative but little persistence, Maria Eleonora made plan after plan for elaborate memorials to her late husband. There was to be a new tomb, then a new chapel, then a new castle, then a whole new city. One French envoy, flattered to be consulted by "this charming woman," recorded his delight in discussing with her "the finer points of every branch of architecture, of Doric and Ionic and Corinthian columns,"[18] seeing in the mother the same advanced aesthetic sensibilities that others would one day see in the daughter. But a perverse talent for leaving things unfinished was equally to be revealed in both, and, needless to say, no stone of the king's great memorial was ever laid.

Christina, meanwhile, did what she could to escape. The means at her disposal were slender, but she exploited them, or so she claims, to the full. Her hours of exercise, and especially of schooling, became her refuge. The mother's weakness was turned to the daughter's profit. "What I endured with her," she writes, "made me turn all the more keenly to my books, and that is why I made such surprisingly good progress—I used them as a pretext to escape the Queen my mother."[19] The indecisive senators had at least been able to agree on the kind of education the child should receive; in fact, prompted by their absent chancellor, the entire *riksdag* had discussed it, and in March of 1635, with Christina already eight years old, they made their conclusions known. Their priorities are revealing. The little queen must learn, states their preamble, "to speak well of her subjects and of the present state of the country and of the regency." Though she must learn something of foreign manners and customs "as becomes her station," she must also "practice and observe Swedish ones and be taught them carefully." She must learn table manners, too, they declared, without, however, specifying whether these were to be homegrown or of some foreign variety.

The men of the *riksdag* were clearly anxious that Maria Eleonora's widely known disdain for all things Swedish should not be inculcated in her daughter. Other foreign errors were also to be strenuously avoided, notably those of popery and Calvinism. The "art of government" was acknowledged to be important for her to learn, but "as this sort of knowledge is learnt rather with age and experience than by the studies of childhood, she should first and foremost study the word of God, and all the Christian virtues,"[20] reading only those books that had been approved by learned men of suitably moral temper. The program for her education was to be reviewed as the little queen progressed.

Gustav Adolf had also been concerned to prevent Maria Eleonora from influencing their daughter adversely. From his campaigns abroad, he had sent back detailed instructions about her upbringing in the event of his own death. The queen was to be excluded from any regency. The child's two governors were to be Axel Banér and Gustav Horn, both senators. As tutor she was to have Johan Matthiae, a theologian and former schoolmaster, and the late king's own chaplain. The two governors were both expert in the use of arms, and both were hard drinkers, but otherwise they were very different men, Banér apparently something of a rough diamond with a penchant for pretty women, Horn more of a courtier, fluent in foreign languages and an experienced diplomat. The tutor, Johan Matthiae, well born and well educated, had studied not only in Sweden's own university at Uppsala, but also in the German lands, as well as in Holland, France, and England. He was a man of calm and kindly temperament, liberal in his thinking, especially in religious matters; in this he reflected, as he had no doubt helped to form, the views of his late king.

Unlike the "five old men" who comprised her regency, Christina's governors and tutor were young, all in their thirties at the time of their appointment, Gustav Horn indeed barely so. Two at least had been Gustav Adolf's beloved friends, Banér even sharing the king's bedchamber before his marriage, and afterwards whenever the queen was absent. Johan Matthiae, too, had accompanied the king on campaign. Christina later described them all as "capable, good men." She appreciated the straightforward honesty of Banér, and admired Horn's foreign polish, but for her tutor she reserved a special fondness. She called him "papa," and he quickly became the confidant of all her little secrets, a steady and reassuring presence in her difficult young life.

Together, the three guardians formed a vital counterweight to the extremities of Maria Eleonora's court, and provided an outlet for the frustrated energies of a bright and active child. But the Queen Mother's continuing obsessive behavior during these years destroyed any chance for a real affection to develop between herself and her daughter. Though Christina claims to have loved her mother "tenderly enough," her respect for her began to fade, she says, when she "seized me, in spite of my tutors, and tried to lock me up with her in her apartment."[21] Three years were to pass before her eventual release, in the summer of 1636, on the return of Axel Oxenstierna from his command of the war in Germany. More determined and less manipulable than his brother senators, the chancellor removed Christina at once from her mother's suffocating embrace and returned her to her sensible aunt. The Queen Mother herself was also promptly removed, and placed under comfortable but tedious guard in the island castle of Gripsholm at Mariefred, some fifty miles from Stockholm. Like her own once imprisoned daughter, Maria Eleonora would do her best to escape.

LOVE AND LEARNING

CHRISTINA WAS ONCE MORE in the safe and steady care of the Princess Katarina. With her tenth birthday approaching, she was now taking her lessons in the company of her two cousins, Maria Euphrosyne, age eleven, and Eleonora Katarina, age nine. The three girls seem to have shared an easy friendship, though Christina did complain—to their father—that her elder cousin was falling behind in her schoolwork; it would be a good idea, she suggested, if he made her work a bit harder.

The late king had left instructions that his daughter was to receive "the education of a prince," and to take plenty of exercise, an uncommon emphasis for a girl of the time. He had, no doubt, seen that, even as a little child, she was physically very active, and perhaps he had wished to distance her from the precious femininity her mother had evinced. Christina was to be trained to only two conventionally feminine habits, modesty and virtue, though in the former, at least, she was to fail spectacularly. But her schooling with the two little countesses suggests that her academic training was not exceptional for a girl of her position, at least by the standards of other European courts. Though she may have been more capable than either of her cousins, they all read the same texts and wrote the same inkblotted exercises.

In later years, Christina's accomplishments were to be the subject of a good deal of extravagant praise, not least from her own pen. It is certain that she was a clever and inquisitive child who enjoyed learning. She welcomed this "pretext to escape the Queen my mother," and claimed that by the age of eight she was already studying twelve hours every day "with an inconceivable joy," though she does not mention that by the same age she was given to wild exaggeration. Seen against the background of prevailing local standards, her schooling was very good indeed, for until the most recent years, education in

Sweden had been deplorable. Only a few years before Christina's birth, with the country at war with Poland, there was not a single diplomat available with enough Latin to conduct negotiations with the enemy. Many local officials, it seems, "could not even write their names."[1] Older men had gone abroad for their education, if indeed they had received any, generally to Leiden, or to one of the German universities. Only the clergy had been schooled at home. A young man might study theology or biblical languages in Sweden, but none of the "modern" subjects of law, history, politics, mathematics, or science; all these Gustav Adolf had introduced as part of his great internal reforms of the 1620s. But even by 1632, at a vital period of the war in Germany, there had been no one capable of serving as secretary to any of Sweden's generals in the field—only theologians were available.

In the light of this situation, it is not surprising that Christina's contemporaries were impressed by her educational accomplishments. The regents, apprehensive of her mother's legacy, were relieved to find her a clever and studious child. The late king had prescribed for her a broad humanist education, progressive in some details, but on the whole a legacy of the great Renaissance tradition in which he himself had been brought up. Gustav Adolf's tutors had been independently minded men, and this in turn did much to shape the education that Christina herself now received. Like her father, she was inquisitive, strong-willed, and eager to learn, but unlike him she had no particular enthusiasm for the "Christian virtues" that were expected to be the basis of all her learning. By her own admission, the only parts of the Scriptures she cared for were the Book of Wisdom and "the works of Solomon"—in short, the most secular parts. She remained unmoved by the Gospels, and her lack of devotion to—indeed, lack of any interest in—the person of Jesus Christ was to remain a curious blank in her dramatic religious development. Nevertheless, for a few years during her girlhood, she was intensely pious, even to the point of bigotry. It was hardly surprising, given the narrow brand of Lutheranism prevailing in Sweden at the time, but it also reflected Christina's own very determined nature. A touch of self-righteousness, untempered by experience, led very naturally to dogmatism. Not least, for a girl who enjoyed confrontation, a staunch Lutheran conviction was in direct opposition to her tutor's own evenhanded views; Johan Matthiae's firm belief was in a future union of all Protestant creeds.

Christina's piety, whatever its cause, did not help her to endure the many long and dreary sermons of the Swedish Church. She hated them, she said, with "a deadly hatred," though one of them did inspire her, at least temporarily, with a solid Lutheran fear of the Lord. Its subject was the Last Judgment, and it was preached every year just before Advent, and hence just before her birthday. It was a reliably ferocious tirade, full of hellfire and brimstone, and, to a sensitive and imaginative child, really terrifying. Hearing it for the first time, Christina turned in frightened tears to Johan Matthiae, who comforted her with the promise that she would escape damnation and live forever in Heaven, provided she was "a good girl" and applied herself properly to her lessons. Christina took the warning seriously, and did her best to behave, but the following year, on hearing the same sermon, she found it somehow less menacing. Another year later, the menace had retreated further still, so far, in fact, that she ventured to suggest to Matthiae that it was all a lot of nonsense, and not just the threats of damnation, but all the rest of the stories, too—the Resurrection of Jesus, and everything. Matthiae was alarmed, and warned her in serious tones that thinking of that kind would certainly lead her down the road to perdition. Christina respected her "papa," and loved him, too, and she said no more on the subject. But the seed of doubt had fallen on fertile ground. By the time she was out of her girlhood, Christina believed "nothing at all of the religion in which I was brought up," and she later declared that all of Christianity was "no more than a trick played by the powerful to keep the humble people down."

Matthiae was a theologian and a Lutheran clergyman, but his views were liberal, and he made the great humanist tradition part of Christina's daily lessons along with the harangues of Roman senators and the dry texts of the Swedish Constitution. Christina was particularly attracted to neostoicism, a revival of ancient Stoic thought in a form compatible with Christianity—the inconvenient materialist beliefs of the Romans, for example, had been modified away. In neostoicism, she found a bridge between the Lutheran world that she was gradually abandoning and the classical deism that she was moving toward. The humanists had not gone so far, but Christina read into them what she needed to see, and for now, a deity unhampered by sect or priest or bible was precisely what she wanted to believe in. Besides, the earnest bravery of the neostoics was a perfect complement to the heroic classical tales that she so loved, and it encouraged her enthusiasm for the bookish, boyish virtues— *mens sana in corpore sano*—of the disciplined Roman republic. In her fifteenth year, Matthiae introduced her to the *Politica* of Justus Lipsius, a collection of pithy classical maxims well suited to her own rather apodictical nature. She was never to lose her taste for maxims; from those of ancient Greece and Rome she progressed to those of modern France, and in later life she wrote some of her own, happily contradicting herself with the courage of each changing conviction.

Christina's religious studies were conducted in German and Swedish, and also in Latin, which she had begun to study seriously. She hated it, and periodically vowed to give it up altogether, but Latin was essential, a written and spoken *lingua franca* used everywhere in Christendom; she would never have been permitted to abandon it altogether. Matthiae wanted her to speak to him only in Latin, but she would not; a brief memorandum on the subject, written shortly before her tenth birthday, reveals the state of the case. In schoolgirl Latin, and using the royal "We," she wrote:

> We hereby promise to speak Latin with Our tutor from now on. We will hold Ourselves to this obligation. We know We have promised this before, and not kept Our word. But with God's help, We will keep it this time, beginning next Monday, God willing. Written and signed by Our own hand. . . . [2]

Whether or not she kept her promise "this time" is not known, though she spoke the language well enough in later life. But if she did not like learning Latin, she did enjoy the history accessible through it, and in this she was spurred on by the Royal Librarian, Johann Freinsheim, an authority on Tacitus who taught her most of her Roman history. He seems to have taught her well, for in later years, the French ambassador noted that she seemed to have no trouble with Tacitus, "even the difficult passages, which I found hard myself."

But it was not the quality of Tacitus's writing that attracted Christina. She loved the stories of the ancient world, loved reading of the heroic exploits of Caesar and Alexander, loved the tales of nobility and virtue and the unending quest for glory. They were for her a world of adventure, where bravery and fortitude triumphed, a world of danger and daring where the strong took all and the wisest man was the unflinching stoic. To her they outshone even her own great father, and in her extravagant expectations of what she herself would accomplish, she identified herself with the ancient heroes. Not content to read about them, she took to writing about them, too: a brief ten pages on Caesar, and a more substantial essay on Alexander, which ran to seven drafts—very little changed from one to the next, however, and all but the first copied out in the weary hand of a court scribe. Alexander remained her greatest hero, and she lived surrounded by his exploits: even the walls of her own room in the castle were hung with tapestries depicting them. Matthiae taught her no Greek, but Christina later persuaded Freinsheim to help her study it, and found a new hero from ancient times in the Persian emperor Cyrus. Christina admired the faultless Cyrus, but did not seek to emulate him. Caesar and Alexander, though rougher diamonds, produced, she thought, a greater light.

Christina believed that she "surpassed the capacity of my age and sex," and learned "with a marvelous facility," but Matthiae records the steadier pace of a bright but not brilliant girl. By the time she was eleven, they had read together "the usual beginner's Latin texts,"[3] and the following year she began systematic instruction in French grammar, though she had learned to speak the language long before then. Living with her Palatine cousins had provided the occasion; they were the first family in Sweden to speak French at home—a decided affectation in the eyes of the other nobles, but it gave Christina an easy familiarity with the language, though, even in an age of unsettled orthography, her spelling was quite unusual. French was to remain her preferred language. Arriving in Stockholm just after the queen's nineteenth birthday, the French ambassador was to note that she spoke it "wonderfully well," and in later years she would use it almost always, even when writing to friends and family in Sweden. And, along with her native Swedish, Christina had spoken German from her infancy, with both her mother and her father; it seems to have been the first language she learned to write.

The modern languages, in any case, were not of great interest to Christina during her girlhood. Her heart was in the ancient world, where all her heroes had fought their battles in field and forum. The classical texts served many purposes; they included literature and philosophy and the history that she loved, but they were also an important part of the young queen's political education, tried and tested examples of *Realpolitik* from which a present-day ruler might take counsel. Christina enjoyed this aspect of her training in the "art of government." She likened the ancient political feuds to games of chess in which the shrewdest manipulator took the prize, and liked to think of herself as a master of "dissembling"—it was one of her favorite words—who could always outwit even the cleverest men about her, including Chancellor Oxenstierna himself. He was now spending several hours each day instructing her in practical politics and statecraft. These hours she relished: the chancellor was a man of vast experience who knew "the strengths and weaknesses of every state in Europe," and she listened, enchanted, to his firsthand stories of battles planned and bargains struck and enemies undermined. "I really loved hearing him speak," she writes,

> and there was no study or game or pleasure that I wouldn't leave willingly to listen to him. By the same token, he really loved teaching me, and we would spend three or four hours or even longer together, perfectly happy with each other. And if I may say so, without undue pride, the great man was more than once forced to admire the child, so talented, so eager and so quick to learn, without even really understanding what it was that he admired.[4]

It is not very likely that the chancellor's understanding failed him now as he contemplated the talents of his young pupil. Axel Oxenstierna was among the most gifted men of his generation, and he had comfortably taken the measure of the likes of Cardinal Richelieu while keeping the upper hand, five hundred miles distant, with his every opponent at home. Christina's noisy insistence on her own cleverness reveals, more than anything else, an uncertain confidence, and a very keen need to measure up to the heroic standards set by her great father. But if it did not satisfy herself, her progress was certainly good enough to please the chancellor. He reported to the Senate with satisfaction that the young queen was "not like other members of her sex. She is stout-hearted and of good understanding," he said, "to such an extent that, if she does not allow herself to be corrupted, she raises the highest hopes."[5] The chancellor did not elucidate the nature of Christina's possible corruption, but his reference to her "allowing herself" to be corrupted suggests that he had observed some weakness or unwelcome tendency in her nature. It was no outward menace that he feared for her, but rather, it seems, the consequences of her own contradictory impulses. He may have been made anxious, perhaps even saddened, by the small deceits she had begun to practice on him through her correspondence with her Palatine uncle, Johann Kasimir. The count had once served her father as comptroller of the national finances, but shortly after the king's death, he had been given a clear hint to resign from his position and betake himself to the country. The regents and senators disliked his German origins and suspected him of harboring too great ambitions for his elder son, Karl Gustav. Christina's "wise and prudent" uncle swallowed the insult and retired without demur, but he kept in touch with her, and she seems to have enjoyed the opportunities for petty subterfuge that his ambiguous position provided, seeing them, perhaps, as chances to display her mastery, such as it was, of small diplomatic subtleties.

As part of her training in statecraft, Christina had studied William Camden's Latin biography of Elizabeth I of England, the Virgin Queen with the "heart and stomach of a King" who had overseen the defeat of the Spanish Armada fifty years before.[6] The Protestant queen Elizabeth was widely known and admired in Sweden, and during Christina's girlhood, memories of her were still fresh in many minds. The English queen's wide culture, her strength of mind, and, not least, her mastery of statecraft, had framed a golden age for her small country, which, like Christina's Sweden, had only recently emerged onto the world's wide stage. It was agreed that a queen like Elizabeth would be a fine successor to the great Gustav Adolf, and her glorious reign seems to have aroused Christina's envy. In a later rant against all women rulers, she avoided mentioning the legendary English queen, but Elizabeth's shadow lingers nonetheless in a series of phrases anticipating the obvious interjection of her name: there have been no good women rulers, or if there have, none "in our present century"; women are weak "in soul and body and mind," and if there have been a few strong women, well, that's not because they were women. For Christina, the capable woman ruler was merely the exception that proved the rule. She took her model of all women from her mother, and declared that, of all human defects, to be a woman was the worst.

> As a young girl [she wrote], I had an overwhelming aversion to everything that women do and say. I couldn't bear their tight-fitting, fussy clothes. I took no care of my complexion or my figure or the rest of my appearance. I never wore a hat or a mask, and scarcely ever wore gloves. I despised everything belonging to my sex, hardly excluding modesty and propriety. I couldn't stand long dresses and I only wanted to wear short skirts. What's more, I was so hopeless at all the womanly crafts that no one could ever teach me anything about them.[7]

Christina's ungenerous attitude toward her own sex had been long fomenting. The hyperfemininity of her unloving mother cannot have helped, but her distaste for all things feminine was mostly, it seems, the result of her own very masculine nature. The late king's instructions that she should have a "princely" education consequently accorded very well with what she herself most enjoyed. Even her dolls, it seems, were the classic toys of little boys. They were "pieces of lead that I used to learn military maneuvers. They formed a little army that I set out on my table in battle formation. I had little ships all decked out for war, little forts, maps. . . ."[8] Whether they were really her own toys, or whether they were inherited from her cousins, Karl Gustav and Adolf, Christina does not say, but her enthusiasm for them was genuine enough; she loved cannon and swords and all things military. She loved being outdoors, too, and loved animals, especially horses and dogs; when Karl Gustav went off to university, he left his hunting dogs in her particular care.

"I can handle any sort of arms passably well," she wrote, "though I was barely taught to use them at all." From this it seems that she must have learned to fence, but if she did not receive much instruction, this is not surprising. Fencing was an aspect of military training, and consequently not something that any girl, even an honorary prince, would be expected to need. Perhaps Christina persuaded one of her governors, both expert swordsmen, to give her a few lessons, or perhaps her two cousins, happy to display their boyhood skills, passed on to her some of their own instruction. Christina did not keep up her fencing, though from time to time in her adult life she would appear sporting a short sword.

Hunting, by contrast, was a noble sport of long standing for both men and women, and fast and furious riding was an integral part of it. Christina loved it all. Whether or not she was "barely taught," she was a very good shot; the French ambassador remarked that she could "hit a running hare faster than any man," though as she herself insisted, "I wasn't cruel and I have never killed an animal without feeling real sympathy for it."[9] She was a very fine horsewoman, too, though she used a lady's sidesaddle, and was probably taught by her governor Axel Banér, himself a superbly skilled rider. Christina admitted that she had been taught to ride "a bit," but in fact she received a good deal of instruction, and she spent many exhilarating hours on horseback in the royal hunting grounds of Djurgården, across the lake from her castle home. In short, she was perfectly suited to the vigorous princely upbringing her father had commanded for her. In the young girl racing on horseback through the forest, the Swedes saw their great king's own active spirit embodied once again:

> Between what I was taught and what I wanted to learn myself, I was
> able to learn everything that a prince should know, and everything

a girl can learn in all modesty. . . . I loved my books with a passion, but I loved hunting and horse-racing and games just as much. . . . The people who had to look after me were at their wits' end, because I absolutely wore them out, and when my women wanted to slow me down, I just made fun of them. Every hour of my days was occupied with affairs of state, or study, or exercise.[10]

It is a rather boastful account, and a touch defiant, but Christina's description of her girlhood self is more or less accurate. She was clever, and generally hardy, though given to sudden illnesses, most apparently emotional in origin, and she did spend her days more or less charging at the world, infuriating and exhausting those about her.

The Princess Katarina died a few days after Christina's twelfth birthday, in December of 1638. Christina had been fond of her kindly aunt, and she missed the company of her cousins, most of whom now returned to their own castle at Stegeborg. Only the youngest girl remained with her in Stockholm, and she stayed for four years, a companion "suitable for my age" in schooling and at play. Both Christina's governors died within these years, and they were not replaced. Johan Matthiae was to remain until Christina was sixteen, and from then on, for all but her political education, she was to be left to her own devices.

WHILE CHRISTINA HAD been poring over Caesar and Alexander, a latter-day hero had been making his way, in less martial mode, through other lands. In 1641, Karl Gustav returned to Stockholm, age just nineteen, with the happy weight of student life and foreign travel on his broad young shoulders. If his portrait is to be believed, he had grown into an exceedingly handsome young man, with dark eyes and dark hair, and fine but manly features. He was well liked among his peers and well regarded by those above him, liberal but not extravagant, courageous, and very capable, a young man full of promise, but with no settled future as yet before him.

It had been more than three years since Christina had last seen her cousin. She was now fifteen, and she found at once that her former easy, boyish talk of fencing and hunting no longer felt appropriate when she was with him. Awkward chatter soon gave way to whispers and sighs and secret glances, as the friend of her childhood metamorphosed into her first love. It became a conspiracy. With chaperones in the way, the two resorted to impassioned notes, delivered by an excited Maria Euphrosyne, cousin and sister to the lovers, or a surprising alternative go-between, Christina's learned old tutor, Johan Matthiae. There need not have been much intriguing. For a girl of her rank, Christina was now of marriageable age, and the match would have been welcomed by Karl Gustav's family—it had in fact been a long-held wish of his mother, Christina's Aunt Katarina. Chancellor Oxenstierna would have been less pleased. He disliked the Palatine family and suspected them of manipulating Christina's affections for their own advancement. But as head of the regency council, he could in any case have forbidden any marriage until Christina had formally attained her majority at the age of eighteen. This was almost three years away; by then the youthful romance would surely have run its course.

The chancellor had miscalculated the strength of Karl Gustav's affections, but where Christina herself was concerned, he need not really have worried. She seems to have enjoyed the subterfuge as much as the romance itself. She wanted to write in code, and though she often enough swore "eternal love" and "faithfulness unto death," she spent as many lines trying to keep the young man calm, and urging him to think of his professional future. "I will wait for you," she wrote, "but for now you need to think about the army. All good things come to those who wait. We can marry once I have become Queen in fact as well as in name"—an event still several years distant. The eager young lover could be packed off to the wars, and the game of love continue to be played without danger of any real involvement.

Sidelined in love, Karl Gustav became the focus of Christina's first attempt at practical politics. An important position had fallen vacant at the court. The chancellor's brother had recently died, and the Senate was debating who might succeed him as High Treasurer and member of the regency council. Christina had proved a keen and able student of politics, and it was thought that, as she was now age fifteen, she might add her voice to those of the senators—her father, after all, had begun to attend *riksdag* sessions at the age of only twelve. The senators suggested she might like to nominate her cousin, Karl Gustav, for the newly vacant position. It was welcome news to the young man himself; he had no other employment, and his family had no wealth beyond what they could earn through the grace of the court. Johann Kasimir was delighted. He had himself once been a member of Sweden's highest council. Now, despite his German blood, his son would take his own place there. They could count on Christina, he knew—but he had reckoned without her paradoxical support.

Excited by this first foray into real politics, she devised a small subterfuge, apparently to persuade the Senate that she was not especially predisposed toward her cousin and his family. In fact, Chancellor Oxenstierna seems to have been her real target. Though she hung on his every word and "never tired of listening to him," she had begun to resent the great man's power; he was her regent, after all, and not the king. The chancellor disliked her uncle, Johann Kasimir, regarding him as an untrustworthy foreigner who had come to the country with nothing and who intended to take from it whatever he could. He disliked the fact that Karl Gustav had his own claim to the Swedish throne—like Christina, he was a grandchild of Karl IX—and he disliked the evident fondness that existed between the young queen and her Palatine family. Christina's own growing jealousy of the chancellor was reason enough for her to strike against him, but the vacant position of High Treasurer provided the opportunity that until then had been wanting. She would win the chancellor's confidence by pretending to stand on his side against the Palatines, and in future would use this trust to further her cousin's interests, and her own—the senators might even appoint Karl Gustav anyway. She set her mind to scoring this first political point, and in so doing managed to harm the very person whom she most wished to help.

To the senators' invitation to nominate Karl Gustav, Christina replied that she could not do so, for the improbable reason that his own father would not approve of the appointment. More sensibly, she added that it was not suitable for her to choose one of her own regents; the matter, she wrote, should be referred to the chancellor. Her cousin was astonished, her uncle dismayed. How could she have declined so valuable a position on their behalf? More than once she was obliged to point out herself how clever she had been. "If I had nominated Karl Gustav," she wrote to her uncle, "the other regents would have thought I was only wanting to plant a spy among them."[11] It seems she had not stopped to consider how useful such a spy might have been to her, almost as useful indeed as the powerful and well-paid post itself would have been to her impoverished cousin. Instead, the noise of her self-congratulation quickly drowned out the sound of his own puzzled disappointment. Years later she would describe the episode as evidence of her capacity for "profound dissimulation, which even in my early youth deceived the most astute people."[12]

The "most astute people" were, of course, the chancellor, but it is not very likely that her ruse persuaded him of any sudden lack of fondness on her part for the Palatine family. Christina's "dissimulation" was of no benefit to Karl Gustav, and indeed cost him a great deal. It cost the chancellor nothing, and left Christina herself, in the eyes of her nearest relatives, with an aura of immaturity, or untrustworthiness.

It is a measure of Christina's naïveté at this stage that she believed she had somehow outwitted the chancellor. It is revealing, too, of her great confidence in her own powers that she regarded the little ploy as an exercise in "profound dissimulation," a capacity to which she would always lay extravagant claim. But above all it is significant that Christina's first attempt at political influence was an attempt to deceive. Just fifteen years old, in a position of extraordinary privilege, with a hundred hardened graybeards awaiting her response, she might have revealed a precocious wisdom or even simple humility. She might have made a bold stand to assist the family to whom she owed so much. Instead, she responded deviously, leaving Karl Gustav to bear the risk.

Christina's ploy did not help her cousin, but, quite by chance, it may have helped her country. The new High Treasurer, chosen by lottery, was the senator Count Per Brahe, a cousin of her father's former love, and a man of immense experience and talent in military and civil affairs. Per Brahe was no doubt better suited to the position than any nineteen-year-old, no matter how handsome, could ever have proved to be. The adverse effects of Christina's clumsy subterfuge had been prevented, quite literally, by the luck of the draw.

Johan Matthiae's reports on Christina's education ended in her seventeenth year, when he left Stockholm for Strängnäs, some fifty miles away. Here, despite his lukewarm Lutheranism, he had been given a bishopric. Christina had been a good pupil, talented and studious, but Matthiae's efforts to educate her "as a Christian prince" in the way of Erasmus must be said, on the whole, to have failed. In her adult life there would be little trace of the humanist virtues her tutor had so exalted. Christina was not without admiration for them, and apt quotations were never to be far from her fluent tongue. But, although in her earnest girlhood she embraced some of their values, it was not in her nature to pursue them beyond these years. She would be seldom stoical, often unprincipled, and generally, at least where her personal affairs were concerned, rational only ex post facto. On the rock of her own ebullient temperament, the fine-wrought vessel of her education was doomed to break apart, nature triumphant over nurture.

—

DURING THESE YEARS of her girlhood, Christina saw her mother hardly at all. Confined at Gripsholm Castle, Maria Eleonora made only one brief appearance in Stockholm, and it seems that the visit was never reciprocated. Christina approved of her mother's exclusion from the regency, regarding it as "a most sensitive mark of my father's love" to have insisted upon it. If her mother had had a hand in ruling the country, she wrote, "she would no doubt have ruined everything, like all the other women who have tried it. But," she added, "though I praise the regents for keeping her away from the business of governing, I must admit it was rather harsh of them to separate her from me completely."[13]

It is hard to say whether Maria Eleonora really missed her daughter; her maternal interest had been erratic, after all. It is certain, in any case, that she was miserably unhappy at Gripsholm Castle. Perched on an island in the sparkling lake, to the Queen Mother's mind it was the bleakest fortress imaginable. For four bored and angry years, she had stewed inside its red-brick walls, her coterie of German ladies-in-waiting simmering about her. Unmoved by the loveliness of her surroundings, or by her daughter's occasional pleas for calm, she had taken consolation in a secret correspondence with King Kristian of Denmark, himself no friend to Sweden's governors. In this she gave full vent to her resentment of the chancellor and his men. Adding insult to injury, Oxenstierna had dismissed her to Gripsholm with the suggestion that she "learn to grow old gracefully." Gradually, with cunning and charm, she laid her plans for a vengeful escape.

She was now age forty, and still, it seems, despite the chancellor's injunction, in full possession of all her womanly assets. Only a few years before, her widow's weeds notwithstanding, she had been described by two French visitors as "the most beautiful, radiant woman we had ever seen. We were," writes one, "quite dazzled by her beauty."[14] The Frenchmen, apparently, were not the only ones to admire Maria Eleonora's "charming features" and her "truly royal figure." Her official captor was now captivated in his turn. Marshal Nilsson, whose army days had no doubt accustomed him to less insinuating prisoners, had been readily acceding to her majesty's wishes: she had such a passion for Homer, it seemed, that she wished to spend her days on the shores of the island, reading the majestic lines of the *Iliad*, listening to the majestic sound of the waves. Maria Eleonora must indeed have been a woman of many charms; after four years of confinement, during which she had evinced no interest whatsoever in classical literature, her improbable ploy worked perfectly. She was soon aboard a Danish sailing ship en route to Helsingør.

In France, delighted tongues whispered that Maria Eleonora and the Danish king were lovers; to join him, she had braved the seas, defying the wrath of mighty Sweden. Kristian does not seem to have appreciated the irony of the rumors; his wife, after bearing him twelve children, had braved his own wrath for the embraces of a German count. The cuckolded king was now age sixty-three; his gallantry toward the lady had been prompted more by politics than by love. He duly received a protest from the outraged Swedes, and sent them a cool apology, but he soon turned his energies to ridding himself of his turbulent guest. Her brother, the ailing Elector Georg Wilhelm, flatly refused to permit her return to Brandenburg, and by Christmas, an exasperated Kristian was applying to the new elector, Friedrich Wilhelm, to take her off his hands.

The young elector was not pleased. Brandenburg had recently been at war in the emperor's service, and Friedrich Wilhelm was now suing for peace with the hard-pressing Swedes. He had no wish to embrace a major diplomatic embarrassment in the person of his volatile aunt. The refugee herself was apparently happy enough to go; indeed, she had little choice, since the Swedes had rescinded all her rights to income and had confiscated the many personal belongings she had left behind her. But her nephew kept her waiting for almost two years, while the Swedes were gradually persuaded to restore her income, and the Danish king descended into desperation. Once in Brandenburg, Maria Eleonora was to remain four years, returning to Stockholm at last to find her daughter fully grown, and a reigning monarch.

acorn beneath an oak

CHRISTINA'S KINGDOM WAS now her own. On a cold November day in 1644, shortly before her eighteenth birthday, she summoned her "five great old men" to give a formal account of the twelve years of their regency. They spoke of the past, and also of the future. From now on, Christina would be queen in fact as well as in name. Once the war was concluded, there would be a spectacular ceremony of coronation to confirm the beginning of a glorious new reign, and in the meantime she was to take into her own hands the governance of the realm.

It was a curious young woman who stood before the regents now. She was rather small, not quite five feet tall, and her habit of wearing flat shoes made her seem even smaller to her high-heeled contemporaries. Her delicate upper body was marred by a pronounced unevenness of the shoulders, the result of her fall in infancy, but her arms were round and womanly, tapering to fine, small hands. Her face was finely made and oval-shaped, framed by straight fair hair, and her forehead was high. Her long, hooked nose led to a small mouth, from which most of the back teeth, it seems, were already missing, narrowing the delicate jaw and emphasizing the small, pointed chin. All accounts agree that her large, blue, close-set eyes were beautiful, lit with intelligence and humor; they revealed pride, too, and often anger, and at times a kind of penetrating stare that seems to have alarmed every recipient into quick submission, but their expression does not seem to have ever been cold. Despite her small stature and fairly delicate build, the young queen's movements and gestures were far from feminine. She walked like a man, sat like a man, and could eat and swear like the roughest soldier. Her voice was deep and gruff, and her temper warm—her servants were no strangers to blows or bruises. She was clever and well read, but she liked best to talk of manly things, and whenever she spoke of military action, she adopted a sort of martial pose, planting one foot in front of the other. Her many unusual traits notwithstanding, she formed an impressive figure, and she left her old counselors broadly reassured for the future of their country.

Not the least anxious observer of the young queen's development had been the remarkable Baron Axel Oxenstierna, whose own premier position of many years' standing was about to encounter its first challenge. Since his first appointment as chancellor in 1612, at the age of only twenty-nine, he had served Sweden with great distinction in every field from military logistics to city planning. A lawyer by training, an outstanding administrator and diplomat, he was also an able politician, and for more than thirty years he had steered a well judged course between Sweden's longstanding adversaries of crown and nobility. His years as chancellor to Christina's father had been a turning point in the life of his country; the two had worked together to transform their homeland from a backward outpost on the cold periphery of Europe to a major power on the continent's center stage. Oxenstierna's considered temperament had provided a perfect complement to the exuberant genius of Gustav Adolf, epitomized in a famous exchange between them: "If we were all as cold as you are," the king had once exclaimed, "we should freeze." "If we were all as hot as Your Majesty is," replied the chancellor, "we should burn." Over decades of service, Oxenstierna had revealed not only his abilities and his strength of mind but also his profound patriotism, a golden thread running through the many antagonisms of his public life, in Sweden and abroad.

His achievement had been phenomenal. By the end of the regency, in 1644, there was no stone of state that he had left unturned, and his rare combination of vision and pragmatism had earned him admiration and respect and, in the areas of Swedish military action, no small fear. To the senators and the men of the *riksdag*, his remarkable partnership with the late king remained a vibrant memory, and in the years after Gustav Adolf's death, Oxenstierna's own powerful aura had only shone the more brightly.

Now, in the small firmament of the Swedish court, there was no longer room for two stars of equal brilliance. In the eagerness and arrogance of her eighteen years, Christina felt it was her turn to shine. She was intimidated by the chancellor's achievements, and mistrustful of his reforms, seeing in them a threat to her own power. Despite his long years of service and his championing of the Vasa dynasty, she convinced herself that he was taking advantage of her inexperience to weaken the crown and advance his own authority instead. Her tutelage, she decided, was at an end. During the years of his guardianship, she had listened to him attentively, but now she would speak and he would listen. She did not seek the fruitful equipoise of monarch and chancellor that had served her father so well. In her mind, this was only history, after all; for the twelve years of her girlhood the chancellor had ruled alone, seconded and supported by his ubiquitous family. But the right to rule was not his at all; he had used it while he could, but he would not usurp it now that she was of age. It was her own right, and she would exercise it.

The chancellor thus appeared less a complement than a foil to Christina's own designs, and his prominent position merely a conspicuous target for her keen and jealous eye. Her concern became to oppose him, and from a willful principle it grew into a habit. His great abilities, his vast experience, and, not least, his own majestic presence, so often remarked upon by contemporaries, all struck deeply at the defensive heart of an uncertain girl not even five feet tall. She responded by perversely attacking the great oak that might have sheltered her own tender growth, developing at the same time an attitude of terrific outward pride, insistent to the point of comedy and even pathos.

Though the chancellor had now formally ceded his place as first power in the land, his position remained immensely strong. He stood supported by his own men, with wealth and patronage at his disposal, and about him a wall of skill and influence three decades thick. He was not without enemies, old rivals for office and riches, and those envious of his family's great standing, but they were not as yet a solid flank to be used in opposition to him, and Christina in any case lacked the experience to manipulate them to that end. She began instead on her own, cautiously, and her plan of attack was simple: the mighty old oak was, above all, a northern oak; it flourished best under its own wintry skies, mistrusting the dazzling sun and the rich soil of the south — most particularly, the soft, sticky soil of France. This soil, in gleeful handfuls, Christina now determined to spread.

In 1635, under the chancellor's leadership, the Swedes had entered into a cautious alliance with France against the Habsburg Empire. It had not been a happy partnership. Both sides were wary of each other, the chancellor disdainful of the French with their devious and frivolous ways, and Richelieu raising his eyebrows at the majestic Swede — "very astute," he thought, "but a bit Gothic." The replacement of Richelieu by his protégé, Cardinal Mazarin, had not improved relations between the two countries. For almost a decade their awkward alliance had remained in place, with the French offering but not always paying subsidies for Sweden's armies, expecting in return a biddable northern ally, and the Swedes accepting the offers, and the money when it was forthcoming, but continuing to make their own decisions, watching their backs all the while. The chancellor's personal experience negotiating in Paris had confirmed his prejudices, and he had not modified them in the ensuing years. The French were unreliable, he believed, and too concerned with fashion, and they ate too much, and none of their fancy food could bear comparison, anyway, with a good stew of sun-dried salmon with plenty of pepper. Though he knew French well, in recent years he had not been heard to speak that capricious tongue; with more courtesy than candor, he insisted that he could not favor any one country over another.

No such scruples had restrained him from unleashing new conflict with a nearer neighbor. At the end of 1643, in supposed outrage at the Danes' involvement in Maria Eleonora's flight from Gripsholm, Swedish forces had invaded and quickly overrun vital coastal areas of Denmark. The Queen Mother's escape had proved a useful pretext for attacking a hostile power whose control of the Baltic trade routes was altogether too strong for Sweden's liking. By the spring, the Swedes had secured access to the routes for themselves, taking an eye in the process from the bold but ageing Danish king. An ancient balance had once again been tipped, this time in Sweden's favor.

The Danish war was the chancellor's war. For him, Sweden's deadliest enemy would always be the Danes, once ferocious overlords, still dangerous neighbors, inevitably competing for domination of the great thoroughfare of the Baltic Sea. The Habsburg Empire, by comparison, was a distant threat, drawing precious men and money away from the northern lands. The French, naturally enough, took the opposite view. For them, the Danish conflict was a peripheral matter, requiring a swift conclusion so that Sweden's men could return to the field against the emperor. To this end, Cardinal Mazarin had dispatched a peacemaker in the guise of a new ambassador to Sweden, a Monsieur de la Thuillerie, who quickly brought the eighteen-year-old queen around to the French way of thinking.

For Christina, it was a golden opportunity to take a stand against the chancellor. The Danes were suing for peace, but Oxenstierna hoped to continue the war until they had acceded to Sweden's territorial demands for the southern peninsula; it was still in Danish hands, preventing Swedish access to the crucial sound. Christina allowed the Frenchman to persuade her that if the Danish terms were not accepted at once, she would be "blamed by posterity" for her "unbounded ambition." To this effect she wrote several times to the chancellor, defensively couching her argument as the wish of the Senate— evidently she had not yet the courage of France's convictions. "Most of them feel quite differently than you and I do," she wrote. "Some of them would give their hands to end the war."

In the late summer of 1645, a treaty was finally signed between the two old enemies.[1] Though advantageous to Sweden, it did not cede all that the chancellor had wanted. To add insult to injury, Christina suggested that a double celebration be held to mark not only the signing of the treaty but also a recent victory of the French army over imperial forces. As the French had just been discovered in secret negotiation with Sweden's Bavarian enemies, the idea progressed no further. But Cardinal Mazarin was satisfied. The young Swedish queen, whether she realized it or not, had begun her steady transformation into France's creature.

None of it was lost on the chancellor. His regard for Christina was now being severely tested, and exchanges between them became markedly cool. Despite her formidable adversary, Christina did not retreat, but as the stubborn days wore into tired months, the strain of her opposition to Oxenstierna began to undermine her health. Within a year of the regency's end, she had fallen seriously ill and was, or so she believed, in danger of her life. She attributed her illness to "the great exhaustion" of managing the affairs of state, though in fact she had assumed little responsibility beyond continuing to attend the sessions of the Senate. The chancellor was still very able and very willing to continue at the helm, had Christina been content for him to do so. Her recuperation once begun, she relapsed into illness again, and then succumbed to a serious case of the measles, but it was emotional distress, then as later, that seems to have caused the greater part of her illness. "I loved him like my own father," she said of Axel Oxenstierna, but like her father, too, the gifted chancellor cast a long shadow over Christina's sense of her own greatness. Inexperienced as she was, delighting in any intrigue, attracted by the sophisticated ways of a foreign people whom Oxenstierna disliked and mistrusted, she burrowed ever more deeply into a self-deluding syllogism, harmful to herself as to her country: the chancellor opposed the French; Christina must oppose the chancellor; therefore Christina must support the French.

It was a simplistic hostility, but it did not relent, and it left her exposed to easy manipulation by the less scrupulous figures about her. Soon after her recovery, she allowed it to govern a second clumsy foray into the country's foreign affairs, at the same time revealing her susceptibility to a particular type of artful and persuasive opportunist who was to feature prominently in her public and private life.

The first adventurer appeared to take his advantage just as the regency was ending, a Monsieur Duncan de Cérisantes, brawler and seducer extraordinaire, former gentleman of Constantinople, future Catholic aristocrat, current Huguenot diplomat-conveniently-at-large. In earlier incarnations he had been known by the prosaic appellation of Mark Duncan, but Christina accepted him at his own aggrandized word, and before long she had dispatched him to Paris, to "assist" Sweden's permanent minister there, the celebrated jurist, Hugo Grotius. Grotius had occupied this post since his appointment by Axel Oxenstierna almost a decade before, and had overseen a long period of cautious alliance between the two states. Needless to say, he did not appreciate the encroachment, and was soon penning outraged letters, complaining that he was being spied on. If so, no good report of him was making its way back to Stockholm. The French disliked Grotius as heartily as he disliked them. A staunch Protestant Dutchman, Grotius could not conceal his disdain for the frippery and popery of Mazarin's court, and he refused to extend the usual diplomatic courtesies to France's "Prince of the Church," claiming that the rank of Cardinal was unrecognized by those who were not Catholic. His dour comportment became quickly comical in the company of his wife, whose advancing years had enveloped her sturdy frame with an excessive *rondeur*. In her youth a heroine of political resistance, Madame Grotius had since spent her time in quiet retirement, so that one refined newcomer to the court was obliged to ask her identity. "Who is that bear?" he asked of the young lady standing beside him. Unhappily, his unknown companion was Mademoiselle Grotius. "It is my mother, sir," she replied.

Inelegance was as good an excuse as any. At the end of December, only weeks after the regency had ended, Christina recalled the minister, awarded him his pension, and shortly afterwards appointed Cérisantes chargé d'affaires in his stead. Grotius was among the most learned men of his day—theologian, historian, the "father of international law," and one of Gustav Adolf's own heroes. His replacement by the conniving Cérisantes was a fall from the sublime to the ridiculous, which left Cardinal Mazarin and his government puzzled and amused. As might have been expected, Cérisantes rendered the Swedes no service; eventually he actually deserted his post. Christina rewarded this by offering him a position in the Swedish army, but, being then on the way to Rome, he declined, and was soon collecting a handsome sum for his noisy public conversion to Catholicism.

Cérisantes had duped Christina, and he provided an archetype for later artful characters who would dupe her in their turn. Always men, always plausibly capable, always of doubtful origin, they were to form an infamous gallery of lovable and not so lovable rogues in her life. She would repeatedly be defrauded by them, repeatedly forgive them, repeatedly refuse to hear a word spoken against them. Their crimes would run the gamut from petty theft to abduction and murder — she would tolerate, indeed defend, it all.

It is hard to see how Christina could have been so readily ensnared by Cérisantes and his ilk. They were none of them subtle characters, and few other people were taken in by them for long. At the start, perhaps, Christina enjoyed the subterfuge, sharing the thrill of deceiving, or supposedly deceiving, her sturdy, straightforward compatriots. Perhaps, too, she recognized in each opportunist the genuine dissembler that she believed herself to be. Christina was very proud, and would not have found it easy to admit to an error of judgment, but her intelligence was considerable, too, and it should not have been easy to deceive her. It would have been hardest of all for her, perhaps, to accept that she herself was not party to the joke but rather the butt of it, that the deceiver's ground had been whisked out from under her, and that she, too, could find herself, bereft and foolish, among those deceived.

CÉRISANTES'S PLACE AS Christina's representative in Paris was taken by a nobler but otherwise no more likely contender, Count Magnus Gabriel De la Gardie. Scion of a prominent Franco-Swedish family, he was in fact a cousin of sorts to the queen — his great-uncle was her own uncle, Karl Karlsson Gyllenhjelm, illegitimate half-brother to Gustav Adolf. Magnus's father was the Grand Marshal General Jakob De la Gardie, who had served as military instructor to the boy Gustav Adolf, and his mother was none other than Ebba Brahe, the beauty who had once captured the young king's heart; Magnus, her "dear and noble son," was the eldest of her fourteen children. In 1645, just twenty-two years of age, he returned to Stockholm after almost ten years of study and travel in Sweden and abroad, including a lengthy and expensive soujourn in Paris. He had rounded it all off with a tour of duty in the Danish war, adding a soldier's dash to his courtly accomplishments.

Christina was delighted with him. He was tall and muscular, handsome, charming, extravagant, the son of her father's old favorite, and, above all, very fluent in the elegant ways of France—in short, perfectly calculated to annoy the chancellor. They became intimate friends, and she soon made him Colonel of her Guard. It was a swift advance for so young and inexperienced a man, and few doubted that Christina had fallen in love with him—some even whispered that they were lovers. It is not likely to have been true, not least because Magnus was himself in love with Christina's schoolmate and favorite cousin, Maria Euphrosyne. He soon made a proposal of marriage to her; she soon accepted.

Christina responded by separating them. In the spring of 1646, she announced that Count Magnus had been appointed Ambassador Extraordinary to France; "extraordinary" thanks were owing to the French, she felt, for their involvement in the Danish treaty. There was in fact no political need for any such appointment to be made, and the chancellor opposed it strongly, adding to Christina's determination with his every objection. Magnus was to go, and he was to go in splendor such as no Swedish envoy had ever before enjoyed, splendor that was to impress even the extravagant French. A carriage of gold and silver was prepared for him; some three hundred persons were to form his personal retinue; his allowance would be enormous. For three months she delayed his departure with fond excuses, so that those about her, "not wishing to cast aspersions on her majesty's conduct," assumed that, despite his engagement to her cousin, Magnus would soon be married to the queen. The infuriated chancellor could only look on, kept company by a sad Karl Gustav, whose promising romance had evaporated into the perfumed air surrounding his rival. Toward the end of July, Magnus finally set out for Paris. Christina took to her room and wept.

She might have wept more bitterly if she had learned what Magnus had to say of her once he arrived at Mazarin's court. At first he spoke of her "in passionate terms," and "so respectfully" that the French, too, suspected that his feeling exceeded that of a normally dutiful subject for his queen. But the matter was soon made clear: Christina was an extraordinary monarch, wonderfully learned, but not very feminine—in fact, not like a woman at all, not in her appearance, not in her behavior, not even in her face—a surprisingly ungallant remark from so suave a tongue. Magnus made full use nonetheless of her continuing indulgence of him, exceeding his huge allowance three times over, referring his debts to the queen without her leave, and perversely raising

Sweden's reputation as a land of some financial resource, while her soldiers remained unpaid in their garrisons and camps. Little wonder that Christina's former man in Paris, the incorrigible Cérisantes, thought it worth his while to protest that he himself had not been reappointed.

The French appointment served a multiple purpose. It gave Christina time to recover from her love for Magnus. Alternatively, it gave Magnus time to recover from his love for Maria Euphrosyne, and to reconsider what the love of a queen might bring in its train; the costly embassy in Paris was an obvious indication. Loyalty to Maria Euphrosyne may have even played a part. In any case, the appointment indicated that it was Christina's voice, and not Axel Oxenstierna's, that was now to be decisive. The link between Sweden and France would, at least formally, be strengthened, though in fact Magnus's inexperience only weakened Sweden's standing in the eyes of the French.

Magnus remained in Paris just seven months, capably discussing French poetry with the court *précieuses*, while political matters passed beyond his ken. In Stockholm, Christina exchanged daily visits with his mother, and together they sang the praises of their absent idol. Magnus's fiancée herself does not seem to have been included in these laudatory afternoons, but she was there readily enough when he returned, "preceded by the sound of his expenses," to celebrate an unrepentantly lavish wedding. Christina managed to upstage bride, bridegroom, and priest: placing the couple's hands together, she declared to Magnus, "I hereby give you the most precious thing I have." Precious things continued to flow in the same direction, so that within a year, while Christina's treasury limped along, Magnus, at the happy age of twenty-four, was believed to be the richest man in Sweden.

MAGNUS WAS MARRIED, and Karl Gustav rejected, but Christina's affections were not long idle. This time they took a different turn, one that kept the gossips as busy as they had ever been with Magnus or Duncan de Cérisantes. The queen's attention was now fixed on one of her own ladies-in-waiting, a quiet young beauty who had been left in her care on the death of her courtier father some years before. Her name was Ebba Sparre, but in compliment to her loveliness, Christina called her Belle.

Apart from their age, the two had little in common. Belle was timid, feminine, and sedentary, with no particular interest in learning or high culture, but she accepted Christina's attentions, and seems to have returned her affection.

They commonly shared a bed, no unusual matter at the time for two young unmarried women, but Christina enjoyed the provocative possibilities of the situation. She drew deliberate attention to it before the prudish English ambassador, Bulstrode Whitelocke, whispering into his reddening ears that Belle's "inside" was "as beautiful as her outside." Her insinuations quickly ossified into supposed fact, and before long it was widely believed that the queen was a lesbian, or possibly, in mitigating afterthought, a hermaphrodite. Her reluctance to marry added weight to the charge—had not the Count Palatine been trailing on his leash, unfed, for years behind her?—and there was plenty of circumstantial evidence to be brought to bear. Her mannish way of walking, her love of hunting, her gruff voice, her flat shoes—to a roomful of courtiers eager for scandal and impatient for an heir, all betokened clear sexual aberration.

Christina did nothing to quench the little flames, declaring in round terms her aversion to the idea of sex with a man. "I could not bear to be used by a man the way a peasant uses his fields," she said. At the same time, it was clear that neither modesty nor timidity had prompted her attitude. Her coarse language, though she herself regarded it as a natural Swedish defect, was the cause of frequent comment. She was fond of bawdy jokes, too, and was not above teasing the maidenly Belle. She led her one day to the chamber of Claude Saumaise, a Frenchman and a favorite of the queen who had absented himself from some scholarly rendezvous on the pretext of illness. They found him sitting up in bed with a risqué book in his hand. Recognizing its title, Christina disingenuously asked Belle to read a passage aloud from it. Belle began confidently, but was soon blushing and stammering, to a loud roar of laughter from the queen, and a quiet smirk from Saumaise.

It was never clear, nor can it now be determined, precisely what the relationship was between Christina and Belle. Though there was plenty of talk about the queen, no one suggested that Belle herself was lesbian. It is possible that Christina abused her position to force Belle into an unwanted sexual intimacy, but it is not likely; had she done so, Belle would probably not have remained the loyal friend that she appears to have remained for the rest of her life. Christina's lesbian tendencies seem to have been genuine; Belle was to be followed by a number of beautiful young women who captured the queen's admiration and, perhaps, her heart. In later life, Christina would often choose paintings and sculptures of naked women to decorate her private apartments. But her interest in women did not preclude an equal interest in men. Karl

Gustav may have been just a youthful romance for her, but she was certainly infatuated by Magnus, and perhaps by Duncan de Cérisantes, and one great love for a man was eventually to come her way. But, surprisingly, given her active and self-indulgent nature, Christina does not seem to have followed any of her passions to their natural conclusion. Where men were concerned, she understood her own reticence to being "used"; where women were concerned, she bought a lot of pictures, and gave a lot of presents, and wrote a lot of flowery letters, but physical love itself she seems never to have sought.

Christina was clearly fond of Belle, and may even have loved her, but she did not refrain from making use of this most innocent friend in her ongoing battle with Chancellor Oxenstierna. For some time Belle had been engaged to his son, Bengt, but Christina persuaded her to break off the engagement, and to marry instead Jakob Casimir De la Gardie, Magnus's younger brother. A story made the rounds that, during the wedding celebrations, the queen ordered all the guests to take off their clothes and dance—at least—in the nude. The story is mere gossip, but that it could even be suggested reveals something of the reputation that Christina had by now acquired.

Belle's own epitaph was not happy. There was no real affection between Jakob and herself, and even after the wedding, she continued to live with the queen. She had three children, but all died in infancy, and within a very few years she became a widow. Thereafter, despite Christina's continuing affection for her, Belle's young life declined into illness and sadness.

Talk of Christina's lesbian tendencies, meanwhile, did not recede. That it was grounded in at least partial truth was recognized, if reluctantly, by some of those closest to her. Her two uncles, Count Johann Kasimir and the Grand Admiral Karl Karlsson Gyllenhjelm, had long hoped that she would marry Karl Gustav. But by the time Christina was twenty, Gyllenhjelm at least had acknowledged that the marriage was unlikely. He urged Christina instead to seize her chance to choose an heir if she would not choose a husband. "If Your Majesty does not marry," he wrote, "you must act in good time to secure the succession for *a certain family*." His reference was to the queen's Palatine cousins: the bridegroom *manqué*, Gyllenhjelm hoped, might yet wear a Swedish crown. In either place, he would be a powerful counterweight to the great noble families, and in particular the Oxenstiernas, who might otherwise mold the monarchy to their own liking, or even dispense with it altogether. Moreover, it was they who had ousted Karl Gustav's German father from his position on the Council. The father's revenge would be rich indeed if the son after all should ascend the Swedish throne, not as the queen's consort but as king in his own right. Christina did not disagree. She was very willing to assume her uncle's attitude, which put a rational face on her own antagonism toward the chancellor, and she wrote to her uncle that there were some, she believed, who would be only too happy to feed Karl Gustav "a dose of Italian soup" to get rid of him once and for all. She made no formal statement about the marriage, but allowed it to be generally understood, by all but the would-be bridegroom himself, that in due course it would take place.

In due course the anxious chancellor challenged her on the subject. The talk had been going on for long enough, he declared. Was there really any substance to it? The queen's marriage was a matter of the greatest importance to the state. The Senate should have a say in it. They should at least be kept properly informed, and not have to wait to hear the latest story from the fish-wives and gossipmongers about the town.

The queen began with a denial, or rather with a confirmation. It was true that she had intended to marry Karl Gustav, but she had changed her mind. She was not going to marry him. She had in fact no wish to marry at all. However, she did intend to make him commander-in-chief in Germany. The chancellor called her bluff. The count was German himself, he objected, or at least his father was, which amounted to the same thing. Command of the Swedish armies could not be entrusted to a foreign hand. The only way his loyalty could be assured was for the queen to marry him. Christina stumbled: she was not going to marry the count, she declared, indeed she was not going to marry anyone. However, if she did marry anyone, it would be the count. In fact, yes, since the chancellor was asking, yes, she was going to marry him. In fact, yes, they were already engaged.

The news was soon out, leaving no one more surprised than the fiancé himself. He had time to take a few elated steps before being interrupted by a private communication from the queen, informing him that the supposed engagement was no more than a ruse to increase his own public standing. If he were generally believed to be her future husband, his appointment as commander-in-chief would be the more readily approved.

He quickly sought a clarifying interview with her, to which she slowly agreed. It took six months to bring it about, and it was not, in the end, the private discussion that Karl Gustav had requested. Instead, Christina insisted that Magnus De la Gardie and Johan Matthiae, her former tutor, should be present throughout. With two other men in the room, it seems, the count was less likely to become passionate or desperate. Here, as on the battlefield, there was a precarious safety in numbers.

She managed one decisive statement. She was not going to be bound by promises she had made as a young girl. At the same time she didn't want to take away the count's last hope, but she was not going to marry him unless reasons of state made it absolutely necessary. If she didn't marry him, she would see that he became her successor, though if she couldn't persuade the Estates to agree to this, she would marry him after all. In any case she would give him a final answer within the next five years.

Karl Gustav's response was robust. He protested his love for the queen, and declared that the succession proposal was of no interest to him. He would accept no consolation prizes. If she would not marry him, he would leave Sweden and never return.

The queen told him not to be ridiculous. He was indulging in romantic fantasies, she said. He should count himself honored that she had even considered him as a possible husband. Even if he died before she made up her mind, it would still have been a great honor for him, as everyone would acknowledge. But she accepted that he was fond of her, and agreed in the end that he could continue to plead his cause—though not in person. He was to declare his love in letters to his father and to Johan Matthiae. They could pass the messages on to her. And he must leave immediately to assume command in Germany. And above all, he must pretend that she had agreed to marry him. This would make it easier for him to succeed her, if she should die.

Karl Gustav's response was human: he became ill, plagued with constant headaches and fainting fits. Christina did not relent, and so, defying the chancellor's anti-German insinuations, he sought consolation in the time-honored Swedish way: he took to drinking heavily, then turned his mind to soldiering.

But from his post in Germany, the young commander-in-chief sent pleading and desperate letters, not to Christina but, as she had instructed, to his father and to Johan Matthiae. If the queen would not marry him, he wrote, he would exile himself from Sweden, seeking a sad alternative fortune at the hands of kinder princes. Some, at least, believed that his suit was not yet lost. He received encouraging letters from Magnus De la Gardie, the friend of his youth and now his brother-in-law, who had much to gain if the marriage could be achieved. "You must risk everything to win her," wrote Magnus. "Remember, fortune favors the brave!" It was easy advice from a man who had never himself risked very much, and Karl Gustav had no need of it in any case. By threatening to leave Sweden forever, he had already risked everything. Apart from his country, his family, the castle at Stegeborg, the promise of wealth, the crown itself, he had nothing else to risk, save his own life, and this he had already risked many times on German battlefields, fighting in Christina's name.

Christina's hesitancy was not the result of callousness. It was not a cat-and-mouse game that she was playing for her own perverse pleasure. There were gains to be made in championing the Palatine family in the teeth of opposition from Axel Oxenstierna and his supporters. Karl Gustav's appointment as commander-in-chief was a slap in the chancellor's face, just as Magnus's appointment to Paris had been. But the hardest slap that Christina could give would have been to marry Karl Gustav. Unlike her, he had brothers and sisters. His own rise would be followed by a train of honors and riches for them all, advancing them at once from dependency to dynasty and demoting the Oxenstiernas to a permanent second place.

Christina hesitated to marry Karl Gustav not because she did not love him, but because she did. It was not the love of a woman for a man, and so it could not be the love of a wife for her husband. Rather, it was the sturdy old love of a childhood friend, of a comrade-in-youthful-arms, of a brother in all but name. Christina saw, as clearly as anyone, how advantageous the match would be to the family that had been in effect her own family, to the uncle who had welcomed her as one of his own, to the girls and boys who had played with her and fought with her and grown to adulthood with her, to the people who had given her her only sense of belonging. Marriage to Karl Gustav would have been a perfect ending to her childhood's only idyll. This she saw as she told him to wait, to keep his hope alive, to do this or that beforehand, to prepare the way. But her own ambivalent nature, and her distaste for the act of sex, made the realization of any marriage impossible for her, and this she saw at the same time. She could not marry Karl Gustav, and so she tried to console him with an army, with a fortune, and at last with a crown.

Karl Gustav's love for Christina was a strong and genuine love, overlain perhaps, but not tainted, by the great advantages that marriage to her would have brought him. Because of it, he endured more than ten years of her ebbing and flowing, endured the prodding of his friends and the sniggering of his enemies. In the years that followed, its urgent flame would fade to the quieter glow of loyalty, of kindness, protectiveness, and patience, but despite myriad gusts of provocation, it was never to be extinguished.

warring and peace

KARL GUSTAV'S DOGGED LOVE was not the only recurrent theme of Christina's early reign. Problems of state recurred, too, on a larger scale and at a faster pace than the young queen could hope to manage them. Pride in her own capacities and resentment of older and wiser heads made the problems worse than they might have been, and hindered their solution.

The first problem was money, or rather, a serious lack of it. It was not all Christina's fault. It had begun nobly enough, years before, with the drive to improve public services. Her own father had set it in train, building schools and hospitals, endowing universities, developing the post office, laying new streets, boosting local industries. In every enterprise he had been assisted by his eminently capable chancellor, who had carried on the work through the years of the regency, creating in the process a proud and beautiful city worthy of its standing as a European capital. To raise the money for such vast reform, Gustav Adolf had sold what belonged to the crown: land, industries, the right to raise revenue. He fully expected to regain what he had sold by way of indirect taxation—the land and the industries and everything else would be more productive, it was presumed, in any hands other than the crown's. His chancellor approved the sales, calling them "pleasing to God and hurtful to no man—and not provocative of rebellion." They seemed to be a way of modernizing the state's finances, replacing the old herring-and-rawhide payments with efficient cash in hand.

For more than thirty years, all the years of Gustav Adolf's reign and all the years of the regency, it worked. But it provided a dangerous precedent for Christina's extravagant temperament, and in time she came to view the crown's assets like the loaves and fishes on the Mount of Olives—miraculously renewable, no matter how many hands dipped into the basket. Moreover, she could not distinguish, or would not distinguish, between the crown's property and what belonged to her personally. It was all endlessly available for public works or for presents to favorites or for libraries or paintings or armies or orchestras. She used it all, sometimes justly, rewarding a soldier's bravery or a civil servant's hard work, but more often at random, and always more lavishly than was needed. She had little understanding of finance, and she made no attempt to learn.

Reserves soon dwindled. The quickest way of raising more money, Christina saw, was to sell noble titles, and she began to sell them by the dozen. The long-established families protested, but Christina ignored them. Always impulsive, seldom with any thought of tomorrow, she had no interest in compromising with them now to ensure their support in the future. When all the old titles were gone, she created new ones, handing them out impartially to the highborn and the low, until steady citizens were heard to complain that a man could now "leap into the highest posts straight from his pepper-bags or his dung-cart."[1] Within a few years, she had increased sevenfold the number of Sweden's earls, swamped the nine old barons with forty-one new ones, and almost trebled the number of noble families. "We now have arms and escutcheons by the hundred," wrote one disgusted courtier. "The court is overrun by the mob they call counts."[2] Worst of all, most of the country's new aristocrats were not even Swedish. Artists and merchants and mercenary soldiers arrived to claim their laurels; they came from the Baltic states, from England and Scotland, from Germany and the Netherlands and, especially, from France. Townsfolk and peasants alike muttered that there were altogether "too many nobles and too many foreigners" in the country. Some at least had paid for their new positions, but just as many received them simply as tokens of the young queen's regard. Extravagance, it seemed, was her credo. "Magnificence and liberality are the virtues of the great," she wrote. "They delight everyone."

But there were many who were far from delighted. For with the noble titles went, too often, noble land, or rather, crown land sold to provide an instant family estate for the new-made aristocrats. It seemed that the number of nobles would keep on growing, that the queen would continue to sell off land, or give it away, until there was nothing left. At the crown's land registers, where titles had once changed more slowly than the pace of generations, the clerks could not cope with the sudden flow of transfers. Serious mistakes were made; some land was sold twice over, and one man, with an entrepreneurial spirit lacking elsewhere in the country, did very well for himself selling land that did not even exist.

As the nobility grew, so the crown's assets shrank. Christina attempted to redress the balance by raising taxes, a measure that was bound to be of limited effect when there were so few people to be taxed in the first place. Worse, the many ennoblements had been continually reducing the numbers liable to taxation at all; nobles paid no tax, and their peasants paid taxes to them, rather than to the crown. It was a simple equation—more nobles, less tax revenue—but Christina did not master it.

The great families themselves, nobles ancient and modern, did nothing to halt the downward spiral. Official rewards and simple plunder during the long years of war had expanded their understanding of the good life, and they now began to emulate their extravagant young queen in a hedonistic parade of new wealth. Once modest to the point of discomfort, their homes and their habits were now thoroughly up to date. They lived as fashionably, and owed as much money, as any of their compeers in France or Italy. Over the years of the regency, palaces and manors had been built in town and country to house their new art collections and their new aspirations to cultured living. Most magnificent of them all was the home of Jakob and Ebba De la Gardie, Magnus's father and mother, which stood proudly in the middle of Stockholm. Adorned in the Italian style with sculptures and fountains, it was named, appropriately, Makalös (matchless). Other magnates tried nonetheless to compete, among them the chancellor himself, whose own impressive red palace stood boldly facing the city's cathedral. Inside the great new houses, tapestries warmed the walls, lovely objects drew eye and hand, and many a looted German grandee looked sternly out from his portrait, while the candlelight danced on the new silk gown of his captor's wife or daughter.

The real problem was that Sweden—isolated, sparsely populated, half-frozen—simply did not produce very much. Although Gustav Adolf and Oxenstierna and Christina, too, had encouraged the potentially valuable mining industry and promoted foreign trade, including the slave trade,[3] it was not enough to meet much more than the people's daily needs. All was consumed in the prosaic traffic of hand to mouth. Except in the leanest years, most simple folk lived better than their counterparts in other lands, but there was no general surplus for the kind of luxuries now demanded in the towns and in the manor houses. Moreover, most Swedes were too used to thinking in terms of farming or soldiering to turn their minds to commerce, and the country owed what modest industrial success it had so far achieved mainly to foreign entrepreneurs, almost all of them Dutchmen.[4] Their influence encouraged some of Sweden's governors to view the innovative and prosperous Netherlands as a possible model for their own economic advancement. A South Sea Company was set up, and an Africa Company, and favorable conditions ensured for adventurous investors at home, but those who might have taken advantage of it failed to do so, and for the huge deficit in Christina's crown revenues, it was in any case too little, and too late.

The queen, whether really at fault or not, was an easy target for criticism. Voices were raised against her, and pamphlets slyly printed, and one summer Sunday, as she knelt at prayer in the castle chapel, a man armed with two naked daggers slipped through the congregation and ran toward her. The two guards standing in front of the queen, despite their spears and battle-axes, were unable to stop him; he knocked them both to the ground, snapping the spear of one before jumping over the other. Their captain, standing beside the queen apparently in pious reverie, had completely failed to notice the commotion. Christina gave him a shove and he leapt into belated action, seizing the assailant by the hair. On questioning, he was found to be insane; he was spared punishment, but was carried off to a madhouse.

The attack lent an urgency to the government's demands that Christina should marry as soon as possible. She was already age twenty; she had not been free of illness; now there had been an attempt on her life. If she should die without heirs, how would the succession be assured? How could they avoid dissension, civil war, foreign interference, a Catholic king? Christina responded wryly, equivocally, angrily, but always without committing herself. From Brandenburg, her frustrated cousin, the Elector Friedrich Wilhelm, continued his suit via envoys and agents, who never in fact managed to see the queen. She was too often strategically absent on hunting trips, and the men she had designated to deal with the envoys—her uncle, Karl Karlsson Gyllenhjelm, and Magnus's father, Jakob De la Gardie—both appeared to be "tending their estates in the country" with annoying frequency. In Copenhagen, the Danish king's second son, encouraged by Maria Eleonora, began to hope for success where his brother had failed; in due course he failed, too.

Though she ignored, and worsened, the country's financial problems, and delayed the question of her marriage, there were other matters that pressed on Christina daily, and which she could not dismiss. Privately and publicly, in court and in government, she encountered the same antagonisms between the crown and the nobles, and between the nobles and the commoners' Estates, that her father had known, and that he had never fully overcome. During his long absences on campaign, almost every year of his twenty-year reign, Gustav Adolf had left the government in the hands of the great noble families, ensuring their loyalty by allowing them to monopolize the best offices almost as if they were their own personal property. This had maintained a long internal stability, but it had worked against able men of humbler background, who would have preferred instead some form of meritocracy such as earlier Swedish kings had had, a "rule of secretaries"—essentially, men like themselves who had made their way up through talent and effort, who could govern the kingdom with the monarch's support, or, indeed, without it. During the years of the regency, without the king's charisma to bind them together, the two sides had diverged more sharply. Many who were themselves of noble birth had become openly hostile to the powerful old families, the Brahes and De la Gardies and the Banérs and the Bielkes and the Sparres and, above all, the Oxenstiernas, who dominated the government and the court. Christina's own uncles, Johann Kasimir and Karlsson Gyllenhjelm, resented and feared them, and she quickly learned to do the same—not without some reason: when an appeal case between the Oxenstiernas and the Bielkes was brought before the Senate, it quickly became apparent that every single senator was related to one or other of them, or to both.

Christina could not dispense with them, and as yet she lacked the skill to undermine them, but she struck out at them nonetheless, confusing her dislike of their influence with her own continuing rivalry with the chancellor. In the first months of 1647, soon after her twentieth birthday, her old tutor, Johan Matthiae, now Bishop of Strängnäs and recently ennobled, unwittingly provided an opportunity for the young queen to test her power.

As the late king had done, and as he had wished his daughter to do, Matthiae supported the idea of a single Protestant Church, uniting both Lutheran and Calvinist creeds. This kind of syncretic thinking was anathema to the adherents of Sweden's rather narrow form of Lutheranism, among whom the chancellor himself was counted—Calvinism, like Catholicism, had been outlawed. From his diocese in Strängnäs, Matthiae had written a book promoting Protestant unity.[5] It had infuriated the chancellor, and at a session of the Senate, he denounced it roundly, calling for the book to be banned and for Matthiae himself to make a formal apology before the five hundred men of the *riksdag*. Matthiae did so, and the Senate and the *riksdag* together then demanded the outlawing of any movement prejudicial to the accepted rites; an old document of 1580, the Liber Concordiae, was to set the terms thenceforth for religious observance in Sweden.

Christina seized her chance. Just as her father had done almost forty years before, she rejected their decision and refused to accept the Liber Concordiae. There was nothing wrong with the bishop's views, she declared; indeed, her own views were the same. The chancellor remonstrated; the queen stood her ground; the chancellor insisted; and the queen burst into tears. The match was a draw, more or less: the book was not banned, but neither was it reprinted, and the chancellor went off to his country house, muttering that the queen was absolutely impossible, that the late king would never have behaved so imperiously, and that the bishop was not to be trusted.

At the Tre Kronor Castle, Christina's angry tears were dried by the kindly old Count Per Brahe, who had taken Karl Gustav's proffered place as High Treasurer. Her majesty was young, he said, and, with the greatest of respect, had much to learn; she would be wise not to place all her trust in a priest—any priest, even a beloved former tutor. And if he might be so bold, her majesty could perhaps exercise a little more discretion in her choice of companions. That Magnus De la Gardie was altogether overstepping the bounds; he needed to learn his place. The chancellor and the senators were experienced men; they would serve her majesty very well, if she could only put aside the pride of youth and trust their judgment.

IN THE NAME of the most holy and individual Trinity: Be it known to all, and every one whom it may concern, or to whom in any manner it may belong, That for many Years past, Discords and Civil Divisions being stir'd up in the Roman Empire, which increas'd to such a degree, that not only all Germany, but also the neighbouring Kingdoms, and France particularly, have been involv'd in the Disorders of a long and cruel War . . . from whence ensu'd great Effusion of Christian Blood, and the Desolation of several Provinces. It has at last happen'd, by the effect of Divine Goodness, seconded by the Endeavours of the most Serene Republick of Venice . . . that there shall be a Christian and Universal Peace . . . between his Sacred Imperial Majesty, and his most Christian Majesty of France, . . . the most Serene Queen and Kingdom of Swedeland, the Electors respectively, and the Princes

and States of the Empire . . . and that there shall be on the one side
and the other a perpetual Oblivion, Amnesty, or Pardon of all that has
been committed since the beginning of these Troubles, in what place,
or what manner soever the Hostilitys have been practis'd. . . . Done,
pass'd and concluded at Munster in Westphalia, the 24th Day of October, 1648.[6]

The peace, like the war, had been years in the making. Since the early
1630s, there had been sporadic attempts to secure it; many smaller truces had
been made and broken. A few individuals had laid down arms of their own accord, then taken them up again as their personal interests had shifted. By the
1640s, Bohemia and the German lands had become, as it were, a vast chessboard where the powers played out their alliances and antagonisms, religious
or political. Apart from the occasional Scandinavian skirmish, all Europe's
wars had become more or less "fused," in Gustav Adolf's phrase, "into a single
war." But in 1645, a Turkish attack on the island of Crete, then in the hands of
the Venetian Republic, had finally concentrated the collective mind of Christendom, forcing the European powers to realize the external peril threatening
their territories and their ideals. "While the Christians squabble among themselves," wrote an anxious Dutch poet, "the Turk is sharpening his sword."[7]
    The Venetians at least had perceived the threat, and had set themselves to
broker a general European peace. Now, foreseeing that assistance from their
coreligionists might be needed in their own struggle, they redoubled their efforts. And so it was that, "having implor'd the Divine Assistance, and receiv'd
a reciprocal Communication of Letters," representatives of the various powers
came together at last in the German province of Westphalia. Christina, as
queen of the all-conquering Swedish armies, was a guarantor of peace along
with France's boy-king, the ten-year-old Louis XIV.
    Even at the negotiating table, it was not considered safe to seat Catholic
and Protestant together. In consequence, the treaties were to be discussed and
finally signed in two separate cities, thirty miles apart—Münster for the emperor and his Catholic allies, Osnabrück for the Protestant powers. An exception was made for the representatives of Catholic France: evidently unable to
stomach Austrian company, or perhaps Austrian food, they assembled with
the Swedes and their Protestant allies in Osnabrück. By early August the main
proposals had been agreed, and on the twenty-fourth of October, the treaties
were finally signed.

Sweden emerged as a determined victor, with major territorial gains, including control of the trade-rich Oder River and the whole of Western Pomerania, as well as huge indemnity payments and permanent representation at the German parliament.[8] Many in Sweden felt cheated nonetheless, maintaining that the war should have been continued until the Protestant cause was victorious, or at least until more money could be exacted. Some of the clergy condemned the treaty from their pulpits, stirring up opposition to it until they were formally forbidden to do so. French gains were particularly resented, the more so as they had been largely brought about by Christina's personal intervention. The whole of the central Rhine area and a dozen Alsatian cities passed into French hands, making a bitter mockery of Gustav Adolf's last warning, only days before his death, that France must not be allowed to gain control of any German territory.[9]

France's star had begun to rise, and its neighbor's long bright day was drawing to a close. In a clear signal of the continuing decline of Spain's Habsburg Empire, the United Provinces of the Netherlands finally gained their independence. With revolts on their hands to the east and west,[10] and continuing war with France, the Spaniards could hardly afford to press for better terms.

For the land of the first brave rebellion, it had all been in vain. There was to be no confessional liberty in Bohemia or Moravia, and no restitution of the lands confiscated from the rebels. To Prague's many exiles there remained two simple choices: embrace Catholicism or stay away. "We are abandoned," one despairing Czech wrote to the Swedish chancellor. "You hold our liberty in your hands, and you are handing it over to our oppressors."[11] In France, too, the boy-king Louis was "oblig'd to preserve in all and every one of his Countrys the Catholick Religion . . . and to abolish all Innovations crept in during the War." Only in the German lands did a partial confessional tolerance prevail, a tolerance for rulers, if not for those ruled. By the principle of *cuius regio eius religio* (whose rule, his religion), German princes might choose their religion, and their subjects might follow suit. After all the years of fighting, there would be no single faith across the Continent. People stopped talking of Christendom, and began instead to speak of Europe.

It was all too much for the pope, who saw in the treaty a certain end to the Catholic hope of a reunified Church, cherished since Luther's first revolt 130 years before. In a furious outburst, he denounced it as "null, void, invalid, iniquitous, unjust, damnable, reprobate, inane, and devoid of meaning for all time."[12] As in Stockholm, so in Rome: France's gains were a source of particular outrage; it had been Cardinal Mazarin who, four years before, had attempted to block the pope's election, and the two had nursed a mutual enmity ever since. Unable to strike at France's heart, the pope fixed on the francophile queen of Sweden as the object of his personal vengeance. Proclamations were pasted up in the imperial capital of Vienna, inveighing against the imposter Christina who had stolen the crown from its rightful Polish owner. The emperor, though in private no doubt agreeing, was readier to recognize that the time for conflict was past. He saw a different writing on the wall, and quietly had the proclamations taken down. In Münster, Cardinal Chigi, the pope's unhappy representative, turned at last from the negotiating table with a resigned "*O tempora! O mores!*"

But if the pope had lost his dream of a reunited Church and Spain had lost its prosperous Dutch provinces, the greatest loss had been sustained by the people of Germany, whose homes and farms and cities had been the main theater of the war. The "great Effusion of Christian Blood" had mostly been their blood; a third of the population, possibly half, had been killed. Weapons had not been the only threat, nor often even the main one. Hunger and disease, including periodic outbreaks of plague, had claimed the lives of soldier and peasant and townsman indiscriminately. Always on the move to the next battle or the next supply area, the armies had carried their disasters with them across the increasingly ravaged land, spreading dysentery, typhus, and worse as they passed.

The treaty brought the Germans peace, but they made no other gains. By the end of 1648, much of their territory was in ruins. The western regions and the three great rivers lay in foreign hands.[13] The deep disruption of war had broken the many vital bonds of ordinary daily life. In some areas, there was no trade at all. Though property could be given back and titles reconferred, the "general Restitution" occasioned by the treaty had no power to recreate "those things which cannot be restor'd." In the bitter aftermath, a once advancing German political culture was dashed into the parochial pieces of smaller rival states. Thenceforth they would all defer to the bold young giant, France.

And in the end, the savage tragedy of thirty years turned to dispiriting farce. When a team of weary riders arrived at last with the emperor's letter accepting the terms of the treaty, it was found to be in code, and their dusty saddlebags contained no key. At length the letter was deciphered, but further delay ensued: in a near parody of baroque formality, it took the next three weeks to agree the order in which the different sections of the treaty should be signed.

IT WAS NOT OUT of pity for soldier or peasant, or concern for trade and treasuries, that Christina wanted to end the war. In later years she would be quick to suggest the use of arms when it was in her own interest to do so. But warfare was quintessentially a man's game, and no amount of little lead soldiers on her schoolroom table could turn it into a game that she could play. Like Elizabeth I of England, she might have "the heart and stomach of a King," but unlike Elizabeth, she also had Axel Oxenstierna, who had been capably directing the war for almost fifteen years. While it continued, he was bound to retain his premier position in Sweden, and bound to detract from Christina's own authority in other matters of government. Her stratagem for the peace conference was thus a perfect complement to her tactics at home. Her aim in both was to undermine the chancellor.

The chancellor did not attend the conference himself. Instead, he sent his eldest son, Johan, now in his middle thirties, through whom he intended to direct the Swedish negotiations. Johan was tall and majestic, but apart from this he could not boast—although he did boast—any of his brilliant father's qualities. He was a headstrong man, inordinately proud, hot-tempered, red-faced, fond of wine, and very fond of women. He arrived in Osnabrück at the beginning of the negotiations to a guard of honor five hundred strong, with a retinue of almost a hundred and fifty servants. Through the three long years of talks, every day was punctuated by trumpet fanfares announcing the rising and the setting of the chancellor's son, and every meal in between. They were seldom blared at the usual times; Johan gave many elaborate banquets and generally slept late into the morning. Exasperated locals rumored that he and his men kept supplies of bitter almonds to chew during the discussions—it was supposedly the only thing that could keep them sober.

Johan was the official leader of the Swedish legation, or so he repeatedly insisted, but there was an unofficial leader as well. Not daring to override the chancellor formally, Christina had sent a second, smaller legation headed by her late father's representative, Johan Adler Salvius. Of modest birth, Salvius was among the very few men in Sweden who had managed to rise through the ranks to a position of national influence. Trained in law, medicine, finance, and the science of war, he had also made a fortune by the shrewd courting of a rich widow. He was now almost sixty years of age, with an impressive record of diplomacy behind him, and he was certainly better suited than Johan Oxenstierna to lead the Swedish legation in Osnabrück. But Christina had lacked the courage, and perhaps, too, the necessary support, to propose him instead of the chancellor's son, and so the two proceeded in parallel, or rather at cross-purposes, alternately amusing and frustrating the representatives of the other powers. Johan was directed to draw out the negotiations until certain conditions had been met; if necessary, he was to threaten a resumption of the war. Salvius was to settle for peace at any price, regardless of the chancellor's instructions.

Fortunately, or unfortunately, the Swedes were not alone in their division of efforts: the French, too, had dual lines of counsel, each with its own spokesman. Both detested both the Swedes, who in their turn detested both the Frenchmen. The Comte d'Avaux relayed a loud disgust of the proud young Oxenstierna, "sitting there on his throne as if he's about to pass judgment on the twelve tribes of Israel." The chancellor sent his son a cool word of advice: "If he writes to you in French," he told him, "write back in Swedish." Mistrust flourished. Christina wrote to Salvius: "The chancellor is being very obliging, but I am wary of Greeks bearing gifts." And, though she took their part against the Oxenstiernas, Christina did not always feel sure of the French delegates, either. "I am very well acquainted with their ways," she wrote. "For the most part, it's all just compliments. But civility won't cost us anything—we can pay them in their own coin." Her own often impulsive intervention, however, ensured that France earned much more than compliments, and it even cost Christina something in a personal sense. She had wanted to have the town of Benfeld as a grand bestowal for Magnus, but the French took it along with the other Alsatian territories. Magnus had to be content with the Benfeld cannon instead—he quickly sold them to the town's new owners.

Despite their internal rivalries, the Swedes and the French between them took the lion's share of the treaty's benefits, and in the end they were happy enough to sit down together at the great celebratory banquet hosted by Karl Gustav in Nuremberg. Among those present was the new-made Count of Vasaborg, Christina's illegitimate half-brother, Gustav Gustavsson, only half-rejoicing. His blood ties to the queen had not been enough to overcome the stain of his long service to the chancellor, and Christina had placed no trust in him, nor had she, or the French, supported his personal claims—he had had his eye on a couple of German dioceses. Johan Oxenstierna attended the banquet, too. After sobering up, he traveled on to Pomerania, its new post-treaty governor.

The Russians, though they had not been among the combatants, enjoyed nonetheless the best of the peace celebrations. After thirty long years, they did not at first believe that the war had ended at all, and it was decided that an extravagant spectacle would be the quickest way to convince them. Consequently, in the border town of Narva, between Swedish and Russian territory, a "joyous day of thanksgiving" was prepared, with religious services and feasting and cannon firing off, and particularly elaborate fireworks that could comfortably be viewed from both sides of the border.

THROUGHOUT THE SPRING and summer of 1648, as negotiators wrangled in Münster and Osnabrück, the Swedes themselves had instigated the last important military episode of the war. Fittingly, and sadly, it took place in the beautiful city of Prague, where the conflict had started three decades earlier. Led by General Königsmarck, with Magnus alongside him, a large Swedish contingent marched unbidden into Bohemia, and by the end of July they had captured the western part of Prague on the left bank of the Vltava River, by the great Hradčany Castle. Prague was the last, symbolic bastion. For years the Swedes had been urged to retake the city by exiled Czech reformists.[14] The great blaze was dying down; its last flare should illumine the poetic recapture of the ancient town where the first match had been struck.

From the Minor Town the Swedes began an artillery bombardment of the Old Town across the river, and for a time it seemed they would take the whole city, but quite suddenly they stopped the attack, and, without pressing their advantage, took to plundering instead. Their orders had been countermanded, and a new, secret instruction received, from the queen herself, that they should occupy the castle and seize all that remained of the famous collections of the Emperor Rudolf II. They did so, resisted only by the castle's unhappy keeper, the too aptly named Miseroni. Evidently the Swedes felt they had fought enough for one day; they simply tortured him until he gave them all the keys. On the last day of August, an itemized inventory of the collections was drawn up and sent back to Stockholm, where Christina received it eagerly.

For more than half a century, the vast collections of Rudolf II of Habsburg, Holy Roman Emperor, King of Bohemia and Hungary, had been legendary throughout Europe and beyond. By 1648, however, most of the best pieces, in fact most all of the pieces, had been dispersed. Victims of their own success, over the decades they had attracted a long succession of admirers, most happy simply to stand and gaze, but some determined to enjoy them comfortably at home. The despoliation had begun only a few years after Rudolf's death in 1612, when some of his jewels were sold by Bohemian rebels needing to finance their war against the Habsburgs. After the famous Battle of the White Mountain, in 1620, a victorious Maximilian of Bavaria had returned to Munich with fifteen hundred wagonloads of items from the collections. Following their own visit in 1631, the Protestant Saxons carried a further fifty wagonloads home to Dresden. Rudolf's collections must have been phenomenal, for the items that Christina received, even after all this plunder, included almost 500 paintings, seventy bronzes, 370 scientific instruments, and 400 "Indian curiosities," as well as hundreds of corals, ivories, precious stones, pieces of amber, vases and other *objets d'art*, thousands of medals, two ebony cabinets, and a solitary live lion. Even so, it was not enough for Christina, who penned a hasty letter to Karl Gustav telling him not to forget Rudolf's library. "It is absolutely imperative," she wrote, "that you get everything onto the water as quickly as possible and send it on here."[15]

It was indeed, for everything had to be on Swedish territory before the last signatures were added to the peace treaties. If not, according to the treaties themselves, it would all have to be returned "to its original owner." Karl Gustav got it all onto the Moldau River with twenty-four hours to spare, amid vast rejoicing. For the Swedes, the Hradčany loot represented the apogee of their takings from all the years of the war. There was enough and more to reward all the queen's soldiers, but the lion's share, and the lion, found their way into Christina's own delighted possession.

THE FORTUNATE RUSSIANS had played no part in the long-drawn-out war or the long-drawn-out peace. They had watched from the periphery as Sweden's armies advanced across the continent, and as they watched, so their anxiety grew. The Swedes were old enemies of theirs; the two had been at war for years during Gustav Adolf's reign, and shortly before Christina's birth her Vasa cousins had still been pursuing their own claim to the Russian throne.

Russia was still a minor power, but Gustav Adolf had feared Sweden's fate "if Russia should ever learn her strength." The fear was mutual, and in the early summer of 1649, the Grand Duke Alexei of Muscovy decided that, since the Swedes had stopped fighting in the south, it would be wise to preempt a resumption of their interest in the east. Accordingly, a delegation of one hundred and twelve diplomats was dispatched to Sweden, bearing greetings from their noble Romanov lord. Their visit was observed, and reported in some detail, by the correspondent of a Swedish-controlled news-sheet in Leipzig.[16]

It seems that, from their ships moored on the lovely waters of Stockholm, the Russians disembarked to be met by an assembly of the usual councilors and secretaries, as well as "three substantial-looking old persons," otherwise unidentified. The following day, in an echo of her very first ambassadorial reception at the tender age of six years, the young queen herself received them at a public audience.

The Russians appeared to have lost none of their magnificence in the fifteen intervening summers. They were dressed very richly in gold-embroidered robes interwoven with pearls, and they processed toward the queen in stately fashion, still bearded, it seems, but without any show of the "wild manners" of which she had once been forewarned. Christina remained "on her royal seat," with a cushion beside her bearing her crown and orb and one of her dozen-odd scepters, lengthened since the last Russian visit to suit her now full-grown height.

The ambassadors had come laden with presents for her, including, as the correspondent reported, nine pieces of gold cloth, each one "twelve ells" in length,[17] tapestries worked in gold thread, three suits of Turkish clothes "and similar things," twenty mink furs "for wearing indoors," a beautiful vessel studded with rubies and turquoises, and—in a wintry echo of the lion looted from Prague for her only months before—three live mink. They brought so many presents, in fact, that it took forty soldiers to carry them all. With them, too, came the more prosaic gifts of letters from the Grand Duke Alexei exhorting "eternal peace" between their two lands, and a rather tardy apology for the several hundred soldiers who had deserted the Swedish army to join the Russians more than thirty years before.

Christina had been expected to leave the city directly after the formal reception of the Russian diplomats, to travel to Fi'holm, a day's journey away in the bright summer weather, for the funeral of Madame Oxenstierna, the chancellor's wife. But the opportunity to spite the chancellor had proved irresistible to her, more so than any Russian gold or rubies or mink, dead or alive. The night before she was due to leave for Fi'holm, she became suddenly "indisposed"; though a large retinue had been sent on ahead to prepare for her arrival, she announced that she would not be able to attend the funeral after all. Her transparent stratagem must have saddened Oxenstierna, or perhaps made him angry; certainly it did not convince anyone else. In Leipzig, it was noted sardonically that, once the day of the funeral had passed, Her Majesty "suddenly became quite well again."

It was a petty act, unworthy of any queen, or indeed of any adult. Determined to dim the chancellor's prestige, she had succeeded only in offending him, and in making herself look foolish. In so doing, Christina revealed how much she had still to learn about strength and self-indulgence, and the difference between the two.

Pallas of the North

I N THE SPRING OF 1649, the fabulous collection of the Emperor Rudolf, pushed and pulled all the way from Prague, was brought ashore at Stockholm, and Christina found herself mistress of one of the finest cultural treasures in Europe. It was a splendid crowning of many smaller efforts of plunder and purchase, the work of more than a century, as successive rulers had brought home piece after piece of beautiful tinder to stoke the Swedes' reluctant aesthetic fires. Christina's father had been the most determined of them, to the extent of leaving two of his best generals hostage in Bavaria for the sake of his newly looted Holbein canvases.[1] The Holbeins, along with works by Lucas Cranach and many other German and Dutch masters, were sufficient in number and in quality to form the basis of a first Swedish national collection, installed during Christina's childhood in the Tre Kronor Castle. Though she was quick to appreciate her father's methods of acquisition, she was slower to appreciate the works themselves; the restrained northern painters held little appeal for her, and she was able to give many fine canvases away without so much as a backward glance.

But whether she liked the paintings or not, they were important to her. A certain level of cultural life was necessary if Sweden's national prestige were to be maintained, or indeed even acquired—there was a vast distance to be covered before the Swedes could compare with most of their northern neighbors, let alone with the richly cultured southern lands of Spain or France or, above all, Italy. Plundering was a quick, but not necessarily cheap, way of building up collections; armies were as costly as marble and canvas, and victory was not always assured. Besides, no one would fight for a sculpture or a painting; booty of this kind was unpredictable, to be seized opportunistically like windfall apples from the highest branches. No monarch could afford to presume upon it, and neither did Christina. Even as a young girl, tantalized by ambassadors' tales of beautiful and brilliant things, she had sent emissaries abroad to seek out books and works of art. One envoy went as far as Egypt, lending his hand in excavations for the remnants of the ancient world. Others scoured the studios and libraries of Europe, unearthing sculptures and drawings and a great many books and manuscripts for the avid young queen, whose plundering streak was strong enough for her to leave many bills unpaid.

It prevented her, at the same time, from building up her collections in any systematic way. Though she did request specific books, to match her developing intellectual interests, her agents scouting for antiquities and works of art bought more or less at their own discretion, often sending back things that were not to Christina's taste; ten paintings by Gerrit Dou, for instance, bought at considerable cost by her agent Silfvercrona in Holland, were soon passed on to Silfvercrona's family. The Prague cornucopia did not change her approach to the northern schools of painting, though it contained many eloquent examples of it, but it did provide a concrete elaboration of those Renaissance ideas that had framed the minds of her own teachers. It was largely within that tradition that Christina was now forming her own view of the world.

The Emperor Rudolf had collected not only paintings and sculptures, but also *objets d'art* and all sorts of curiosities, sublime and ridiculous, inanimate and live—the lion now brought ashore for Christina was the lonely representative of a once great menagerie. Caravaggio canvases and Dürer woodcuts had overlooked displays of tools and shells and bits and pieces, including nails said to be from Noah's Ark and a jawbone supposedly belonging to one of the sirens of Homer's *Odyssey*. Rudolf had acquired many spectacular pieces, but not primarily so that they might be admired. Instead, the thousands of individual items were all intended to be understood together as a single entity, a kind of "encyclopedia of the visible world," revealing the harmony of the whole created universe. The myriad items were almost like the words of a lost language; if enough of them could be collected, the links between them might be discerned, and the language of the universe finally understood.[2]

This "pansophist" idea underlying the emperor's great collection had been part of the received wisdom of his day, and it had not yet given way to the ideas of the empirical scientists, who instead were learning to think of the natural world as a vast series of discrete phenomena. Though she was well versed in pansophist ideas, and she was to look to other aspects of pansophism to guide her own spiritual path, Christina had no wish to build a collection in this grand Renaissance way. She accepted what arrived, and she was on the whole delighted with it, above all with the many Italian paintings—Titian, Veronese, Tintoretto, Polidoro, Correggio—which formed so rich a part of the takings. These, all looted from Prague, were her first exposure to the Italian Renaissance masters, and she responded at once to their rich colors and their vibrant emotional energy. She was soon writing about them with the nonchalant air of a connoisseur:

> I shall send you copies of some of the Italian paintings that have come into my hands since I had the good fortune to take Prague. . . . In fact the whole Prague gallery is here now. There are really a lot of paintings but only thirty or forty of the Italian ones are originals. I don't count the others. There are some by Albrecht Dürer and other German masters whose names I don't know. Everyone likes them very much—everyone except me, that is. I swear I'd give them all for a few Raphaels, and even then it would be doing them too much honor.[3]

Christina's letter reveals a great deal of confidence in hearsay, or perhaps in her own prejudice. Though she owned several of his drawings, she had never so far set eyes on a painting by Raphael, nor indeed had any Raphael ever been seen in Sweden.

The booty included a considerable number of sculptures, among them, by way of late and paltry vengeance, several garden bronzes made for her father's nemesis, Count Wallenstein. Prominent among these was a Neptune worked by the mannerist master Adriaen De Vries. The mighty sea-god had been embellished for the count with a pair of sirens, a pair of tritons, two griffin-heads, two lion-heads, four horse-heads, and two ducks. Christina does not seem to have cared much for it, or indeed for any of the sculptures; within a few years they had all been traded or sold or simply given away. She had been captured, heart and mind, by the Italians. Their dramatic sensuality had struck a chord in her own extravagant nature, and it drew from her an emotional response that was to have loud repercussions. The Italian paintings belonged to another world, to the vibrant, flowering world of the sunlit south, and in them Christina saw all the color and drama that she had not found in her plain and austere homeland.

—

CHRISTINA HAD BEEN doing what she could to enliven the gray palette of Swedish cultural life, mostly with splashes of imported foreign color. They came in the elaborately dressed persons of French scholars and their satellites, more or less brilliant, fleeing their country's civil wars of the Fronde. The French Parliament had rebelled against the principle of absolute monarchy, forcing the boy-king Louis XIV to flee his capital in the middle of the night with his much-resented regent, Cardinal Mazarin. Led by the Prince de Condé (*le Grand Condé*), the French nobles were now fighting to drive the Italian Mazarin permanently out of France.[4]

Along with their learned tomes and their silk suits, the French scholars had brought with them a loud cosmopolitan disdain for the provincial. Though Stockholm had undergone a good deal of recent change, its fine town palaces and modern streets did not impress the sophisticated newcomers. One of them was moved to declare it "more fun than Switzerland, anyway," but on the whole they had nothing good to say about Christina's chilly northern capital, finding themselves by far the most interesting aspect of it. The Swedish courtiers responded with the same resentment that their fathers had shown to Maria Eleonora and her haughty Berliners, and with more reason. Christina had showered the new arrivals with personal gifts purchased with public money, and feelings ran high against "this crowd of parasites" who were seducing the young queen away from the wholesome ways of her own people. Their feathers and laces and bows and buckles were found so offensive that sumptuary laws were passed in an effort to return the court to the somber tones of earlier days. The French ignored them, provoking a disgusted comment from one sober Swede. "The further they are from their homeland, the more insolent these people seem to become," he declared. "It seems they travel with no other purpose than to mock other peoples, to insult their customs and to break their laws, and to parade their own pride and extravagance through the world."[5]

Christina quickly aligned herself with the Frenchmen in their disdain for her own country. It may have begun defensively, the reaction of a sensitive and unusual girl who might otherwise have become an object of ridicule herself. But it soon became a source of strength in her ongoing opposition to Chancellor Oxenstierna. It gave her a group to call her own, an alternative to the great old families who surrounded him. Thus, whether consciously or not, the queen's circle of exotic favorites developed into a substantial political force.

Magnus, though homegrown, remained at their center, as yet unfazed by the new competition. He moved easily among them, aided by his natural flair and the elegant ways he had acquired on his travels, and his example contributed in no small measure to a swift refining of manners at Christina's court. Elaborate French ways were steadily supplanting the grave, ceremonious traditions that had long characterized the Swedish courtiers—at least before the start of their drinking bouts. Among the younger men in particular, indignation evaporated before the simple wish to be part of it all, and besides, no one wanted to be counted among the country bumpkins. Everything was affected, from hats to handshakes, but not all the changes were frivolous; the young queen found her own position altered, too. At the beginning of her reign, Christina had had no more than a single chamber to call her own, and had been frequently besieged at her bedroom door by town and country folk seeking an audience with her. Now, a small antechamber was arranged, and order of a kind imposed on the usual mêlée of petitioners. There was an improvement, too, in some of the smaller comforts of daily life, "this damnable Swedish cooking" notable among them. Though the chancellor missed his honest salmon stews, the Frenchmen found an unexpected ally in the English ambassador, relieved to find the menu expanded at last from the "boiled, roast, or fried cow" that he had too often been served before.

Though the French now dominated Christina's court, not all of her visitors hailed from those fashionable shores. There were a goodly number of Germans educated in Holland, and Dutchmen educated in German towns, and a Dane or two, and others from the earnest lands of the north. Among the first had been Christina's Royal Librarian, Johann Freinsheim; he had been laboring steadily and quite happily until the arrival of the thousands of books and manuscripts brought back as booty from Prague. He had balked at the work of cataloguing all these, and at first the job had passed to Isaac Vossius of Leiden, whose own father's library was already in the queen's possession. Vossius, like his Latinist friend, Nicolaas Heinsius, was eventually felt to be more useful out in the field, buying more books, and Christina sent them both off with a vast budget, at least in theory; in fact, Heinsius found himself paying the bills more often than not. Back in the library, Vossius had been replaced by a rather famous figure—physician-in-chief to Louis XIII, former librarian to Richelieu and Mazarin, machiavellian political theorist and presumed atheist, Gabriel Naudé. Naudé brought a substantial recommendation along with him in the form of thousands of books from Mazarin's library, which the Fronde had dispatched, like the cardinal himself, to a dozen different cities.

But the librarians were vastly outnumbered by erudite birds of a different feather. Most of the early arrivals were philologists, specialists in languages, particularly biblical languages, and these Christina decided to study, not in order to deepen her understanding of the Bible, but in order to pursue her interest in the occult. Christina's occult was not magic or witchcraft or the "black arts," though traces of them all did remain in it. It was a part of much of the serious learning of her day, indeed a part of natural science, battling for predominance with the new empirical methods that in the end would prove so fruitful. Occult learning was an older way of investigating the material world, a legacy of the Renaissance, and prominent among its priestly caste were the philologists. Their knowledge of ancient languages gave them access to the esoteric writings of bygone ages, writings that supposedly contained secret knowledge about the nature of the world. The very words and letters themselves, it was believed, concealed the spirit of the universe. Understanding them brought understanding, and also control, of the natural world, for whoever could decipher the alphabet of creation could write with it as well, turning stones into bread, and evil into goodness, and lead into gold. Galileo, the revolutionary, had claimed that mathematics was the language through which the "book of the world" could be understood, but for adherents of the occult, the key was not mathematics but the written word. Like the myriad items of the Emperor Rudolf's collections, the words of the ancients together held the ultimate truth.

The study of ancient languages consequently appeared to be of vital importance, even for everyday living. Christina began to make plans for a school of theological linguistics at the new university in Dorpat,[6] and soon she had gathered to her court some of Europe's best philologists. Typically, she had not waited for any of them to arrive. Eager to begin her study of biblical languages, she had seized on the first possible tutor—the Royal Librarian, Johann Freinsheim. Freinsheim had already been drafted to teach her Roman history, and his duties were further expanded, and his leisure hours reduced, by Christina's demand for regular lessons in Greek. There was a lesson every day, and every night she sat up late, reviewing what he had taught her—five hours' sleep, it seems, was all she needed. Despite this, she does not seem to have made very rapid progress, but in Greek at least she was persistent; in other languages that attracted her, such as Hebrew and Arabic, she dabbled, but made no progress.

Dabbling, in fact, was Christina's forte. She dabbled in philosophy, and dabbled in history; she dabbled in astronomy and alchemy. She dabbled in music and dabbled in dance, and in all areas, with her quick mind and her excellent memory, she picked up enough to make a strong first impression on everyone who met her. Still in her early twenties, she presented already a rich façade of learning that sparkled with extra gems of gossip from a dozen different courts. She had read about everything, and heard about everyone, and a judicious mixture of boasting and teasing ensured that her visitors were quickly apprised of those facts. They lent her a daunting air, a perfect match for her sense of her own innate majesty.

One of those most favorably impressed was the French Resident and later ambassador in Sweden, Pierre-Hector Chanut. He had arrived in Stockholm on the very last day of 1645, and he quickly became one of Christina's most enthusiastic eulogists. An experienced diplomat, forty years of age, a devoted Catholic husband and *père de famille*, Chanut was nonetheless captivated by the clever young queen with her "sweet smile and her big blue eyes." The very day after his arrival, he sat down to pen a few besotted lines to one of his many correspondents:

> She speaks French as if she had been born at the Louvre, she has a quick and most noble mind, a soul wise and discreet, and she has a certain air about her. Her every pastime is the Senate or her study or her exercise. She speaks Latin very easily and she loves poetry. In short, even without the crown, she would be one of the most estimable people in the world.[7]

Christina had just turned nineteen at the time, and perhaps, at least on formal occasions, she was still somewhat reticent; it is otherwise hard to match the description of her "soul wise and discreet" with what is known of her, both before and after this meeting. But the *coup de foudre* was mutual: her studious habits answered Chanut's own bookish tastes, and he, of course, was French, and could also claim a personal link to the world of serious scholarship: his wife's brother was the translator of the celebrated philosopher Descartes, and Chanut himself was a good friend of the great man.[8] The queen and the ambassador soon became firm friends, and over the years Chanut continued his infatuated letters from Stockholm, evoking teasing replies from his amused friends. The queen was "a marvel," it seemed, clever, cultured, virtuous, devoted to duty—and what a memory! She was pretty, too, he wrote, protesting nonetheless that he had not "taken the liberty of examining her beauty very closely." Encouraged to do so, he was obliged to admit that her features were "not very regular," though still, he insisted, they were "highly expressive." The queen, in any case, he wrote, had "no interest in her own feminine allure, and she will not permit the slightest allusion to it." From Paris, a daring Cardinal Mazarin hazarded a mention, anyway, acknowledging receipt of her portrait with a flattering verse; the queen's noble soul and her lovely features, he declared, "must do battle with each other for the highest honor." Susceptible to flattery only where she flattered herself, Christina was on this occasion not deceived. Hearing the translation of the cardinal's verse, she displayed a rare modesty: knowing she did not really deserve the compliment, she blushed.

Ambassador Chanut was an early recruit to Christina's "academy," a formal, regular meeting of all the scholars and artists, where they discussed such matters as interested the queen. What is the difference between spiritual and sensual love, she asked. And of love and hate, which is worse if misused? Chanut recorded that Christina seldom gave her own opinion on the subject at hand. Instead, she would wait until the discussion had finished, then provide a neat summary of all the different arguments, and the conclusion, if any had been reached. Her reticence did not stem from intellectual deference; rather, she did not want to expose her own ideas to the possibility of contradiction. "I never could stand being corrected," she had once remarked to Chanut. Christina had to have, in every sense, the last word.

The foreign scholars, in any case, set little store by the outcome of the discussions. Used to more meaty fare, they found them trivial, and did what they could to avoid them altogether, feigning illness and arranging tactical sojourns in the country. Despite their lack of interest, through their learning and their very numbers, they dominated the meetings, to the disgust of the local luminaries. Writing to a friend, one Swedish scholar expressed his uneasy mixture of resentment and respect toward them:

> It is true that for the natural sciences, for an infinity of languages, and for reading every author in existence, I cannot compete with them. But as for appreciating a poem, one doesn't really have to know the whole of the Greek Anthology by heart.[9]

Whether to pacify the natives, or to advance her own reputation, Christina now began to talk of a separate academy exclusively for Swedes. Cardinal Richelieu had founded the Académie Française in Paris, after all; with Christina at the helm, Stockholm could surely soon boast the same. It might well have, had Christina been steady enough to stay on deck, but her quicksilver attention darted elsewhere, and for the moment the Swedish Academy progressed no further than a few plans and papers.

THE SWEDES' ANTAGONISM did nothing to foster unity among the foreign scholars themselves, whose rivalry flourished in the cold, hard soil of exile. The philologists looked down on the philosophers, the artists looked down on the scientists, who looked down on the theologians, the Protestants looked down on the Catholics, who looked down on the freethinkers, who looked down on both of them, the Dutch looked down on the French, and the French looked down on everybody. Most of the tensions, profound or petty, were produced by the smallness of the Stockholm circle, but some had arrived with the scholars, on the same cantankerous winds and tides. No one caused more trouble within Christina's community of luminaries than the talented French physician, Pierre Bourdelot—also cook, perfumer, dancer, and zither player extraordinaire. A former employee of the Prince de Condé, he had been recommended by Claude Saumaise, and the two made a fine pair of irreverent libertines in stalwart Stockholm, the Protestantism of the one and the Catholicism of the other being of equal amusement to both. Though Saumaise was a really brilliant scholar, Christina had always been most appreciative of his boisterous self-confidence and his earthy sense of humor, and she was delighted to find the same qualities in the new arrival. She appointed Bourdelot her physician-in-chief, and he was soon in constant attendance upon her. Malicious tongues whispered that he was not a physician at all, but his medical training was genuine, and he cured the queen of her many longstanding complaints by the simple expedients of lighter food, more rest, and— evidently a new idea for her—regular baths.

Christina called Bourdelot her "lovable ignoramus." He was in fact a cultivated man who for some years had maintained an excellent academy of his own, with Pascal and Gassendi in regular attendance, but he could not compete, nor did he try, with the vast learning of some of the others now assembled at the court. He disliked their superior ways, and was quick to make fun of them in the jolly fashion that Christina most enjoyed, mimicking their accents and throwing snowballs at them. Bourdelot had done very well for himself, but he may have felt some resentment toward the coterie of highbrows who so disdained him—he had started life as a barber's son, after all. The king's daughter was not always comfortable among them, either. Despite her great self-confidence, she was at times intimidated by them, feeling "that awe" as she later wrote, "that everyone feels when confronted by something greater." Her father had been a great man, but she had hardly any memory of him. Chancellor Oxenstierna was a man of vast ability, and she had responded to him with reverence, then caution, then outright hostility. There had been clever and capable people about her, but now, for the first time, she was surrounded by a large number of scholars and savants of the first rank, many with decades of learning behind them. Though she could hold her own in the learned discussions, she had little of the practical scientific experience or the literary skill that most of the scholars, including those of her own age, could take for granted. Even where she most prided herself—in her knowledge of languages—almost all of them had comfortably surpassed her. Christina had been brought up with three languages, Swedish, German, and French, and she spoke them all equally well. She read Latin easily, too, though she did not write it well, and she understood some Dutch and Italian and bits and pieces of other, related, languages. Enthusiastic admirers among the new arrivals made all sorts of extravagant claims for her: she knew Finnish and Arabic, they declared, and any number of other languages—"Seven!" said one. "No, eleven!" said another, "and the swear words, too." The claims were not well founded, and even if they had been, they paled by comparison with the eighteen tongues, ancient and modern, Eastern and Western, that Bochart knew, or the twenty-six known by the philologist Ludolphus. Even Bourdelot spoke good Latin.

Whatever the reason, Christina now threw in her lot with the physician, encouraging his jokes and even on occasions joining in with them. The dignified Ambassador Chanut was spared, but not many others, it seems. One day, fed up with listening to Meibom's talk about Greek song and Naudé's about Greek dance, Bourdelot suggested that the elaborate descriptions of the two learned gentlemen might be improved by demonstration. Christina agreed at once, and the two were obliged to appear before the entire court in a song and dance routine revived, supposedly, from the ancient world. The middle-aged Naudé swallowed the humiliation without riposte, but Meibom, a notoriously prickly character and thirty years younger besides, took advantage of his next meeting with Bourdelot to punch him squarely in the nose. Though he was obliged to leave Sweden directly thereafter, Meibom departed with his pride intact.

But at court, his bruised nose notwithstanding, Bourdelot kept the upper hand, and he did not hesitate to compromise even the queen herself, when necessary, in order to gain his point. His favorite target was the prim orientalist Samuel Bochart. Bourdelot loathed the humorless polyglot pastor with his three doctorates in theology and his games of shuttlecock—apparently his sole amusement, which, it must be said, Christina enjoyed as well. The loathing was mutual. Bochart despised his compatriot's ingratiating ways and his frivolous attitude toward religion; he had once referred to Bourdelot as "original sin itself." Bochart's masterpiece was his immense and immensely learned *Geographia sacra*, a study of the age of the world from biblical sources. It was on the strength of this work that he had been invited to Christina's court, and she had now arranged a private meeting to discuss it with him. An hour or so before the meeting, Bourdelot declared quite suddenly that the queen's health required urgent preventive care. As her physician, he said, he could not advise delay, and he administered her directly with an enema. Bochart and his biblical geography were simply obliged to wait. Christina seems to have taken it remarkably well, under the circumstances; she later observed to the disappointed pastor that, anyway, the world was like a woman: after a certain stage, it wouldn't do to investigate its age too closely.

She carried on heaping her favorite with riches and honors, with Bourdelot protesting disingenuously that it was all "really more than I deserve." At the few official meetings that the queen now bothered to attend, he accompanied her, standing by her side as the business of the day—the business of state—was discussed. Afterwards, they would retire together to her private apartments, where Bourdelot behaved in the most familiar way, sitting down while the queen was standing, and now and then even relaxing with his feet up on her sofa. Like all her favorites, he was soon rumored to be Christina's lover, and it was even said that she had become pregnant by him, and had "found the remedy along with the cause"—in other words, that he had procured an abortion for her, with the aid of the French surgeon Surreaux, and Madame Wachtmeister, wife of one of the queen's generals. Surreaux was said to have received the huge sum of thirty thousand riksdaler for his services; Chanut's physician, Du Rietz, was supposedly requested to assist him, but refused.

It is probably impossible to know whether there was really any truth in the talk. Like other unconventional women of her rank, Christina was ascribed many lovers, both male and female. Her love for Belle was widely known, and she was said as well to have taken advantage of Charlotte de Brégy, Saumaise's niece, during her bright, brief sojourn in Stockholm, forcing the lady "to perform immoral acts." Two daughters were reputedly born to Christina, both fathered by Magnus. And there is a letter from Bourdelot to Georges de Scudéry, which mentions his fondness for the queen, and adds ambiguously that she has "begun to taste."[10] None of it is conclusive, but if it was only gossip, it must be said that Christina herself provided plenty of material for it. With Bourdelot, she sat up or drove out at all hours, and received him alone in her bedroom; at times, he even stayed through the night. It is likely that they enjoyed no more than a close and easy friendship; certainly, though they were very often together, their feelings for each other seem to have been affectionate rather than passionate. Christina was in her twenties and Bourdelot fifteen years older, but they were, in a way, like a pair of high-spirited students, absorbed in their own amusements, keeping clever secrets behind closed doors, talking about forbidden subjects, poking fun at the dour professors.

THE PROFESSORS, FOR their part, were beginning to find it all rather tiresome. It had been convenient to escape the trouble in France, and they had been paid well, on the whole, and it would be amusing to regale their friends about life at the North Pole, or near enough. But there was nothing they could really *do* in Sweden, or at least nothing that they could not do more comfortably elsewhere. The queen had disappointed them. She treated them as mere specimens of learning, valuable objects that she had captured to bolster her own reputation. Her interest in science and scholarship had proved superficial, even frivolous. She had ignored the calculating machine that Monsieur Pascal had so carefully sent to her. She had given no money to Menasseh ben Israel for his grand new edition of rabbinical writings. She was not trying to do anything spectacular herself—even her alchemy equipment was beginning to rust. The professors breathed a sad, collective sigh, and began to think of leaving.

Christina's response was to pay them more, or to replace them with others of their kind. They stayed on while it suited them, and she basked in their reflected glory, effectively buying a reputation as the new Pallas, the Pallas of the North.[11] On closer acquaintance, the dazzling royal mind had turned out to be not brilliant, not original, but only clever in a rather ordinary way. They were disillusioned, and Christina may have been disillusioned, too; she had imagined herself seated in splendor among them, impressing them all, as she had once impressed her teachers, with her intelligence and curiosity and all the facts and figures she had learned. She was sensitive enough to see their disappointment, and she turned from it at once. Other eyes must reassure her by the reverence of their gaze. Newer voices, not yet familiar, must confirm her superiority, her sovereignty, her majesty.

The newest eyes, and the darkest, and the longest-lashed, belonged to a handsome Spanish diplomat, in whose honor Christina had founded a new order of merit, the Order of the Amaranth, named for the legendary flower whose petals of royal purple were said to bloom eternally. Wicked tongues whispered that the new order's emblem, an entwined AA, signified two lovers, Antonio and Amarantha, a name Christina slyly began to accept.

Christina bestowed the order on everyone whom she wished to please or to patronize. The gossips may have been deterred by its principal require- ment: a vow of perpetual celibacy. Prospective members who were already married—in practice, almost everyone but Christina herself—were obliged to swear that, if widowed, they would not remarry. The English ambassador, Bulstrode Whitelocke, though three times married, was one accommodating recruit. "Three times married!" declared the queen. Then how many chil- dren did his excellency have? His excellency had three children—"one per wife." "By God!" said the queen. "You're incorrigible!"

The motto of the Order might have suited Whitelocke, and others equally steady of purpose, but it was a singularly inept reflection of Christina's own volatile nature: *Alltid Densamme* (always the same), read its golden letters. In fact, her inquisitive and responsive nature left her constantly at the mercy of the latest new idea. Her need to dominate made things worse, and she wasted time and money that she could not spare on project after project that never came to fruition. She planned a vast reform of education in Sweden and Fin- land; her father had begun the work, the chancellor had continued it, but she vowed to outdo them both. The French had their Académie, and proud insti- tutions for painting and sculpture; Christina's Swedish Academy was to sur- pass them all with ribbons flying, leaving the French and everyone else admiring and gasping in its wake. She was alight with grand ideas, but none of them could sustain her interest for long, and she lacked the stamina even to see them brought to fruition by others. Few ideas lasted beyond their first, fine, careless rapture, and too often she found herself with unpaid bills for fireworks that had long since fizzled out.

Christina was a creature of impulse. Lacking a kernel of self-confidence, she lived in a constant swirl of defensive responses to the people around her, swamping them with gifts, lying to them, lashing out, undermining them, withdrawing from them completely. Only her image of herself remained the same: brilliant, powerful, authoritative — even, in the face of the clearest evidence, tall. It was no more than the thickest layer of bravado, and it concealed an interior world of fearful fragility. Christina almost always managed to convince herself of the truth of her own illusion. "To attack me," she once wrote, "is to attack the sun." She was surprised and hostile toward those, like the chancellor, who did not accept her at her own estimation, but she would reserve her bitterest revenge for those who attacked her sense of personal greatness.

TRAGEDY AND COMEDY

HE SUNDRY STORIES of the scholars who came to Stockholm have long since faded beside that of the greatest of all those whom Christina lured to her court. He was a physician, and, like many of the others, a refugee from France's troubled times. He was a mathematician, too, and a pioneer of the new scientific learning that was to sweep away the old world in a succession of mighty strokes. Though most of Christina's scholars admired his work, and some knew him personally, it was through Ambassador Chanut that she came to know—and, in a sense, to kill—the most famous of them all, the great French philosopher, René Descartes.

Descartes had been living in the little village of Egmond, on the northern coast of Holland. Here he had retreated after years of increasing difficulty in France and in the Dutch cities, working in the shadow of the Inquisition, pricked and poked by the smaller demons of Calvinist bigotry. Secluded in Egmond, he had at last been able to work in peace, seeing only friends, and refusing to publish anything new.

Descartes's translator, Claude Clerselier, was Chanut's brother-in-law, and through him the ambassador himself had become a close friend of the great man. Descartes was not a freethinker, but he had earned the displeasure of the Catholic authorities by his rationalist analysis of matters they regarded as their own preserve. Chanut himself, though a dutiful son of the Church, admired Descartes's philosophy and was proud of their friendship, and from the beginning of his stay in Stockholm he had maintained an eager correspondence with him on metaphysical and moral questions, relaying his enthusiasm for the philosopher to the queen, and his enthusiasm for the queen to the philosopher.

Christina was soon drawn into it. Convinced that she would be enraptured by Descartes's ideas, Chanut requested his brother-in-law to send a copy of the as yet unpublished *Metaphysical Meditations* to present them to her. Descartes hesitated. He had had some experience with clever young women. The Swedish queen, he felt, was no doubt like the young Princess Elisabeth of Bohemia. He had been corresponding with her already for years. She was more interested in moral questions. No doubt Queen Christina would be, too. On the other hand, he was not eager to begin a discussion of such subjects with so prominent a person. The Church in France would certainly get to hear of it: "If I publish anything about morals," he wrote, "they'll never give me a moment's peace."[1] He decided, philosophically, to write about the natural world instead.

Christina's response to his ideas was soon made known to him through the ambassador's own letters. Descartes had written that he considered the universe to be so vast as to have no definite limits, or at least none that could be perceived by man. Chanut recorded the queen's response:

If the universe is so vast as you say, then Man himself can be of no great importance within it. He and the entire earth that he inhabits can be no more than a tiny and insignificant part of the whole. If this is so, it is just as likely that the stars are inhabited, or the planets peopled with better and more intelligent beings than Man himself. Man can no longer believe that the universe is made for him, or that it can serve his purposes at all.[2]

Descartes replied, through Chanut, "I am not inferring that there are intelligent beings in the stars or anywhere else. . . . I leave these sort of questions undecided, without affirming them or denying them," but he conceded, "I am not sure that Man is the final purpose of creation."[3] The ambassador then conveyed a question that Christina had raised in her academy: What makes us love a person, she had asked, before we know his true merits? Descartes's surprisingly personal reply arrived in due course:

Love [he wrote] is a disposition of parts of the brain, although it may derive from the objects of the senses. . . . These pass through the nerves to reach the brain . . . and they leave a sort of imprint, so that the next time we encounter a similar object, we respond to it in the same way. . . . When I was a boy, I fell in love with a girl who had a bit of a squint, and for a long time afterwards, whenever I saw someone with a squint, I felt the passion of love. . . . So, if we love someone without knowing why, we can assume that that person is somehow similar to someone else whom we loved before, even if we don't know precisely how.[4]

Happily, it seemed, the philosopher had not been doomed from his boyhood to fall indiscriminately in love with every squinting woman he encountered. "Once I recognized what was happening," he added, "I was able to cure myself of it."

In the autumn of 1647, Christina asked Chanut to include in his next letter a direct query of her own. It was indeed a moral question: What was Monsieur Descartes's opinion on the nature of the Sovereign Good—not in a religious sense, but "in the sense that the ancient philosophers have spoken of it."[5] Despite his unwillingness to write of moral matters, Descartes replied quickly, adding, in a note to Chanut, "I hope that what I write will be seen by no other eyes than Her Majesty's and your own." To the queen, he responded as she had wished: the ancient philosophers, he wrote, "being without the light of faith, knew nothing of supernatural blessings," and therefore in his reply he would consider only "what good we might have on this earth." For an individual, he wrote, physical well-being and the blessings of fortune were not always at one's command, nor was the knowledge of good. There remained only the will to do good, the "firm and constant resolution to do exactly what one judges to be the right thing, and to use all the strength of one's mind to determine what that is. Then, although one can still act wrongly, one is at least assured of having done one's duty." And he added, "As I am sure Your Majesty sets more store by her virtue than by her throne, I am not afraid to say here that it is only virtue that deserves praise. All other good should be merely esteemed. . . . Only virtue is obtained by the right use of free will."

Christina was later to say that she owed to Descartes her first thought of rejecting the Lutheran faith of her fathers and embracing Catholicism, and it is perhaps here, in this letter, that the first seed of courage or justification was sown. Ironically, despite his own adherence to the Catholic Church, Descartes's insistence on "the will to do good," "to do exactly what one judges to be right," was very far from the unquestioning acceptance of its dictates that the Church enjoined upon its flock. Little wonder that he had been called a skeptic, and even, as he complained to Chanut, an atheist, "just because I tried to prove the existence of God."[6]

Perhaps the philosopher's November letter had set the queen to thinking. If so, she remained thoughtful for a very long time. In February, Descartes wrote to Chanut to express his misgivings about the letter; it was not a good explanation, he feared; he might have reworked it to better effect. And, he added, "I am really very eager to know what Her Majesty will make of it."[7] In May, still without a reply, he wrote again to the ambassador. "I think it cannot have pleased her, because although she has read it, or so you say, she still has not told you what she thinks of it. But you say that she intends to look at it again. No doubt she will like it better on a second reading."[8] In the event, it was not until the following February that Descartes received the queen's cursory reply, requesting a copy of his *Principles of Philosophy* to read. He acknowledged the letter swiftly. "Madam," he wrote—

If a letter was sent to me from heaven, and I saw it descend from the clouds, I could not be more surprised, and I could not receive it with more respect and veneration, than I have received that which it has pleased Your Majesty to write to me.[9]

And he concluded by declaring that he could not be more zealously or more perfectly devoted to obeying the queen's every command, "even if I had been born a Swede or a Finn." On the same day he wrote to Chanut, relaying a few hints for the queen about his *Principles*, and revealing, perhaps, a touch of pique at her delay of "several"—in fact fifteen—months in replying to him:

Of course you are quite right. It is enough to wonder at that a queen, perpetually engaged in affairs of state, should have recalled, after several months, a letter that I had had the honor to write to her, and that she should have taken the trouble to reply at all, let alone to reply sooner.[10]

The queen's attentions, having been too little, were suddenly too much. Having kept Descartes waiting for more than a year, she now suggested that he come to Sweden to wait upon her. Descartes was startled and dismayed, and he replied to the invitation immediately with two letters, both addressed to Chanut. The first, which was full of courtesies, was in fact intended for the queen's eyes:

I have so much veneration for the rare and lofty qualities of this princess, that I regard the least of her wishes as an absolute command. In consequence, I will not pause to consider this journey; I am simply resolved to obey.[11]

But in the second letter, Descartes revealed to his friend what he really felt about the prospect of the visit:

I am sorry to give you the trouble of reading two letters at the same time, but I thought you might want to show the other one to the queen. I have reserved for this one what I do not think she needs to see, namely, that I am myself surprised at how very little I wish to undertake this journey. It is not that I do not wish to be of service to this princess . . . if I could really believe that my journey would be of some use to her . . . but I have learned by experience that even among people of good understanding, even when they are really eager to learn, there are very few who can take the time to comprehend my ideas fully, and I certainly cannot expect this of a queen, who has so many other claims upon her time. . . . [12]

He did not want to go. Stockholm was too far away, and too cold, and the journey would not be an easy one, and he was "getting lazier and lazier all the time." He had no interest in Sweden "with its rocks and ice and bears." He wrote to Freinsheim in Uppsala, asking whether a Catholic would really be welcome among the Lutherans. The Dutch had proved less tolerant than they had appeared; so might the Swedes. Freinsheim reassured him, declaring besides that there would be no lack of furs and fires to protect him from the cold, and promising to arrange a dispensation from the usual tiresome routines of court etiquette. Descartes was not encouraged. He was only recently back in Egmond, anyway, after a very disappointing journey to Paris at the command of the king himself, or at least at the command of his regents. They had sent all sorts of promises and guarantees of what he might expect at the court, all written formally on parchment, and in the end, he had even had to pay the postman for delivering the invitation.

It was the most expensive and useless piece of parchment I have ever had in my hands. . . . None of them wanted to know anything more about me than what I looked like, and I came to the conclusion that they just wanted to have me in France like some sort of elephant or panther, on account of my rarity. . . . I am sure it would not be the same where you are, but after all these unhappy journeys of the past twenty years . . . I could be set upon by robbers, or be shipwrecked and lose my life. . . . If you are really convinced that this incomparable Queen still wants to study my work, and she can take the time to do so, I shall be more than happy to be of service to her. But if it is only a question of curiosity that will not last, please make some kind of excuse for me, and spare me this journey.[13]

It was not that Descartes doubted the capacities of the young queen. Apart from Chanut's endless praises, his predecessor in Stockholm, the former ambassador, Monsieur de la Thuillerie, had also spoken "very flatteringly" of her. Nor was the philosopher averse to female students: great abilities might be found anywhere, he maintained. He had long insisted that all his own work be translated from Latin into French, so that "even women" could read it. Above all, Descartes's correspondence with the Princess Elisabeth had convinced him that "persons of high birth, regardless of their sex, can surpass other men in learning and virtue even if they are very young."[14] In fact it was Christina herself who had made him hesitate, and given him reason to fear that he would be no more than just one more "elephant or panther" in her collection of scholars and artists. Her tardy reply to his letter implied a lack of real interest in his ideas, or perhaps simply a lack of time to study them — in either case, he felt, a visit to her court would be pointless.

In the event, Christina made the decision for him, not by any royal command, but by the simple expedient of sending a small militia down to Egmond to collect him. They were headed by Admiral Herman Fleming, son of one of Sweden's greatest heroes, but, although Fleming was an officer and a gentleman, Descartes was not eager to entrust himself to his hands. He sent off an urgent message to Chanut. "What shall I do?" he wrote. "The Queen has sent one of her admirals to get me. What sort of thing should I expect from a Swedish admiral?" Chanut urged confidence, Fleming kept smiling, and so the philosopher was captured.

They set off on the first day of September 1649, and after six hard weeks of travel by land and sea, with the ship's captain dazzled by Descartes's knowledge of astronomy, they arrived in Stockholm, just in time to see a few last rays of sunlight, and to watch the tardiest autumn leaves drifting along the slowly freezing river. Christina was impatient to meet the new arrival, and she arranged to receive him the very next day. She waited in some excitement, preparing herself for the encounter, imagining his noble mien, his commanding presence, and all the distinguished fixtures and fittings that must naturally accompany so great a mind. Into this heightened expectation, the great man toddled at last, shortish, fattish, mildly spoken, uncomfortable in his new pointed shoes, thoroughly unprepossessing. His hair had been curled in honor of the occasion, but it did not help him. Christina was sadly disappointed. A great man should look like a great man, or so she thought. What was the point of being great at all, if no one was likely to recognize it?

Descartes was in need of visible promotion, and Christina decided to arrange it. His grandeur would certainly be increased by a title—no doubt there would be one or two available, and if not, a new one could easily be created. He would need more money, too, and a suitable establishment; he could not lodge with the ambassador indefinitely. If nothing could be found for him, something would have to be built, something big, something fitting. In short, Christina promised everything to her trophy philosopher—everything, that is, but the chance to philosophize.

At first, Descartes did not mind unduly. He was delighted by her knowledge of French, and encouraged by her cleverness, and quite overwhelmed by her generosity. He felt sure she would be willing to help his talented young friend, a fellow royal bluestocking, the Princess Elisabeth. The princess, daughter of Bohemia's deposed "Winter King," a longtime refugee from the Thirty Years' War, was at the moment in particularly difficult straits. The queen was clearly in a mood to give, and Descartes began to drop hints. The princess was a marvelous woman, he said, modest, virtuous, beautiful, devout, and so brilliant! In the study of metaphysics, he declared, he had never met anyone so gifted. It was quite the wrong thing to say to Christina, who did not appreciate competition, even from distant refugees. While Descartes continued his panegyrics, she sent instructions to her mother, then traveling toward the princess's current abode, that she was on no account to receive her.

If Descartes had arrived too early, with the fierce winter still before him, he had also arrived too late, for if Christina's attention had ever been fixed on philosophy, it was by now engaged elsewhere. Her study of Greek, for one thing, was absorbing a good deal of her time. Descartes was dismissive of it, and of all the occult ideas that lay behind it. The ancients had nothing to tell us, he declared. Studying their writings now was not science, he said, but history. No one could claim to be right, simply by quoting old authorities or finding secret links between one text and the next. Nothing could be known to be true, he said, unless it was proved to be so. Christina took no notice, and plowed on with her conjugations. She had got him to her court, and he was welcome to stay in an ornamental capacity, but as for all his new ideas, she had read a summary of them which Chanut and Freinsheim had provided for her, and for the time being, that was quite enough. She suggested that he absent himself for five or six weeks, in order to get to know the country better. Descartes, astonished, felt no inclination to trudge around Sweden's bad country roads with the winter setting in. He opted to stay put, but the queen's alternative proposal did not please him any better.

Her birthday was approaching, and in her honor a grand court ballet was to be performed in the castle's new ballet hall. It was a handsome space, reconstructed from the apartments of an unhappy group of dislodged civil servants. Affairs of state were all very well, but a colorful spectacle, her majesty had informed them, was "necessary and useful at every court"; they could take their notes and their niceties elsewhere. Christina was excitedly involved in every aspect of the new production, and she stunned Descartes by suggesting that he take part as well. Pleading too little capacity, perhaps, or too much dignity, Descartes managed to escape performing in the ballet, and also composing the music, but, at the queen's insistence, he was obliged to write the libretto for it. In the wake of the Westphalia treaty celebrations, still ongoing in Stockholm as elsewhere, it was to be entitled *La Naissance de la paix*—The Birth of Peace. It was supposed to include both heroic and comic elements, but Descartes indignantly injected a loud and serious note into the proceedings by transforming the usual cast of lovable rogues into a troupe of maimed soldiers and refugees. The festivities of peace, he felt, should not be so joyous that the horrors of war would be forgotten. The libretto was not a masterpiece, and Descartes made several subsequent attempts to destroy it, but each time it was rescued by an undiscerning, or mischievous, Chanut. Despite its limitations, the ballet was received well—too well, Descartes may have thought; he was soon "requested" to produce another work for the stage. This time it was to be a play, and there were to be no maimed soldiers and no refugees: an Icelandic princess would be the heroine; there would be a lover, and there would be a tyrant, and there would be a dramatic escape in attractively rustic disguise. Descartes started the work, but did not finish it. He turned instead to a third request from the queen, namely, that he prepare the statutes for her long-promised Academy for Swedish scholars, using Richelieu's Académie Française as his model. He did so, and with great care, including among the Academy's rules the injunction that "everyone must listen to everyone else with respect, without showing disdain for what is being said." In the event, it did not matter, for Christina's Swedish Academy never saw the light of day.

It was all very discouraging, and Descartes wanted to go home. His apprehensions had proved only too correct. He had served his turn as an exotic elephant in the land of bears, and there was nothing more for him to do. Freinsheim, having eagerly encouraged him to come to Sweden, had taken flight himself to thaw out quietly in warmer climes. Chanut had been a genial host, and had even climbed a little mountain with him to take some atmospheric readings to send to Pascal in Paris, but the queen was too busy— though not with affairs of state—to pay him any attention. Perhaps Chanut mentioned something about the matter to her; in any case, she finally realized Descartes's despondency, and bestirred herself to spend some studious time with him. With her Greek and her ballets and the Italian paintings and the Senate and all the myriad matters of the court, she had little of it to spare, so little, in fact, that the thrice-weekly philosophy lessons had to be scheduled for five o'clock in the morning.

It was January 1650, the coldest month of the coldest year of an exceptionally cold century. It is possible that Christina suggested this inhuman hour to put an end once and for all to the talk of philosophy lessons; no doubt she had heard from Chanut of Descartes's habit, cultivated since his boyhood, of spending every long morning tucked up in bed in his "stove," a well-heated, not to say overheated, room where he could read and write and philosophize, whatever the weather. He had found Holland more than cold enough, and had comforted himself in anticipation of the Swedish winter by reflecting that "they have better measures against the cold up there."[15] But in the queen's library, at least, they did not. It was in fact not heated at all, and, moreover, while in the presence of her majesty, Descartes was obliged to remain bareheaded. No amount of fur or philosophy could protect him from the fierce cold; lack of sleep and the sharp disruption of the lie-abed habits of four decades made their own cruel contribution, and by the end of the month, Descartes had fallen ill with influenza.

Chanut had succumbed as well, and for some weeks the two looked after each other at the ambassador's well-heated residence. Christina sent along one of her personal physicians, but Descartes, trained in medicine himself, declined his help. He had no faith in Swedish physicians, he said—overlooking the fact that the man in question was Dutch—and he resorted instead to a medicine of his own devising: hot drinks of brewed tobacco. They did not help him. His condition worsened; he developed pneumonia, and at the beginning of March, he agreed that the queen's physician might attend him after all. He allowed himself to be bled, warning that there should be no wasting of "good French blood," but if the treatment could ever have helped him, it was now too late. On the eleventh of the month, "content to withdraw from this life, like a true philosopher and a true Catholic," he died, not dramatically, as he had once feared, at the hands of robbers or by shipwreck, but quietly and prosaically, from the flu.

Christina was shocked by Descartes's death, and she determined, perhaps not without a trace of guilt, that the attention she had failed to pay him while he lived should be fully accorded him now. She made plans for a magnificent funeral for him, to be held as soon as she could organize it. He was to be laid to rest in the Riddarholmskyrka beside all the kings of Sweden. His tomb was to be of marble, inscribed with the noblest references to the noblest mind in the noblest of all the sciences. In the meantime, as he had not been a Lutheran, the great man was hastily buried in Stockholm's cemetery "for the unbaptized," and his grave marked with a wooden plank. For a week or two, the grand memorial plans absorbed Christina, but soon they were forgotten, and two years later, on his pilgrimage to Descartes's grave, the young orientalist Pierre-Daniel Huet was shocked to see the paltry memorial, rotting away in the wind and rain. His playful nature quickly got the better of him, however. With a glance behind him, he took out his penknife, and to the solemn words *Beneath this stone*, he added a mischievous *made of wood*.

CHRISTINA HAD TAKEN CARE to provide a musical accompaniment to all the philosophizing and the talk of belles lettres. She began by demoting her mother's German musicians in favor of people of her own choosing. Maria Eleonora had done a good deal to develop Stockholm's modest court band of lutes and trumpets into an impressive professional ensemble. Christina could not get rid of them, but she did her best to lessen their standing. She sent for musicians from France, and once they arrived, prevented them from joining the various ensembles already in place at court. She promoted a rival English group as well, instrumentalists in the entourage of Ambassador Bulstrode Whitelocke. "Her Majesty would often come to me," the ambassador recorded, "and discourse with me of her musicke."

Christina admired the English music that Whitelocke's ensemble performed, and asked him to obtain copies of it for her. But she liked even more the music composed for the newly fashionable French-style ballets. From this music, and the ballets themselves, a few tentative new works sprouted at her court, operatic in style, a mixture of masque and song, entertaining if not memorable. But there was also vocal music of fine quality. At the behest of Alessandro Cecconi, already snugly ensconced in Stockholm as a petted favorite of the queen, a large company of Italian singers had arrived, with a troupe of actors in tow, and Christina's attention was soon turned to them. They had the usual repertoire of cantatas and madrigals, but also some very different works in the emerging Italian opera style, including one, if reminder were needed, portraying the frenzied grief of Maria Eleonora following the death of her husband. The music at least was new to Christina, and she was enraptured by it. The French musicians quickly fell from grace, though their masques and ballets continued to be staged. The queen signaled their demotion by allowing them to join forces with her mother's Germans, while the Italians, and Cecconi in particular, continued their rise. Christina came to trust Cecconi implicitly—he had taught her all her best oaths, after all, in French as well as Italian—and before long she sent him off to hunt out banned material from the backstreet booksellers of Florence.

Despite her love of music and her expensive encouragement of all kinds of music-making, Christina does not seem to have been a musician herself. No doubt she had learned to play as a girl, the lute at least, and perhaps a keyboard instrument; she also received dedications of music for viola da gamba and baroque guitar, suggesting that she may have played or particularly liked these instruments. A girl of her rank, even a tomboyish, horsey girl, would certainly have had some lessons, especially as her father and mother were both excellent lutenists. Magnus, as a boy, had spent an hour at his lute every day, and even Chancellor Oxenstierna played passably well. Christina may have lacked the patience or the application to become a good instrumentalist, but it is more likely that she was simply not very good at it. Christina did not like to be where she could not shine. If she could not quickly dominate, she withdrew.

With dancing, and the grand theatrical gestures required for the French ballets, the story was very different. She loved it all, and commanded fabulous performances at every opportunity, whether she could afford them or not. She took part herself, always in a starring role, usually as a queen or goddess, often as Diana, pursuing the chase and eschewing marriage with equal vengeance. She brought in specialists from Paris to advise her on the staging and to make the elaborate costumes, and to design the extraordinary machinery required to make the seasons change and the waves move and the gilt sun rise in splendor. It was, in a way, Christina's element, extravagant, spectacular, with a clear central figure to whom every other character deferred. It was an art of fantastic excess in an age of the same, and to bring it off successfully, for the audience, at least, may have required a touch of distance, or of irony. It is not a touch that Christina could have brought to any grand portrayal of herself, even in allegory, and she may have taken it all a little too seriously. The sophisticated Frenchmen, watching her dancing and posing, chuckled into their lacy sleeves, but they took care that she should not hear them, and she herself never realized the melodrama of her own performances.

Loving music, but not playing, loving to dance, but dancing badly, Christina had yet to come into her own in the difficult world of the fine arts. Her real gift was visual, and as yet it had only barely started to blossom. Sweden itself had not many master painters, though the best of them, Jakob Elbfas, had for some years taught her himself. The hundreds of paintings from Prague, especially those of the Italians, had opened her blue eyes wide at last, and several foreign masters were now invited to Stockholm. From time to time, curbing her restlessness, Christina agreed to sit for them, and many beautiful paintings resulted from the gifted hands of Sébastien Bourdon, David Beck, and Pierre Signac, and from other, lesser, artists. They painted all the prominent people of the court: Karl Gustav, and Belle, and the chancellor, and Johan Matthiae, and Maria Euphrosyne, and—several times over—Magnus.

HOLLOW crown

WHILE THE SCHOLARS and artists had been coming and going, one consistent problem had stood fast at Christina's court. The queen was now in her early twenties, and still she showed no sign of accepting a husband. Karl Gustav had been packed off to the army, and, after fifteen years of unencouraged wooing, the young Elector Friedrich Wilhelm of Brandenburg had finally given up and chosen another princess. In January 1649 came menacing news from the east: Christina's Catholic cousin, Jan Kazimierz Vasa, stepbrother to the late King Wladyslaw, had been crowned Poland's new king. Despite being a cardinal, Jan Kazimierz had secured the throne after agreeing to marry his stepbrother's Italian-born widow, Maria Ludowika, an ambitious and popular queen, much interested in the perilous politics of her adopted country,[1] and still young enough to bear many heirs. With a new Vasa marriage in Catholic Poland, the Swedish succession was now a matter of urgency. Christina, repelled by the "handing over" of the royal widow, found herself pressed ever more earnestly to settle the issue of her own marriage once and for all. But she could not bring herself to marry, and for months, even years, she had been slowly coming to a resolution of the dilemma at the heart of her young life.

Despite her proud claims of physical strength and great stamina, she had not been well for some time. She had known illness in childhood, but since her formal assumption of the throne five years earlier, the problems had increased in frequency and in degree. She suffered excessively from menstrual pain, and was prey to frequent headaches and fevers, insomnia, fainting fits, and even heart palpitations. The fevers may have been malarial (the ague), but the other symptoms seem to have had a more emotional cause. No one was able to help her: her several doctors between them could suggest only doses of brandy heated with peppercorns. Disliking all forms of alcohol, she cannot have welcomed even this mild remedy, and in any case, it had no effect. Her malaise continued, and in the end, she decided to help herself.

In the cold and gloom of a January day, Christina made her way to the *riksdag* hall, and there, to a startled and disbelieving audience, she announced that she wished to nominate a successor to the throne: her cousin, the Count Palatine Karl Gustav. She had not made the decision impulsively, she said. It was the fruit of long and careful consideration. But she needed, and now she asked for, the *riksdag's* agreement to it.

A great roar of consternation grew from the benches. What could her majesty mean? What need was there to nominate a successor? Her majesty had already agreed to marry Karl Gustav. It was on that understanding that he had been appointed commander-in-chief. The marriage would surely be blessed by children, and if not, her majesty's consort would naturally succeed her. Her majesty was clearly unwell; she must still be suffering from the effects of her recent fever. The whole idea was preposterous. They would certainly not agree.

Christina departed, and shortly afterwards, in her private apartments, a deputation of *riksdag* men arrived to remonstrate further with her. Her majesty was young, they said, and understandably nervous, but there was really no need for this latest idea. The count was a fine young man, a good soldier, and a loyal subject. He would make a fine husband, and a fine consort for her majesty. All would be well once the marriage had been settled.

In an echo of her earlier confrontation with the chancellor, Christina at first responded diffidently, then suddenly turned on the deputies, declaring that she did not intend to marry at all. "I am telling you now," she said, "it is impossible for me to marry. I am absolutely certain about it. I do not intend to give you reasons. My character is simply not suited to marriage. I have prayed God fervently that my inclination might change, but I simply cannot marry."[2]

The men of the *riksdag* departed, shaking their heads, but reassuring themselves that it was all just the apprehension of a young girl, admittedly rather an unusual girl, that Karl Gustav, although admittedly German, was sound enough, and clearly very fond of her majesty, that in due course the two would marry, and all would surely be well. In February, they issued a formal refusal of the queen's proposal.

A sudden gust of chill wind from the west obliged them to reconsider. In March, the court received a visit from Lord Patrick Ruthven, the Scottish

"General Reduving" who had once served, along with many of his country-men, in the armies of Gustav Adolf. Lord Ruthven came on desperate busi-ness, seeking Swedish help for the English king imprisoned by Cromwell's parliamentary government. He came too late: within days of his arrival, the court learned the astounding news that Charles I had been executed. Weeks before, in Whitehall, he had been led to the block, fortified by "bread and a little red wine," and wearing two shirts so that he would not shiver and be thought a coward in the bitter winter afternoon. There, before thousands of silent and bewildered spectators, the English king had "bow'd his comely Head down."[3]

The civil war in England had been followed with interest and some ap-proval in Sweden, where Charles's absolutist stance had not been well regarded; the chancellor for one had been heard to state in round terms that the Stuart king had brought his troubles upon himself by his own intransi-gence. But Charles's execution destroyed any real Swedish sympathy for Cromwell's parliamentarians. Christina herself was reportedly so affected by it that she determined to contribute personally to the punishment of the regi-cides and the reestablishment of the Stuart monarchy. Her indignation passed quickly, however, and she was in fact the first of Europe's monarchs to send formal greetings to the new government in London. Cromwell modestly ac-cepted a translation of it from one of his more learned accomplices, his Latin being, by his own admission, "vile and scanty."

From Poland in the east to England in the west, the Swedes saw rebellion. To the south, too, tremors shook the once sure powers of France and Spain. Catalonians and Portuguese and Neapolitans had taken up arms against their Spanish Habsburg overlords, and the civil war in France showed no sign of ending. Comparisons were soon being made in Stockholm, and talk of for-eign uprisings began to be followed with less eager interest and much more anxiety. The need for stability now seemed paramount. If her majesty could not be persuaded to marry quickly, the succession should at least be assured. At all costs, the Swedes must avoid having to elect a new monarch until the country's financial crisis had been settled; an election would certainly bring dissension, and might even lead to civil war. And so, in the same month of March 1649, the *riksdag* reconsidered Christina's proposal, and finally agreed to accept Karl Gustav as her formal successor to the Swedish throne. To most of them, it seemed a formality in any case; the two would surely be married, they felt, once the queen had overcome her girlish reluctance. And Karl

Gustav would have no hereditary rights; he would be replaceable, by election, if that need should arise in the future. Reassured for the moment, the men of the *riksdag* turned once again to their own concerns.

But to Christina, the acceptance of a formal successor appeared as a major victory. She believed that she had finally succeeded in balancing her strong desire to rule with her antipathy to marriage: while she lived, she would remain on the throne, but she would have no need to marry or bear children; that distasteful duty could be left to Karl Gustav's eventual bride.

A NEW YEAR DAWNED: 1650, a year of bitter weather and failing harvests. Sweden's farms were producing less than they had done for fifty years, and less than they would do for fifty to come. By the spring, Stockholm's bakers were fighting one another at the gates of the town, desperate to get their hands on the scanty supplies of flour arriving from the countryside. Anxiety fed wild rumors: it was said that the queen herself had no money for firewood, and at times no food in her larders. The rumors were false, exaggerations no doubt of delayed payments to tradesmen or reduced courses at formal dinners, but the people's hardship was real enough. Bad weather was partly to blame, but bad management contributed, too: the most productive land had by now been sold or "donated" away, and of the little revenue that it did produce in these slim years, even less made its way back to the crown coffers. Of the Stockholm bakers who did get their hands on some flour, many found that the loaves they baked could not be sold, or at least could not be paid for: the crown's servants had been too long without wages.

Christina, Nero-like, passed the desperate months preparing for her coronation. She was now age twenty-three, and had been a reigning queen for more than five years, and still she had not been crowned. Celebrations had been planned and delayed and planned again; the continuing war, lack of money, disagreement about suitable forms, and for a long time, her own disinterest, had prevented any formal decision being made. The six months of preparation initially deemed sufficient had been doubled, then trebled, then extended indefinitely while courtiers rushed from city to city and funds were scraped together or borrowed to pay for the great event.

The date was finally fixed for the twentieth of October.[4] It was to be a splendid day, a day that would announce to the rest of Europe that Sweden was no longer a cultural backwater, that its artists as well as its armies could hold their own in a European field. Impressing the neighbors was not Christina's wish alone: despite a great deal of grumbling about escalating costs, the Estates and the Senate, and the chancellor, too, agreed that it was time for Sweden to reveal another, more refined face to the world. In Stockholm, the magistrates voted hefty sums to add to the general splendor, though the *riksdag* clergy, condemning anew the exchange of ancient crown land for quick cash revenues, jeered that Christina had better be crowned Queen of Swedish Tolls and Excises.

In March, she fell ill once again, and for many weeks she remained so, recovering only with the weak spring sunshine in the last days of May. She set off for Nyköping to visit Maria Eleonora, and so did not see the arrival of the great ship containing the fireworks—and a decorative cardboard castle—for her forthcoming coronation. As the festive load arrived in Stockholm, another ship sailed out of the harbor with a different and more extraordinary cargo: a letter from the French Ambassador Chanut to Cardinal Mazarin in Paris, relaying the ambassador's astounding belief that, despite all the preparations for her coronation, the Swedish queen did not intend to remain long on her throne.

Chanut's belief was well founded, for he had received the information from the queen herself. For some time, she had been sighing over the difficulties of governing, whispering to him of her dissatisfaction with Lutheranism, hinting at a possible abdication. Her evident distaste for marriage had added to its likelihood, and by the spring of 1650, the ambassador was well and truly convinced. It was "a tremendously bold step," he thought, but her majesty would not be without other resources. Concealed within her own soul she harbored "treasures of happiness and joy" that would be forever at her disposal; the retreat she was preparing was "greater," he wrote, "than all the kingdoms of this earth." For the devout ambassador, the queen's concealed treasures were all religious; a potentially Catholic heart, he was sure, beat in her unhappy Protestant breast.

It was no doubt the question of religion that had prevented Christina from confiding in any of her determinedly Lutheran compatriots. She had many intimates with whom she might otherwise have shared her thoughts: Karl Gustav himself was still in Nuremberg, leading the Swedish delegation at the long series of discussions that followed the peace treaties, but her cousin, Maria Euphrosyne, was close to hand, and so was Magnus, and Johan Matthiae, and Belle, and even Salvius, recently returned from diplomatic service abroad.

But she may have felt that outsiders would be more impartial in the matter—they had less to gain, after all, and also less to lose, if she should abdicate. In any case, it was Chanut, and the emperor's recently arrived ambassador, the Conte Raimondo Montecuccoli—significantly, both Catholic—who were first to learn of the step she was now considering. Montecuccoli, sizing up the queen's likely successor with an earthy Italian eye, reported that Karl Gustav was "a very easy and friendly fellow, and a great drinker, all qualities much appreciated in these northern countries."[5]

While Cardinal Mazarin and the Habsburg emperor were kept abreast of developments by their dutiful ambassadors, no one else had the least idea that the queen was thinking of renouncing her throne. In July 1650, the *riksdag* assembled to discuss the country's parlous financial situation, and for two months they talked of alienations and resumptions of land, of ancient rights and modern needs. It was late in September when the queen finally interrupted them with a new subject for discussion: Karl Gustav, she declared, must be made hereditary prince of the realm. To Christina, it was an interim subject, a stepping-stone to the real matter of her intention to abdicate, but to the men of the *riksdag*, it was a shocking idea, and altogether without reason. They had already accepted Karl Gustav as the queen's eventual successor; in the event of her death, the crown would pass to him, and in due course, some other king could be elected, some Swedish prince of their own choosing. Karl Gustav was a soldier, and might again see military action—in theory, at least, an election might not be so far away. It would all be very different if he were made hereditary prince, if his heirs were entitled to claim the throne after him. It would mean the end of the Vasa dynasty, and a German king on the Swedish throne. The patriotic men of the *riksdag* issued a staunch refusal.

Christina staged a tactical retreat, but she did not give in. The *riksdag*'s unity on this new matter had not overcome their older divisions, and the young queen now revealed how much she had learned since her first clumsy interventions in the great affairs of state. The three commoners' Estates wanted a resumption of crown lands; the nobles' Estate insisted that the status quo must be maintained. The commoners wanted to increase their rights and decrease their obligations toward the nobles; the nobles wanted to maintain their power. The commoners' hope was the nobles' fear, and Christina played on both. She took none of their causes truly to heart, but she was sufficiently astute, and now sufficiently experienced, to exploit them for her own purposes. To the commoners she promised resumptions of land and relief from the most pressing of their obligations; to the nobles she promised protection against the commoners' more extreme demands. In return she demanded, from all of them, that they accept Karl Gustav as Sweden's hereditary prince.

Within a fortnight, she had achieved it. On the ninth of October, the *riks-dag* capitulated. At their last session before the coronation, the queen sat triumphantly beneath her silk canopy, with the new prince, just returned from Nuremberg, seated in unsought glory at her right hand. All parties believed that victory was theirs. No one thought to raise the matter of the financial crisis.

CHRISTINA'S CORONATION WAS a fabulous affair. But although the Swedes may have earned some cultural standing by its magnificence, in the event, as with so much else connected with her reign, it was the French who gained the most.

From France came Christina's coach, her throne, her wonderful robe and all her other coronation clothes, her coronation canopy of velvet and gold and silver, a marvelous saddle for her favorite horse, and liveries and gifts for hundreds of people. All had been made for her in Paris, and much had been years in the making. The coronation robe, arriving via Amsterdam, had lain in careful storage at the home of an ambassador for more than a year, awaiting the completion of other, less magnificent preparations. It was an extraordinary creation of "violet-brown" (purple) velvet, lined with ermine, trimmed with pearls, and laden with circles of solid gold crowns.[6] It was twelve feet long, and with the weight of the crowns was difficult to manage; it obliged Christina to ride in a coach rather than on horseback as her ancestors had done. It seems that even she was unnerved by the enormous cost of the robe. Many of the bills from the unidentified Paris tailor were quickly lost, or hidden, or destroyed. Later monarchs, facing their own financial troubles, would turn to the robe to help them out, removing the gold crowns and melting them down for reuse.

Christina's crown at least required only a modest outlay. It was the same crown that her mother had worn for her coronation as Queen Consort thirty years before. Unlike her own jealous mother-in-law, Maria Eleonora had handed over her crown quite willingly, though arches were now added to indicate the higher standing of her sovereign daughter.

Very tardily, Christina sent to The Hague for an erudite book on emblems and heraldic devices; her own had yet to be decided. The throne room at Uppsala was looking the worse for wear: Christina commissioned thirty-five paintings from Jacob Jordaens in Antwerp, who packed off a series of classical scenes, conveniently to hand in his studio.[7] Christina's private apartments in the castle, she felt, should also be redecorated for the occasion, and the bare cathedral walls must be covered with tapestries depicting the great events of Swedish history, the uprising against the Danes and Gustav Adolf's victories and other grand, inspiring scenes. The order was sent for scores of them to be woven especially, but it was much too late, and the tapestries that were obtained retold instead the usual tales of gods and nymphs and hunting. The arches at least were to be original, splendid triumphal arches through which she would drive—her coronation could certainly not proceed without one or two, at least. In the event, three triumphal arches were deemed to be necessary, all huge, all magnificent, all in the ancient Roman style. The city magistrates agreed to pay for two of them out of the public funds—in fact the public deficit—and the senators met the cost of the principal arch by raising a special tax on every horse, or rather horse owner, in the kingdom.

Until now, it had been assumed that the coronation would be held in the cathedral church at Uppsala, where, by tradition, Sweden's kings had always been crowned. But the affair this time was on too grand a scale, too many people had been invited, and the little cathedral was too small. Worse, Uppsala had not lodging enough, or not lodging good enough, for the hundreds of expected grandees. Stockholm would have been easier for the master-builders and their men: their workshops were there, and their families, too. But most of the materials were already by now in Uppsala, and the work had already started. The astrologers muttered that tradition must be observed, that it would be courting disaster to change the time-honored ritual. Christina professed herself indifferent, with an aside that she would prefer all the same to be crowned in Stockholm, if only to prove that she was not superstitious. In April, with only months to go, it was still assumed that the celebrations would be held in Uppsala. In May, a final decision was made; the work was stopped, and men and materials were transferred back to Stockholm.

The shortage of time was now acute, and it was clear that it would be impossible to complete the three arches by October. The first was simply left unfinished. The second was built, not in stone but out of wood, and painted by a fluid French hand.[8] Efforts were now concentrated on the principal arch. It was to stand near the water, at an elegant angle to the bridge at Malmtorget, but even at this eleventh hour, the proposed site was not quickly approved. General Lennart Torstensson, ferocious, gouty hero of the Thirty Years' War, had already started work on his own new town palace in the same corner of the square; it was to balance, or outclass, the just-completed palace of Field Marshal Lillie, in the opposite corner. Torstensson was not about to change his plans now for any cardboard arch, no matter what the occasion. In consequence, a huddle of little stalls and lean-tos was swept away to make room for the arch, and in early June 1650, with less than five months remaining, the work began in earnest.

It seems that the architect, Jean De la Vallée, had in fact been working on the design of the arch for some time. He was certainly able to produce it with amazing rapidity, and it bore a strong resemblance to the great arch erected for the Emperor Constantine near the Coliseum in Rome. The beautiful marble of Constantine's arch, however, was not to be matched in Stockholm. Despite Axel Oxenstierna's enthusiasm for a permanent monument, Christina had to be content with a wooden structure draped with painted canvas, and decorated in stucco and wax. Shortage of time permitted no other solution, though the senators managed to delay things further by entrusting the work first to the court's old master-joiner, and then to their master-carpenter. A predictable furor ensued, but the two eventually set to work together on the vast pile of materials: out of hundreds of dozens of wooden planks, five thousand yards of canvas, eight hundred pounds of wax, three thousand pounds of various resins, a ton of linseed oil and half a ton of olive oil (inferior grade), sixty barrels of plaster, and assorted bits of iron and stone, they and the artists constructed a magnificent edifice some eighty feet wide and a hundred feet high, with twelve columns, and so cleverly painted that from a distance "it looked as though it was made of stone."[9] Even the painted flags of the German armies captured in the recent war seemed to be fluttering in the wind. It was topped by twenty-four monumental statues, allegories in wood, the work of the old Swedish master-carver Jost Schutze; though in the classical style, it seems they were not without traces of "Germanic exuberance." Among those who painted the arch was Christina's old drawing-master, Jakob Elbfas; in gold letters fifteen feet high, he proclaimed the queen's virtues, but most of his skill went into twenty huge tableaux, depicting, at Axel Oxenstierna's command, Sweden's military victories during the years of the regency. Not only an homage to the young queen, the great arch was also a monument to the chancellor's own achievements.

Christina spent the days before her coronation at the De la Gardies' grand new country house of Jakobsdal,[10] and from here she made her formal entrance into the city, wearing a magnificent gown embroidered all over with gold thread and pearls and precious stones. Like her other coronation garments, it was a Paris creation in the latest style, and like them, it had had to be remade for her in Stockholm—the French tailor had not made allowance for Christina's uneven shoulders. She drove back into the city in an open carriage upholstered in red velvet and silk with silver and gold embroidery, a gift from Karl Gustav. It was drawn by six white horses, and beside it walked sixty young noblemen, all dressed in yellow. Her own horse, unridden, trotted along with them sporting a beautiful pearl-encrusted sidesaddle of purple velvet, and a bridle, reins, and stirrups of enamel and silver gilt, studded with diamonds.

Christina was followed by Maria Eleonora, who was herself followed by six camels, complete with howdahs, a team of reindeer from Lapland, and twelve extravagantly saddled mules. Enthusiastic nobles trailed behind, one in a festive carriage pulled by two brightly caparisoned cardboard elephants. Townsfolk and diplomats watched from the windows along the route—unresolved disputes about precedence had forced the removal of the diplomats altogether from the proceedings, though a place was kept in the church for Chanut. Every detail was noted by a special French correspondent, invited by the Swedes with a view to impressing all the fashionable readers in Paris.

On the twentieth of October, a day of rare autumnal beauty, Christina stepped once again into her gold-and-velvet coach. From the Tre Kronor Castle, she drove in grand procession to the Storkyrka, to be met at the great portals by the Archbishop of Uppsala, champion of the commoners' Estates, with her grandfather's gold annointing horn in his hands. The church was hung with the bright new tapestries of landscapes and hunting scenes and classical myths, and through these they now processed, Christina in a simple dress of white silk beneath her fabulous robe, and her acknowledged heir, Karl Gustav, walking after her, in a crimson cap with a tall spiked crown on top of it. Magnus carried the royal standard of blue silk; his father, the great general, now blind, was led behind him, and a vast train of Oxenstiernas and Brahes and Sparres and other De la Gardies followed, all the prominent men of state, and all the favored scholars and artists of Christina's court, and Maria Eleonora, unforgiven and unrepentant.

Before the altar stood a throne of chased silver, a coronation gift from Magnus.[11] More a beautiful armchair than a throne, it was small and elegant, with a wooden frame. Despite his earlier ungallant remarks about his manly queen, Magnus's gift was decidedly feminine, with slender, curving arms and legs, and backed by two proud and pretty little figures, Justice with her sword and Prudence with her mirror. Here Christina now seated herself for her formal consecration as Sweden's queen. The archbishop presided, while five other bishops uttered prayers alongside him. After anointing her with holy oils, the archbishop placed the crown upon her head, and invested her with the royal regalia: the ornamented sword, the gold scepter and key, and the golden orb, which must have challenged many a royal geographer—the northern countries had all been engraved back to front by mistake. At the end of the ceremony, as if there might have been any doubt about it, a herald proclaimed, "Queen Christina has been crowned, and no other person!"

They processed grandly out of the church, into the fading day, and Christina stepped into a second carriage, draped in silver cloth, while medals of gold and silver were thrown to the crowd around her. She arrived at the castle as night was falling, saluted by eighteen hundred guns; it took a deafening two hours to fire them all off. For the guests, a vast banquet followed. Too large for any of the castle's halls, it spread to a bevy of smaller rooms, while out in the streets, the townsfolk feasted on roast oxen stuffed with turkeys and geese, and public fountains flowed with wine. Fireworks burst above the lovely waters of Stockholm, and many a noble glove took many a roughened peasant hand to dance in the torchlight through the autumn night.

The festivities continued through the late dawn; in fact they went on for weeks, with masques and ballets and feasting and jousting. Magnus, taking his turn among the young blades to tilt a lance at the "Turk's head," fell heavily from his horse, but by evening he had recovered sufficiently to preside at another huge banquet. Through the shortening days of November, a series of animal combats was staged, in gladiatorial style. The most dramatic, at which Christina herself was present, featured a bear and a lion—not, apparently, the one from Prague—a calf (which did not last long), a horse, and a buffalo. An excited crowd assembled, and the coliseum atmosphere intensified as the animals were released. The lion chased the calf, the bear chased the lion, the buffalo butted the bear, and the horse kicked it. The calf fled, the lion turned on the bear, and a satisfying struggle ensued. The bear's roars were louder than the lion's, so the lion was recaged, and the horse and the buffalo rounded up. The bear, which had certainly had the worst of it, salvaged its dignity by washing itself off in a pool in the middle of the arena. It was agreed that the lion had made a very poor showing.

When the noise had died down in Stockholm, echoes of it could be heard for months as nobles across the country vied with one another to produce the grandest dinner or the brightest fireworks. Like their new-crowned queen, most of them paid on credit.

THE CORONATION HAD been a gigantic diversion, welcome even to the chancellor, but in November 1650 the *riksdag* reassembled, with the business of the country's finances still before them. The nobles were anxious and the commoners eager, both expecting that the resumptions of crown lands were about to begin. But Christina had achieved what she had set out to achieve; Karl Gustav was Sweden's hereditary prince, and she had no need now to inveigle the different parties of the *riksdag*, or to keep her promises to any of them. To the nobles' relief, she announced there would be no resumptions after all, nor would she accept any limit to further alienations of land or other assets if that should be her pleasure; the queen's royal prerogative was not to be infringed. The dismayed commoners were obliged to abandon the focal point of their joint efforts, and they quickly fell back into their three separate Estates, leaving the nobles triumphant by the grace of the queen's duplicity. She threw them little tidbits of consolation: a few long-sought privileges were awarded to the clergy, the peasants' labor service was decreased, and the burghers gained a few solid provincial posts, as far as possible from Stockholm.

There was no real reassurance for the nobles. They had retained their exalted place at the nation's table, but they had had to eat from the queen's own hand. Thenceforth the Senate and the great families would have to defer to her. It was not a question of procedure or of statute; there had been no constitutional change. Christina's splendid, public coronation had increased her prestige and her confidence, but her triumph was above all a question of will, and it was now clear that, when required, she could summon it in abundance. She had recovered all the personal authority that she had lost by nominating a successor, and, more importantly, she had greatly strengthened the power of the crown in its longstanding feud with the great noble families. It was a remarkable achievement, and she was very proud of it. And in her victory, she decided to be magnanimous; quite suddenly she became much friendlier to the chancellor. His influence was still great, and now she had need of his support for her newest, most incredible plan.

Christina had so far used much of her power, paradoxically, to divest herself of it, pushing through Karl Gustav's acceptance first as her successor, and then as hereditary prince. She had enjoyed getting the upper hand over the question of resumptions, too, but on the whole, the business of government bored her. There were too many compromises, too many views to consider, too many financial problems, too many Oxenstiernas. Ruling was one thing, so long as it meant commanding. Governing, by contrast, held few attractions for her.

Why need she carry on? Karl Gustav was already her heir. She was still quite young enough to be pressured into marrying, if not Karl Gustav himself, then some other prince. Friedrich Wilhelm had carried on his suit for fifteen years, and then there had been Karl Ludwig of the Palatinate, and Leopold of Habsburg. It was never going to end, as long as she was queen. If she could lay the crown aside, now, once and for all, she would never have to marry. It was the only way she could be sure of avoiding the most repellent of all possible futures. Had not Luther made the fate of Protestant women clear? Let them bear children unto death! he had cried. No, this she would not do. She would not bear children at all. She would not be used by any man "as a peasant uses his fields," and besides, what kind of child might she have, with her sicknesses and her ambiguous attractions and her silly, hysterical mother? "I could as easily give birth to a Nero as to an Augustus," she had said.

She wanted to rule, but without oppositon, without compromises, without a husband. She wanted to be queen, but queen without a dynasty. If she gave up the crown, what would she lose? Even as the queen, especially as the queen, she could not do as she wished. There was so much work, so many problems in the towns and countryside, so many councillors pushing her this way and that. And there would always be the question of marriage. That could go on for twenty years more.

In April 1651, she confided to Magnus that she had made up her mind to abdicate, and toward the end of June, she wrote to her cousins with the same astonishing news. Appalled and disbelieving, Karl Gustav sent his brother off with a letter to Johann Kasimir; Adolf was to explain what Christina had said—it was "too horrifying," wrote Karl Gustav, to be repeated in writing. Christina had meanwhile sent Magnus as her own emissary to her uncle, but as Magnus stood to lose more than anyone but the queen herself, he failed to reassure him. Christina then sent Johan Matthiae to Johann Kasimir, but being hardly more reconciled to the business than Magnus had been, he had no more success. All of them felt that the decision would surely rebound against them once their enemies at court got hold of it. They feared it would be seen as their own doing.

Karl Gustav sagely decided to absent himself altogether. He made a brief sortie into Stockholm to attend the funeral of General Lennart Torstensson—who had had no time after all to enjoy his splendid new house near the coronation arch—then set off for his favorite retreat on the Baltic island of Öland. Christina used his few days in the capital to try to convince him to marry an unsuspecting German princess who was paying a visit there. She felt her decision would be more readily accepted by the Estates if Karl Gustav could be presented as already on the way to producing a new generation of Swedish royals, even if they were German.

For six weeks or more, the news supposedly remained a secret within the two families, though the chancellor certainly heard of it, and Chanut continued his reports to Cardinal Mazarin in Paris. In early August of 1651, Christina informed the Senate of her decision, which required, she told them, "not advice but assent."[12] They responded at once with an absolute refusal, and followed it with a collective letter, in fact written by the chancellor, reminding Christina that it was her duty to remain as queen, sacrificing her own interests if necessary for the good of her people. Though they did not draw attention to it, the good of the senators themselves was also at stake: a new king might change his mind about the resumptions, and retrieve the crown lands after all. They had agreed to accept Karl Gustav, but they could not afford to see him brought to the throne at such short notice; they would need time, Christina's lifetime, to bring him around to their way of thinking. Besides, the queen had only just been crowned—it would take years to pay for that coronation, let alone a second one. Perhaps it was all a ploy to conceal something else that she really wanted. Her grandfather, after all, had frequently used the threat of abdication to force agreement to some plan or other, and even the chancellor had stooped to it once or twice.

Christina could find no allies among her own people for her latest, most outrageous plan, and this surprised her. The proposed royal family themselves could not be enlisted; they feared a backlash that would put their friends at risk and perhaps even make the land ungovernable. Puzzled, she wrote to

Karl Gustav, her oldest ally and now, if she could manage it, her greatest beneficiary—would he not come to Stockholm and help her push the plan through? He would not. On the last day of September, the senators presented Christina with another firm letter, a petition, in fact, urging her to abandon all thought of abdication and offering her greater help if she should be finding the burdern of government too heavy. Christina was impressed by their united stand and pleased by their apparent devotion to her rule, but she gave them no more than a temporizing answer: she was not well, she said, and anyway she could never marry; Karl Gustav would be a far more suitable monarch. The senators departed, unsatisfied, and in the middle of October she summoned them again, determined to gain their assent. The meeting this time lasted more than five hours, with both sides adamant until the very end, but the twenty senators eventually won the day. A few weeks later, Christina announced that she had reconsidered the matter: she would not abdicate, after all—at least not for the time being. The senators, jubilant to have retained their queen and, in consequence, their lands, ignored or forgot her warning suffix: her plan to abdicate had not been abandoned, but only postponed.

As so often had happened before, the emotional strain at length demanded physical payment. One evening, as she sat at supper with Magnus beside her, Christina suddenly collapsed. She gasped a farewell to him, and lapsed into unconsciousness. For an hour she lay still, her pulse so weak that she seemed to be dead—no one thought to try to revive her. When at last she showed a faint sign of life, her French doctor, Du Rietz, was hastily called. She recognized his face through a blur of pain and dizziness. "I didn't think I'd ever see you again," she murmured.

THE TALK OF SUCCESSION and abdication had done nothing to solve the country's financial problems. Through all the meetings and debates and correspondence, Christina had maintained her own priorities. She had achieved what she wanted, or most of it, but she had done little to answer the commoners' pressing grievances. Rebellion still seemed all too likely. "All that's lacking is a leader," growled one angry man. The chancellor's son Erik declared that when the nobles got home from the *riksdag*, the peasants would break their necks. Whether he believed it or not, his father admitted that he was, for the moment, in no hurry to return to his house in the country. At court, a growing faction muttered discontentedly against the queen.

At the end of 1651, the festering crisis was brought to a head by one of Christina's own most favored servants, Arnold Johan Messenius. He was a learned and outspoken man from a famously learned and outspoken line; within his family, imprisonment had become almost as strong a tradition as scholarship. His father, Johan, a celebrated historian, had been a mentor of Johan Matthiae during his student days. Subsequently imprisoned for nineteen years for seditious writings, he had managed nonetheless to continue a married life of sorts with his equally outspoken wife; the boy Arnold Johan was born and brought up in his father's prison cell. On reaching adulthood, Arnold Johan had soon received a fifteen-year sentence of his own for the same offense. His criticisms had been directed against Axel Oxenstierna, and when Christina attained her majority, anxious to wield her own authority and only too happy to discomfit the chancellor, she had pardoned Messenius, then ennobled him, enriched him vastly, and appointed him Historiographer Royal.

With the end of the year approaching, Karl Gustav had left with a party of followers to enjoy the hunting season on the island of Öland. While there, he received an anonymous letter exhorting him to depose the spendthrift queen and her irresponsible ministers, including the chancellor and Magnus and his father, who were, or so the writer claimed, leading the country to ruin. If necessary, he urged, the queen's life itself should be sacrificed. Karl Gustav lost no time in informing Christina, and despite her extravagant support of Messenius, she immediately suspected that the letter had been sent from him; his criticisms of her government had for some time been all too loudly in evidence. Messenius was arrested, but quietly, and not at once—it was hoped that caution might also entrap his accomplices. Karl Gustav, though alarmed enough to have informed the queen, does not seem to have regarded the threat as really dangerous. Fearing her response, he wrote to Messenius and the other suspects, urging them to flee the kingdom. They did not, and as the plot slowly unraveled, it became clear that it involved some of the greatest men in the land.[13] It seems they had wanted to send a warning to the queen rather than to call for any real action against her, but her shock was nonetheless immense. Pride dampened her outrage—it could not be admitted publicly that dislike of her rule ran so deep, and so close. A second arrest was made, of Messenius's twenty-two-year-old son, Arnold, once Karl Gustav's page, but the other suspects remained free, too powerful or too popular to be charged. Christina exclaimed at the two men's "ingratitude"—how could they have behaved so, after all she had done for their family? Messenius at first denied his involvement, as did his son, but when confronted with the instruments of torture, both confessed: the father had known of the letter, the son had written it.

Their trial involved no independent court of law. Too many great names had been implicated, and a legal hearing would have shone too bright a light on the government's many failings. The Senate itself, and Christina, heard the confessions and passed sentence. As Karl Gustav had feared, Messenius and his son were condemned to death.

They were permitted to meet on the morning of their execution, and they made a dreadful farewell, the young man on his knees, weeping at his father's feet. Messenius was beheaded, but his son was taken to be broken on the wheel, his limbs beaten to pieces as the wooden frame revolved. The body was left exposed at a crossroads, while one brave pastor shouted from his pulpit against the queen's injustice.

In a final act of loyalty or penance or bargaining, Messenius had bequeathed his marvelous collection of books to the queen, begging her to ensure that his widow and daughters would be provided for. She took the books, but ignored the women, who quickly declined into a miserable poverty. At court, the biggest birds remained, their feathers ruffled, but unplucked. Karl Gustav, despite his sympathy for the culprits, received a handsome reward from the queen: a portrait of herself, surrounded with diamonds.

Relations improved between the queen and the chancellor. A common cause had for once united them, and they remained for some time on better terms than they had been since the years of the regency. The rapprochement may have softened Christina: three years later, as she rode toward the ghastly crossroads, she was heard to express some regret for young Arnold Messenius, and she ordered his body to be taken down for burial.

THE ROAD TO ROME

LAGUE HAD BEEN THREATENING Stockholm and Christina had withdrawn to the country. In the autumn of 1653, the danger over, she returned to her court, empty now of all that had held her interest. Most of her scholars had departed, driven away by boredom and snow and, above all, by the end of the civil war in France. The boy-king was on his throne again, with Cardinal Mazarin securely perched behind him. Christina wandered alone through the chilly rooms of the castle. All her pretty swallows, so carefully fed, had flown south. The high summer of her reign was drawing to a close.

She had done her own work to hasten its end. Her talk of abdication had sown seeds of change in many otherwise conservative heads. Her lavish way of life and fitful rule had encouraged public discontent in a time of hardship. But, above all, she had by now been involved for more than two years in secret discussions with Jesuit priests from several different countries. They had talked about the old gods and the one God, about the nature of the world and the nature of mankind, about reason and free will and celibacy, and about the possibility of a Lutheran queen abandoning her faith and becoming a Roman Catholic.

It was a perilous journey to be making in Sweden. Catholicism had driven Christina's uncle from the throne; the battle against it had cost tens of thousands of Swedish lives, including the life of her own father. As monarch, she was officially head of the Swedish Lutheran Church. The practice of Catholicism was illegal, and punishments could be severe: during her own childhood, Swedish Catholics and foreign missionaries alike were arrested and sometimes tortured; two had been executed shortly before her birth. Banishment from the country and the confiscation of property remained the standard penalty. Swedish subjects were forbidden to attend the masses held in the private chapels of foreign Catholic diplomats, and even these masses were sometimes interrupted by zealous Lutheran officials.

Christina's own attitude, apart from one burst of bigotry early in her girlhood, had always been relaxed. Her friendship with Ambassador Chanut had dispelled the purplest myths about Catholicism that she had assimilated in the Protestant north, and her own doubts about Lutheranism, and indeed about Christianity as a whole, had encouraged her tolerance. Saumaise and Bourdelot and others from free-spirited France had persuaded her that a greater intellectual liberty would be possible within the Catholic fold. Christina was not yet convinced, but she was interested. Her agents were still hunting down copies of banned books for her, including Porphyry's third-century tome *Against the Christians*, a Jesuit priest's "incautious" book on atheism, and the infamous though apocryphal *Three Imposters* (namely, Moses, Mohammed, and Jesus). She had long outgrown any loyalty to Swedish Lutheranism, whether as a personal conviction or as a state religion. She was temperamentally drawn to the strongly hierarchical organization of the Catholic Church, which corresponded perfectly with her own ideal of the perfect secular state — effectively, an absolute monarchy. She spoke disdainfully of the "weak" forms of government, oligarchy and elective monarchy, that Sweden had known in the past; in later years she would describe herself proudly as the country's "most absolute" ruler. In any case, the Catholics had all the best cities — above all, Rome. With most of her foreign birds flown, there seemed to be nothing to look forward to but more ice and snow. And now a new offer of marriage had arrived, this time from the Emperor Ferdinand pressing her to marry his eldest son. Christina was becoming desperate.

Isolated from her French friends, harassed by the business of government, threatened by pressure to marry, Christina had come to view the Roman Catholic Church as a haven from all her trials and troubles. The Roman Catholics, it seemed, could protect her from it all. Apart from anything else, they valued the celibate life — unlike Luther with his "let women bear children unto death." From this point of view alone, Catholicism was "a beautiful religion" — she had said so herself before. And it was broader, too, better for someone clever like herself, not so narrow as those carping Lutheran pastors. Bourdelot had said so, and Saumaise, and even Descartes had preferred to stay within its fold.

And, it had to be admitted, the Jesuits were certainly clever fellows. They knew everything, it seemed, astronomy and Hebrew and law, and they were somehow never at a loss for an answer, even when she went right up close to them and looked them straight in the eye, they were never daunted, they always came back with some silky reply. Christina felt there must be something in it, if men like that could be part of it all. She was enraptured by the thought of Rome, the thought of living there in the warmth, among the paintings and the beautiful squares and palazzi. Stockholm was nothing by comparison; it was boring, and sterile, and there was no new thinking, no vibrancy, no *life*. People's thoughts just froze solid, like the water in wintertime — Descartes had said so himself. The best idea that anyone had there was simply to leave. All the best people left. Even Descartes had left, in his way.

Christina's path, then, was obvious. She must leave, too, and not wait until she died of the flu or the plague or just — just expired of absolute ennui. And if she were to leave, she must of necessity relinquish her crown.

Abdication was thus inevitable for Christina. Catholicism, however, was not. There were many possibilities for a wealthy noblewoman without the restraint of husband or child. She might have traveled, made a "grand tour" as Karl Gustav and Magnus had done; as yet she had never ventured outside her own country. She might have lived a scholar's life like the Princess Elisabeth, or established herself privately in a handsome country manor, and husbanded her lands in quiet prosperity. She might have become a public advocate for religious toleration; it was her instinct, after all, and her father's prestige alone would have carried her voice far. But none of this was for Christina. For all her curiosity and cleverness, for all her robust days of riding and hunting, she did not have the patience or the tenacity required for a life of sustained purpose. Her nature was restless and impulsive, and surprisingly dependent. Her constant bravado notwithstanding, she took no steps without a confidant, and usually had several. Christina had her role to play as the daughter of the great Gustav Adolf, but there was much in her of her mother, too. She drew her sense of self from those about her, and looked to others to affirm her own desperate need for greatness. She could not simply abdicate and retire to private life. She wanted to give up the crown, but she needed to remain the queen. She wanted to be at the center of things, and that center, or so it seemed to her, was Rome.

Her first contact with the Jesuits had occurred in the spring of 1651. It had come about by chance, through the unwitting mediation of the Portuguese ambassador, José Pinto de Pereira. Pereira had been in Stockholm since shortly before Christina's coronation. He had come to negotiate terms for

Sweden's trade in Portugal's West African colonies. Chancellor Oxenstierna had handled most of the discussions, but when more pressing matters drew his attention elsewhere, Christina stepped in to take his place.

Pereira's supposed secretary was one Antonio Macedo, in fact a Jesuit priest, and it was he who led the discussions with the queen. It seems that this was part of a ruse on Pereira's part to persuade the Swedes that he himself was something of a simpleton—overconfidence might lead them to make mistakes that would be to Portugal's advantage. The talk was in Latin, which Pereira did not understand well, or so he pretended. It did not long concern West Africa. In theory, Christina wanted a leading hand in all the affairs of state, but in practice she had neither taste nor flair for government. She had never been interested in anything to do with trade, and she did not suddenly become so now. But she liked Macedo, and he knew a good deal about things that really did interest her. For one thing, they had a friend in common: Christina's former ambassador to Lisbon was a Catholic convert and a Franciscan friar; he knew Macedo well, and there was no doubt plenty to say about the old life and the new life of Brother Lars Skytte. From here their talk seems to have flowed quite naturally to the great questions of philosophy that had underlain Skytte's conversion.

Christina was particularly interested in the Church's attitude to the new scientific thinking. How could Catholic teaching be reconciled, she asked, with what Copernicus and Bacon and Galileo had said? What about the Inquisition? Its reputation was terrifying. What kind of intellectual liberty could be expected with such a force against it? Macedo demurred. He was not an expert, of course, but he could assure her majesty that, though the outward forms had still to be observed, a Roman Catholic might think freely enough in private. Galileo's work had been permitted, indeed encouraged, at times even financed by Cardinal Princes of the Church. His mistake had been to challenge the Church openly. But he must advise her majesty to seek further elucidation. There were many fine scholars within the Society of Jesus. One, or several, could certainly be sent for. Her majesty had only to command.

Her majesty decided to do so. Macedo records that, one late summer day, she drew him aside "into the most remote rooms of her private apartment," and there whispered into his ear the greatest secret of her life, and of his. She was thinking of abjuring the Lutheran faith, she said, and converting to Roman Catholicism. Macedo was the first and only Jesuit she had known; he was an able man and a man of integrity, and she felt she could trust him to carry this dangerous information to Rome, and send back to her "two Italians of your Society" with whom she might further her knowledge.

Macedo set off rejoicing—*tutto giubilante*, he records—and decided that he must request a period of leave from his ambassador. The secret could not be entrusted to him. Macedo would say that he wanted to spend a bit of time in Hamburg. Pereira was suspicious. Whether he had understood the Latin conversations or not, he had begun to think that Macedo was in some way betraying him. Why did he want to go to Hamburg? Why did he want to leave at all? He would certainly not give him leave to go.

Then, replied Christina, Macedo must go without the ambassador's leave. In the middle of August, she packed him off with a letter for the Jesuit General in Rome, hurrying him out a back door of the castle and onto a little barge. Macedo records that he passed the night "upon a rock" in the middle of the harbor, waiting for the ship that was to carry him away. Pereira sent agents after him, and he was twice arrested in Germany; supposedly he had stolen a number of sensitive documents. But Christina had given her own instructions to Macedo's pursuers: "If you find him," she had said, "pretend you haven't." And to ensure a quick passage through his many ports of call, she had provided him with a passport letter signed in her own hand. With such a guarantor, he could not long be detained by any authorities, and by the end of October he was safely in Rome. Christina had enjoyed the intrigue, but she does not seem to have thought very carefully about the consequences of adding her personal signature to Macedo's passport—perhaps, at this stage, she did not much care. But a Jesuit priest, a fugitive from his own embassy, traveling to Rome in the service of the queen of fiercely Protestant Sweden, could hardly pass unnoticed. Though Macedo's lips were sealed, Christina's own signature hinted at the great secret, and within a few weeks, rumors of her imminent conversion had taken root.

The letter itself, which Macedo had carried for her to Rome, made no direct mention of it, for the risk of interception was too great. It spoke only of more general things, but its tone, and indeed the very fact of its existence, would have been enough to raise the most acute alarm had it been known in Sweden. It was addressed to Father Goswin Nickel, acting General of the "illustrious Company" of Jesuits, Gustav Adolf's "Black Pope":

I would count myself most happy [Christina wrote] if I could be assured, by some persons of your Order and your nation, that you consider me, unknown to you though I am, worthy of your friendship and your correspondence. I have charged Father Macedo to make known to you what it is that I hope from you. . . .[1]

It was a tremendous affair. Father Nickel decided to seek the advice of the pope's own secretary of state, Cardinal Fabio Chigi. Chigi had been papal representative at the Westphalian peace negotiations, and consequently had some knowledge of the Swedes and their clever young queen. It seemed to him that her letter was sincere, and an encouraging reply was soon dispatched: two Jesuit priests, Malines and Casati, both well schooled in theology and the natural sciences, were to make their way forthwith to Stockholm. Christina would have been gratified to see that their mission had been marked *altissimo segreto*—top secret. In a move that she herself had suggested, they were to travel disguised as two wealthy tourists, Malines—with a new beard and hair grown unjesuitically long—as Don Lucio Bonanni, and Casati as Don Bonifacio Ponginibbio. In case their letters should fall into unfriendly hands, the queen herself was given the alias of Signor Teofilo Tancredo—"Monsieur Godloving This-I-Believe"—a convincing enough pseudonym in Italian, at least. The new dons were under careful instructions to remain on good terms with each other, to avoid all political discussions while in Sweden, to listen to the talk of heretics without losing their tempers, and to take care that their secular attire did not encourage them to worldly pleasures. The queen was interested above all, they were informed, in Greek literature and ancient philosophy—they should brush up their own knowledge accordingly. They dutifully did so, and toward the end of November they set off for the north.

While the flurry of priests and papers continued in Rome, a second Jesuit journey was beginning quite separately in Copenhagen, sparked unwittingly by Christina's old favorite, Claude Saumaise. In the same month that Father Macedo had set out on his little barge, Saumaise had set out from Stockholm to return to his post as rector at the University of Leiden. Passing through Copenhagen, he happened to dine one evening at the residence of the Spanish ambassador, Bernadino de Rebolledo, and over the wines he began to talk of his soujourn in Sweden, and of that country's remarkable queen. Rebolledo remarked that it was a pity her majesty's evident mental powers could not be bent to an acceptance of the True Faith: surely some clever priest must be able to persuade her. The clever priest was directly at hand, at least as far as he himself was concerned, in the person of the Jesuit Father, Gottfried Francken. Saumaise, himself a kind of Calvinist unbeliever, no doubt smiled at the ambassador's suggestion, but he made no comment; Rebolledo said no more, and the subject was to all appearances dropped.

Francken, however, a determined Dutchman, decided to follow it up. At sixty years of age, he was no longer young, but he was an experienced teacher of theology and philosophy, and ardent in the service of his faith. He was a hardened missionary, too: returning to Holland after years of study and work in Catholic lands, he had been thrown into prison by his unimpressed Calvinist compatriots. Now he was in Denmark, officially still a missionary priest, but in fact living comfortably in Rebolledo's embassy quarters. It was not enough for an old warrior. Diplomats' confessions and the flowery worries of their ladies were simply no substitute for real work in the field—and what a victory it would be to convert the queen of Sweden! A sortie to the north, Francken decided, was just what he needed.

Christina encountered the priest at Nyköping Castle, where she was visiting her mother in the early weeks of the autumn. Francken had come supposedly on a minor diplomatic mission for Ambassador Rebolledo in Copenhagen. Their first meeting was formal, and Francken records that the queen spoke to him in an "arrogant and sneering" tone. If this was an instance of Christina's famous capacity for dissembling, she seems to have convinced Francken, at least, for he abandoned his plan at once and asked leave to return to Copenhagen. It was granted, and he returned to his quarters to pack up his things directly. But no sooner was he in his room than he received a message from the queen, saying that he should not prepare to leave Sweden but should present himself the next day for a secret interview with her.

Closeted with him on the morrow, she came directly to the point. Was Francken really a Catholic priest? He was. Was he in fact a Jesuit? He could not deny it. Though it had been only a matter of weeks since she had smuggled Father Macedo out the back door of the castle, Christina supposed that Francken must have been sent to her by the Jesuit General in Rome. If he was not a diplomat, she asked him, what then was his real mission in Sweden? He replied that he had hoped to teach her something of "the old religion." Had he come to turn her into a papist, then? Did he not realize he could lose his head for this? Francken was prepared to take the risk. The queen's tone softened. She assured him that he was in no danger, and that she was anxious to learn what she could of his faith. The old priest, as yet unaware of Macedo's separate mission, conveyed what he could in secrecy and haste, before beating a triumphant path back to Copenhagen.

From the Danish capital, Francken sent a detailed report of all that had transpired—not to the Jesuit General in Rome, but to his more immediate superiors in Flanders. He asked to be replaced at Rebolledo's embassy, so that he might give his full attention to the matter of the queen's conversion. This arranged, Francken returned briefly to Sweden to deliver an encouraging letter from Ambassador Rebolledo, along with a worthy essay on morals that Rebolledo had written for the queen. Christina forbore to read it, pleading, perhaps truthfully, that her Spanish was not good enough. Francken departed, leaving Christina to await her deliverers alone, through a winter "so cold that the streets of all the towns are desolate, no creatures stirring in them for many months, all the inhabitants retiring to their stoves."[2] When the springtime came at last, Stockholm life unfolded, and stretched, and stood up, and went outside.

BUT AS THE SPRING of 1652 opened out into summer, Christina began to grow anxious. For months she had heard nothing from Macedo or the Jesuits in Rome. She did not know that Father Nickel had been replaced as General of the Order, that his successor had died, and that a third General was now in office, and that each change had necessitated fresh authority for the mission to Stockholm. In consequence, she had been glad to see Father Francken again, but, though he had been fully prepared for a lengthy stay in Sweden, the queen had not detained him long. Within a week or two she had sent him back to Copenhagen, armed with a letter of her own to Rebolledo, in which she requested the ambassador's help in forging "a bond of friendship" between herself and the Spanish King Felipe. Rebolledo dutifully dispatched a letter to that effect to the king's chief minister, then packed Francken off to Sweden again. He stayed just a few weeks, and it was only now, on his third visit to her, that Christina mentioned to him her dealings with Macedo and the Fathers in Rome. Francken was dismayed. He had not realized there was a rival suitor for the queen's religious affections. He remarked unhappily that there were scholars and priests enough in the Low Countries—her majesty had really had no need to seek instruction from as far afield as Rome. Christina consoled him by requesting him to write an essay for her on the immortality of the soul. He did so, only to be packed off yet again, not back to Copenhagen, but to the Spanish ambassador at The Hague.

He carried a letter requesting the ambassador to negotiate a possible retreat for the queen in the Spanish Netherlands; the governor, the Archduke Leopold Wilhelm, brother of the Emperor Ferdinand, was of necessity to be brought into the plan. Perhaps as a kind of apology for having kept him in the dark about Macedo, Christina also asked Francken to find some learned priest in Flanders who could come to Stockholm to instruct her. If so, it was a sadly backhanded attempt, as the newcomer was to take the role that Francken himself had thought to play. In the event, the chosen priest, Philip Nuyts, was a most unwilling replacement. Not much younger than Francken, he was in poor health and begged to be excused from the assignment for fear the rigors of the north would prove too much for him. His hardy Dutch superior dismissed his concerns, assuring him that the change of climate would do him good, and that anyway he would be out before the next winter. Father Nuyts arrived in mid-April, disguised as a merchant, complete with cape and sword. The temperature was well below zero, and he was suffering from a fresh spring frostbite. Happily for him, the queen kept him hardly a month. Her own spies had got wind of his mission—he was not in Sweden "just to steal hens," it seemed. Alarmed, he burned all record of his discussions with her majesty but later reported that they had mostly talked of how quickly he should be gone. He went almost as quickly as the weary Father Francken, who had only just arrived back in Stockholm on his fourth visit when he received instructions to go home to Flanders. His misfortunes had not come singly: there were in fact two letters, one from his superior in Antwerp, and a second from the Jesuit General in Rome; the General had at last learned of the separate missions and was adamant that nothing was to get in the way of his own envoys, Malines and Casati. Francken, sadly disappointed, and Nuyts, greatly relieved, began to make arrangements for their departure.

Christina had now made three separate appeals for help: to the Jesuits in Rome, via Macedo; to the king of Spain, via Rebolledo; and now to Leopold Wilhelm in the Spanish Netherlands. She had considered a fourth as well. Her old friend Bourdelot had suggested that she look to the French for help, and this had seemed to her an excellent idea. She had been a good friend to France, after all, supporting their claims at the peace treaties in the teeth of all that the chancellor, and indeed her late father, had said and done. She took the precaution of packing up the several thousand books she had acquired from Cardinal Mazarin's library, and sent them back to France, with a message that she had only ever intended to keep them safe for His Eminence while the war lasted. She then confided her plans to Chanut, but his response dismayed her. He was aghast at the idea that she might seek a kind of luxurious refuge financed by the French. Sweden and France were longstanding allies, and the war with Spain was still ongoing; Mazarin was not likely to risk offending the Swedes by taking in their renegade queen. So far the French had had no formal hand in Christina's journey to Catholicism; Chanut, though a pious Catholic himself, and privately overjoyed at the step the queen was about to take, was anxious to preserve his nation's neutrality in the affair.

Chanut's reluctance had been something of a shock to Christina, and had brought home to her how difficult, and how lonely, her road to Rome might be. But now all seemed to be working out as she had hoped. Francken and Nuyts had been and gone, and their fellow Jesuits in the Spanish Netherlands would no doubt speak on her behalf to Leopold Wilhelm and the ambassador there. That would take care of some of the practical problems: they would provide a residence for her, and no doubt arrange an allowance from the emperor. As for the religious question itself, that was being taken care of, too. Malines and Casati had finally arrived in Stockholm after an arduous and stormridden journey, including a stage by gondola, and were already instructing her majesty in the tenets of the Catholic faith. From Vienna, one of the queen's own spies sent a warning that there were "no fewer than four Jesuits" believed to be at large somewhere in Stockholm.

The Italian priests spent almost three months in the city in their guise as wealthy tourists. "Don Lucio and Don Bonifacio" made regular visits to the queen, and their reports indicate the questions that concerned her. Was there really any difference between good and evil? she asked. Was the soul really immortal? Was it absolutely necessary to pray to saints and keep statues of them and revere their relics? Could she not practice Catholicism in secret and remain outwardly Lutheran? Her great-uncle King John had tried to do so. And how, in the end, could faith be reconciled with reason?

The two Jesuits did not deceive her. Yes, there was a difference. Yes, the soul was immortal. Holy Mother Church had good reasons for all its practices. No, Her Majesty could not be both Catholic and Lutheran at the same time. His Holiness had not smiled on the attempts of her majesty's revered predecessor. Her Majesty must recognize that the articles of faith were above reason, and yet not contrary to reason. Her doubts were no more than the promptings of the devil.

Christina confessed that, in her search for the truth, she had been examining each of the tenets of religion in the light of natural reason—a Catholic practice she may have learned from Descartes. She had concluded, however, that the mysteries of faith were not susceptible to investigation in this way. She assured them that she had not lost her belief in the existence of God, though she had to say that the various religions seemed to her to be no more than a political invention designed to keep the common people in their place. She had investigated them all, Christianity and Judaism and Islam, and had at first decided that a simple acquiescence in the forms of her local religion was the wisest and easiest course of action. But it was hard, she said, to live without some real personal faith, for what other foundation could there be for life? A true religion must exist somewhere, she felt, and that must surely be Rome.

"Free will is in itself the noblest thing that we can have. In a sense, it makes us equal to God."[3] So had the great philosopher once written to the searching young queen. Five years later, she had found her reply. "The use of our own free will," she wrote, "is the noblest sacrifice we can offer to God. Reason will not persuade us of the truth of Christianity. We must submit blindly to the Roman Church. It is God's only oracle. To believe in more is superstition. To believe in less is infidelity."[4]

It is a convenient conclusion, with the apodictic ring that Christina had always so enjoyed in the ancient maxims. Perhaps she believed it some of the time, but she certainly did not believe it consistently. In one sense, she was giving up the search, "submitting blindly" to another point of view, enshrining it in apparent inevitability. But at the same time, it suited her to choose the Roman Church. It was, after all, in Rome.

Whatever Christina's rationale, the two Jesuits had no need to press their case further. Casati departed to oversee the arrangements for the queen's reception in Rome; Malines remained as a moral support. Christina's conversion was now only a question of time.

WITH THE RELIGIOUS QUESTION settled, the practical arrangements remained, and these, for the present at least, depended on the Spanish. Late in the summer of 1652, a special envoy from His Most Catholic Majesty, King

Felipe IV of Spain, presented his credentials to Her Majesty the Queen of Sweden. In marked contrast to the somber, stately costume of the Spanish court, she was dressed simply in a gray skirt, a man's jacket, and a black velvet cap, with no jewel or lace about her, but only a black ribbon around her neck in the fashion of the Swedish sailors. Had the envoy known it, she was very relieved to see him. Ambassador Rebolledo had clearly done his work. Her passport to freedom had been delivered at last.

The special envoy was Don Antonio Pimentel de Prado—General Pimentel, in fact, a tall and handsome soldier of commanding disposition, despite being, apparently, completely bald. Christina was delighted with him, anyway, and only disappointed that the king had not sent him with full ambassadorial rank; the matter of her conversion might have merited that much, she felt. Pimentel made his bows and his addresses in the elaborate manner of his countrymen. He had come, he said, to return the courtesies that her majesty's own envoy had recently paid in Spain, and to negotiate a commercial treaty between their two noble nations. His stay would be brief, he said, for the treaty was largely a matter of formalities, and there was little to be discussed.

The queen replied that the envoy of His Most Catholic Majesty was welcome to her court, indeed very welcome. The treaty would no doubt be swiftly signed. This she would leave to her ministers. In the meantime, did Don Antonio have no other message for her? There was something else, perhaps, that might be discussed later, between the two of them, in private? The Spaniard demurred. Her majesty was mistaken. Don Antonio had relayed everything just as he had been instructed.

It was the truth, but not the whole truth. Pimentel did have a secret mission, but it would not have pleased Christina had she known of it. He was to assess Sweden's military strength, and to determine whether the queen had any plans to marry. A new Swedish alliance could tip the balance of power adversely for Spain; the forces, military and otherwise, of His Most Catholic Majesty were well to be prepared. As far as the queen's conversion was concerned, Pimentel had no instructions at all, nor did he have the least idea of the plan that was afoot. Christina found this hard to believe, and for some weeks did her utmost to spirit him away alone with her at every opportunity. Accepting at last that he knew nothing of the plan, she realized the implications with alarm: the Spanish king was skeptical of her sincerity; he had made no arrangments for her; the practical work was still to be done. She decided that Pimentel must not leave Stockholm until her material support was assured. She must have a direct and constant link to Felipe. Pimentel must succeed where Rebolledo had failed.

Christina's hinting, her conspicuous inclusion of Pimentel at every last court occasion, and her myriad attempts to arrange personal meetings with him, led many people to believe that the two were lovers. The scandalous little flames were fanned above all by Peder Juel, the Danish Resident in Stockholm. Juel was generally known as a reliable man, not given to embellishing his reports; moreover, his principal correspondent, the deeply religious Danish chancellor, was a very model of piety and probity. In consequence, Juel's sober reports were assumed to be true. Within a month of Pimentel's arrival, he was writing in disgust that the queen of Sweden was a thoroughly debauched woman, and her court "an assembly of dissolute and profligate nonentities."[5]

Pimentel, at least, was probably perfectly innocent. For six months Christina kept him in the dark about her plans, hinting at him, closeting him away, until he must have begun to wonder himself whether her intentions toward him were entirely honorable. Certainly he made several attempts to get away, without waiting for permission from Spain. His first ship outward was blown back to Sweden by a fierce contrary wind. He returned to the court, where tongues wagged that he had made the whole thing up in order to return to his royal mistress. Deeply embarrassed, he prepared to depart by an overland route. Christina herself assisted his preparations, which no doubt reassured him in one respect at least. But his departure was prevented by a direct order from Spain; the envoy was obliged to remain in Stockholm "until further notice," continuing his secret observations of the Swedish fleet. Frustrated, Christina seized on a Dominican friar, happily to hand — one Brother Güemes, not long arrived from Rebolledo in Copenhagen. She sent him off to Spain instead, and a downcast Pimentel settled in to await his orders.

His chaplain, Manderscheydt, at least, was busy. Now that her majesty's intentions were clear, he had plenty to do through the long summer days, and set about with a will to instruct her in the many articles of Catholic faith. Christina was a readier pupil than most, though in other respects quite typical: apparently her clothes were sometimes splashed with ink and her linen often torn. The chaplain also recorded a telling description of her, once away from her theology books:

There is nothing feminine about her except her sex. Her voice and manner of speaking, her style, her ways are all quite masculine. I see her on horseback nearly every day. Though she rides side-saddle, she holds herself so well and is so light in her movements that, unless one were quite close to her, one would take her for a man.[6]

Manderscheydt, another Jesuit, was as indiscreet as his fellow priests had been cautious, and throughout Stockholm it was soon generally accepted that the queen was receiving instruction from the Spanish papists. Alarmed Lutheran dignitaries complained of the subversive religious activities that were "almost out of hand in the city." As usual, Christina herself did nothing to stem the tide. She spent countless hours in Manderscheydt's company, addressing him always, with great respect, as *mon père*. It was not at all to her advantage to do so. She had as yet nothing to live upon after her abdication, though a goodly number of valuables had by now been spirited away to Antwerp. Perhaps, with Pimentel and his chaplain nearby, and Güemes on his way to Spain, she was feeling more confident about her future life. She enjoyed, too, the chance to provoke the stolid old pastors who had wielded such bleak authority in her youth. She was acting rashly, and against her own interests, but if she did stop to think about it, she carried on regardless.

THE SWEDISH PASTORS were not the only ones to feel the effects of Christina's loosening ties to her homeland. The influence of old friends was waning, too, and none more so than that of Magnus. Though Bourdelot had long ago replaced him at the center of the queen's affections, Magnus's luck had still been running high. He was a lively young man, and now very rich, and at court he had his enthusiasts, among them the famed lyric poet Marc-Antoine de Gérard Saint-Amant, known to all since his sojourn in stodgy Poland as "Big Saint-Amansky." The attentions of Saint-Amant and of his Stockholm friends seems to have blinded Magnus to Christina's own declining interest in him. At any rate, his effrontery remained undimmed as his star began to wane. He picked frequent quarrels with Bourdelot, who was quick to reciprocate. Christina's patience was far from inexhaustible, but for many months she allowed her favorites to carry on bickering, until at last she had had enough. She reprimanded Magnus sharply, but she also let Bourdelot go.

The good doctor was in fact more than ready to take his leave. He had

been in Stockholm for almost three years, and, though the queen had treated
him wonderfully well, he was in need of a wider world. Saumaise was long
gone, and there was scarcely anyone now at court who did not hold some kind
of grudge against him. Besides, the Fronde had run its course. Paris was safe
again, and much might be extracted from the newly reinstated Cardinal
Mazarin, for the French were eager to remove Bourdelot from Christina's circle of influence. He had too many enemies in Stockholm. The Swedes were
turning against France, and France had her hands more than full already
with the Spanish war.[7] Bourdelot's continued presence could only push them
further along that unwelcome path. He must be persuaded to leave.

Bourdelot could afford to be demanding, and so he was. His price for leaving was high. There was to be no public dismissal; officially, he was to travel to
Paris on a special mission for the Swedish Crown. Once in France, he was to
take possession of either a valuable bishopric or a substantial abbey in Berry,
with all its lands and rents—his pragmatic atheism was evidently not to be allowed to stand in the way. The ingenuous Chanut supported Bourdelot's demands, declaring that many wicked tongues had spoken unjustly of him. For
her part, Christina was happy with the pretense. It was not to be suspected that
she had felt obliged to let him go. No one was to think she had consulted anyone's wishes but her own. She ordered the usual formal leavetaking—a round of
morning farewells accompanied by German wine and peppered toast soaked in
vinegar, which cannot have encouraged Bourdelot to reconsider his decision.
Magnus did his best to avoid his turn as host, but Christina obliged him to take
part. The chancellor capitulated without demur, and Karl Gustav's family
padded the pockets of the doctor's traveling coat with diamond-framed portraits
of themselves. Christina warmed his passage, too, with thirty thousand ecus, a
table service in solid silver, an elaborate carriage with six fine horses, and, her
frequent personal token, a heavy gold chain. He passed through Leiden on his
way to Paris, and there a resentful Isaac Vossius caught sight of him, seated
proudly in his new carriage, "laden with riches and curses."

In Stockholm, flush with partial victory—Bourdelot had gone, at least—
Magnus decided to press his luck. He had fabulous riches, it was true, and
lands, and honors, and jewels galore, and a royal wife—all courtesy of the
queen, including his wife, or so the queen maintained. But all those were old
victories. Magnus needed a new token of affection. And he was young still,
only thirty—a young man, he felt, must have some ambition. Magnus decided it was time he became a prince.

As it happened, there was a Swedish principality available, but it was in the gift of the Queen Mother, and the Queen Mother was not inclined to give it. Magnus began to look further afield, and soon saw a vacancy in the German lands. He had begun to make the arrangements, when Christina intervened. Creating princes was her prerogative, and hers exclusively. The count's station was already high; he was not to set himself above it.

Magnus suspected a different motive. Bourdelot had gone, but a new favorite was already emerging, the recently ennobled Count Klaes Tott, even younger than Magnus, and with nothing of his own, though handsome and soldierly, and very dependent on the queen's kindness. Tott must be eliminated if Magnus were to regain his premier place. Attack, he decided, would be the best form of defense—or rather, a concealed attack that would look like defense, or at least, an attack on Tott that would look like an attack on himself, and which the queen would have to defend. Thus, unsagaciously, Magnus approached Christina. There was an intrigue afoot, he declared, a conspiracy, against himself. Not daring to name Tott in person, he lighted upon two other courtiers at random, identifying them as the plotters. The charge could not be sustained for a moment.

Christina fell into a fury. She raged at Magnus for daring to try to trick her, and cursed his ingratitude and presumption. One of the accused courtiers challenged him to a duel, but he declined, supposedly disdaining to cross swords with a member of the lower nobility. It was the last straw for Christina. Laying great claim to physical courage herself, she was disgusted, she said, to find it lacking in her own cousin, whom she had so singularly honored for so many long and loving years. Magnus had disgraced himself. Christina could not bear to have him in her sight. Let him betake himself to his country house, and stay there. He was not to return to the court again. She would hear no pleading. He was banished—and he could resign all his commissions before he left.

Magnus withdrew, only to renew his pleas in a groveling letter, which elicited from Christina a swift and savage reply:

From now on [she wrote] I can feel nothing for you but pity. You have undone all the goodwill I had for you, and you have proved yourself completely unworthy of it. If I were of a mind to regret anything, it would be that I ever formed a friendship with so weak a person as yourself. . . . You have made it obvious that I should never have done so, and that is a secret that I was resolved to keep all my life. . . . Now I just want to forget you. If I speak of you at all, it will be only to reproach you.[8]

In a foretaste of a similar letter on another, grimmer, occasion, Christina had the letter translated into Latin and distributed to all the courts and cities of Europe. Everyone must see her strength. Everyone must know that she would brook no tricks and no treachery.

As the letter itself reveals, Magnus's real fault was to have made Christina look foolish by having favored him in the first place. It was enough to turn her implacably against him. Despite submissions and remonstrations from family and friends, and even from the chancellor himself, who had long despised Magnus, she would not relent. Her fury is evident in the response she made to Karl Gustav's personal appeal for his banished brother-in-law. Why was he pleading on that coward's behalf? Did he not realize that it was Magnus who had turned her against him, that if it had not been for that fork-tongued villain, she would certainly have married him? She had several times been on the point of accepting him, she said, but Magnus had persuaded her against it. It was a remarkably cruel thing to say, and it was a lie as well.

Karl Gustav retreated sadly. Magnus's family gave up. In Paris, Bourdelot was heard to remark that if the queen wanted a reign of peace, she was going the wrong way about it. Magnus did his best to turn his gaze away from the grand and glittering life that he had lost. For many months he was seen wandering about in town and country, sorrowful of mien and garbed in unwonted sobriety, conspicuously engrossed in *The Consolation of Philosophy*, a famous tale of fortune's ever-turning wheel. If he looked at it at all in private, it was only to copy out extracts for his unphilosophical friends. And if Magnus was consoled by the worthy tome, no one else was persuaded.

ABDICATION

TWO YEARS BEFORE, in the autumn of 1651, the chancellor and senators and men of the *riksdag* had been congratulating themselves on warding off the abdication. They had recalled the young queen to her senses, and to her sense of duty. The talk of laying aside her crown had ceased. Christina had warned them that, though she might postpone her plan, she had not abandoned it, but this they had chosen to ignore. In fact, with no clear response from the Jesuits or the Spanish king, she had had to make a tactical retreat. Now, with Güemes on his way to Spain, some guarantee of material support, she felt, would surely be soon forthcoming.

As the winter of 1653 set in, Christina made her final preparations. Pimentel, informed at last of the queen's real intentions, found an agent to act for her in the Spanish Netherlands; here she would stay, after leaving Sweden, until all had been arranged for her in Rome. In Stockholm, she appointed a group of new senators, including Jakob De la Gardie, Magnus's brother and husband of her own Belle. Swedish territories had expanded during the Thirty Years' War, she declared; more senators were needed to administer and represent them. But she had chosen the new men with great care; all would be likely to support her plan.

Early in the new year of 1654, the senators were summoned to Uppsala Castle, and there Christina announced her intentions. The senators returned a formal refusal, but it seems that, by now, most of them had accepted that the queen would not be dissuaded. Only the chancellor persisted, drawing up a petition suggesting that she share the burdens of her great office with her acknowledged heir, Karl Gustav. But Christina wanted no compromise. Within a few days, she had their agreement. She would make a formal act of abdication, she said, when the *riksdag* met in the spring.

All that remained to be agreed was the size of her apanage. Christina stipulated the amount herself—200,000 riksdaler per annum, a comfortably royal sum. It was to be drawn from a number of estates within Sweden itself, including the town of Norrköping and the two large islands of Gotland and Öland, the latter Karl Gustav's favorite hunting retreat. She would have an Estonian island as well, and lands in Mecklenburg and Pomerania that her father had claimed as trophies of war. Hers, too, would be the Baltic town of Wolgast, where long ago Maria Eleonora had kept hysterical vigil over the king's body, waiting for the frozen sea to thaw. It would be more than enough, provided it was paid, and for this her conversion plan must remain a secret, at least until the papers were signed and she was safely out of the country. Even more important to Christina herself, however, was her continued status as a sovereign. She was to remain a queen, though a queen without a realm, and on this she insisted. The members of her court were to be subject to her, and she herself was to be subject to no one, no matter where she should be. She would be answerable "only to God." This was the first article of her abdication agreement, before any mention of her apanage, before any reference to her successor, and it was vital to her. Her sovereignty, her right to rule, she believed she carried within herself. For Christina, it was a personal quality. It had nothing to do with the state or the crown, and she could never be divested of it.

It was not ideal to have two "your majesties" in the land, and some felt that it might endanger the country's political stability. But there was far greater opposition to the financial settlement. Some of the lands to be ceded to the queen had already been given away. They would have to be formally repossessed by the crown—a clear precedent for further expropriations and a subject of furious controversy throughout the country. Public opinion swayed against the queen: the settlement was felt to be too much. But by playing a double game with the Senate and the Estate of the nobles, Christina was able to conclude the business to her complete satisfaction. She managed to persuade each party that the other had agreed to sign, and thus both parties resumed negotiations accepting the settlement as a fait accompli. The senators tried to add a stipulation that she remain in Sweden after the abdication, and "spend all the money in this country," but Christina flatly refused. So, after three weeks of haggling and grumbling, the agreement was signed.

There was no real guarantee, however, that it would be honored. Nothing could compel the Swedes to pay—nothing apart from national pride, perhaps, and respect for the memory of her father, and either one might persuade them to abandon her completely once she had been received into the Catholic Church. As the weeks passed, and no news arrived from the Spanish king, doubts rose once again in Christina's mind, and her actions began to take on a desperate color. Out of hand, she dismissed the Portuguese ambassador, declaring that his king was a usurper, and that his country rightfully belonged to Spain. As a "symbolic gesture," she sent out roughnecks to pounce on the ambassador's servants in the streets and give them a good beating; for a time she even planned an invasion of Portugal. The senators drew the obvious conclusion: the queen must have some ulterior, pro-Spanish designs. Spain was still at war with France, and France was Sweden's ally. The queen's actions, sighed the exasperated chancellor, were "the most grave assault on Sweden in the last forty years," but Christina would not be bridled. Not content with harassing the Portuguese, she demanded that the war subsidies owing from France to Sweden should be paid to her personally. When it was pointed out to her majesty that there was in fact no money owing from France, she tried to sell the French a fleet of warships "to meet immediate expenses" following her abdication.

Christina sent for Ambassador Whitelocke, and announced that she wanted the English to buy Sweden's colonies in West Africa, the profits to accrue to her own purse. Whitelocke, taken aback, made a diplomatic reply, and took his leave, but on the following day, he was summoned again. The queen had been considering her abdication agreement. She had decided that a clause must be added to it, a secret clause, known only to herself and the ambassador and his master, "that gallant man," Oliver Cromwell, Lord Protector of England, who had cut the head off her "dear cousin" Charles not so long before. The clause would guarantee her apanage: if the Swedes failed to pay it, Cromwell was to nullify the trade agreements between England and Sweden.

The ambassador was not normally permitted to sit in her majesty's presence, but perhaps, on this occasion, he sat her down, metaphorically at least, and proceeded to explain a thing or two about international relations.

England was at war with the Dutch. Baltic trade was a matter of huge contention between them. The trade agreements with Sweden had been a painstaking and important victory for the English. They were not going to be jeopardized by any clause, secret or otherwise, with a powerless former queen with no connection whatsoever to England. Only desperation, or an impossibly inflated sense of her own personal importance, could have prompted Christina to think of it in the first place. Whitelocke extracted himself as diplomatically as he could, and took his leave again, with a sigh of relief, or exasperation, or disbelief.

None of this, needless to say, not the sale of the warships or the African colonies nor any of Christina's other wild schemes, was approved or even discussed by the Senate or the *riksdag*. Now, as so often before and afterwards, she refused to distinguish between the assets of the state and her own personal property. In the ship that had carried away Cardinal Mazarin's books, she had already smuggled out a hoard of paintings and manuscripts and other valuables that the Jesuits had promised to keep for her in Rome, all valued, it is said, at some half a million livres—the price of a good fleet of warships—and she had coerced the emperor's ambassador, the unwilling Conte Montecuccoli, to take medals and jewels for her out of the country as well. The income agreed for her would be enough to keep her more than comfortably for the rest of her days, provided it was paid. But she was not dealing honestly with her countrymen, and she feared that, in return, they would not deal honestly with her.

In the midst of Christina's growing anxiety came news from Leiden of the death of Claude Saumaise. At sixty-five, he had been older by far than any of her other friends, but he had been dear to her, and she had fully expected to see him again. Saumaise, an unbeliever to the core, had refused the last rites, but had left instructions for his widow to destroy a number of his no doubt heretical writings, and she had duly done so. Hearing of this, Christina dispatched a furious letter to Madame Saumaise. It cannot have comforted her much:

> Saumaise's death [the queen wrote] is a grievous blow to all rational minds, but can you imagine how I feel about the irreparable loss of his papers? You know how highly I valued his genius. You know I loved him as a father. Do not look to me for consolation. It is right that you should grieve. You should spend the rest of your days bewailing his death and the crime of homicide you have yourself committed by destroying his work. You have killed him a second time over, and I will never forgive you for it.[1]

Though Christina did not relent, she did offer, with typical generosity, to pay for the education of Saumaise's young son. At the same time, with typical avarice, she did what she could to claim his well-stocked scholar's library.

—

ON A BRIGHT SPRING day in the middle of May 1654, five mounted soldiers rode through the royal town of Uppsala, with kettledrums and trumpets beating and blaring before them. They were sent to proclaim an extraordinary meeting of the *riksdag* in the great hall of the castle on the following day. All the *riksdag* men were to appear by eight o'clock in the morning, "upon pain of half-a-dollar mulcted for every default."

It would be an early start for the nobles among them, at least, for the night before they were to attend a wedding at the castle, and the ceremonies were not to start until midnight. The Baron Horne, "of ancient and noble family," was to marry the Lady Sparre, a kinswoman of Christina's Belle, and one of the queen's own servants. The bride came decked in diamonds, and the bridegroom with gold and silver lace on his suit of white silk. They made their vows by torchlight, then marched a solemn round with trumpets sounding, and then they set to dancing.

But at eight o'clock the assembly met promptly enough in the great hall, with the whole *riksdag* in attendance, and ambassadors and other dignitaries watching from the upper gallery. One after another the *riksdag* men filed in: sixty peasants, led by "a plain, lusty man in his boor's habit," twice as many burghers, two hundred nobles, and sixty members of the clergy, "grave men, in their long cassocks and canonical habits, and most with long beards." They were followed by the guardsmen, and the gentlemen of the court, two by two, and then the senators, from humblest to highest. Finally came the queen, with her yellow-liveried pages in attendance. She made her way to the front of the hall, and sat down, whereupon all the many soldiers and servants who had trooped in so solemnly, only minutes before, now trooped out, leaving only the men of the *riksdag*, with the great doors closed behind them.

The queen stood up and beckoned to Chancellor Oxenstierna, who approached her "with great ceremony and respect." They spoke a few moments together, the chancellor returned to his place, and the queen sat down again. The chancellor had been expected to address the assembly, and announce the formal purpose of its meeting, but it seems that in this brief conference he had begged her majesty to be excused, "by reason of an oath I had taken to my king, to endeavor to keep the crown on his daughter's head."[2] Christina was thus obliged to make the introduction herself, and so she did. She rose up "with mettle," and stepped to the front of the dais, and "with a good grace and confidence spake to the Assembly."

This was a strange meeting, she said, and the reason for it "astonishing to many," but the step she was about to take had been long in her mind. She had but one purpose, which was "to give into the hands of my most dear cousin our most dear country and the royal seat, with the crown, the scepter, and the government. And if in these ten years of my administration I have merited anything from you," she said, "it shall be this only which I desire, that you will consent to my resolution, since you may assure yourselves that none can dissuade me from my purpose."[3]

Several good men tried to do so nonetheless. From each Estate in turn, the marshal spoke, asking her majesty to reconsider, to think of her people, of her duty, of her father, of the will of God. The archbishop of Uppsala took first place with an elaborate oration, "which was somewhat long," and concluded with "three congees," some handkissing, and then three more *congés*. The marshal of the nobility and the burghers' marshal did likewise. The last place was left for the peasants' marshal, "a plain country fellow, in his clouted shoon," and he now stepped forward, "without any congees at all," to add his forthright plea:

> O Lord God, Madam [he declared], what do you mean to do? It troubles us to hear you speak of forsaking those that love you so well as we do. Can you be better than you are? You are Queen of all these countries, and if you leave this large kingdom, where will you get such another? Keep your crown on your head, and continue in your gears, good Madam, and be the fore-horse as long as you live, and we will help you the best we can to bear your burden.[4]

When the marshal had concluded his speech, Ambassador Whitelocke recorded, he "waddled up to the Queen without any ceremony, took her by the hand and shook it heartily, and kissed it two or three times; then turning his back to her, he pulled out of his pocket a foul handkerchief and wiped the tears from his eyes," before returning to his place.

It was all to no avail. Christina did not want to be Sweden's fore-horse any longer. She was resolved to give up her crown, and she wanted only the *riksdag*'s assent to retire "from so heavy a burden." Her pages were called, and she withdrew from the hall, and all the men of the *riksdag*, dismayed and disbelieving, withdrew after her.

Late in the evening, the wedding festivities were resumed. It was Maytime in the north, and it would be light through half the night. At one in the morning, the dancing began, and Christina remained through it all. It was too much for Whitelocke, who had been up till all hours the previous night, He took himself quietly home, wondering, he said, "that the Queen, after so serious a work as she had been at in the morning, could be so pleased with this evening's ceremonies."[5]

THE AMBASSADOR DID NOT attend the solemn ceremony of abdication, which took place in the great hall of the castle at Uppsala a few weeks later, on the sixth of June. He was in Stockholm by then, preparing his journey home to his unanointed master, but late in the night he was visited by Karl Gustav Wrangel, field marshal and senator, who had ridden the forty miles down from Uppsala to relay the proceedings of the day.

They had begun early. At nine o'clock in the morning, the queen walked into the great hall of the castle, with a train of servants in attendance, and draped in her velvet coronation robe, her crown upon her head. At the front of the hall stood a long table with five velvet cushions laid on it, and on the first four of these lay the royal regalia that she had received at her coronation: the ornamented sword, the gold scepter and key, and the golden orb with its back-to-front engravings. Reaching them, the queen paused a moment, then turned to the assembled dignitaries, and spoke. It was a brief address: she was resolved, she said, to resign the throne and government of the kingdom, and she had come now to execute this design. To her "most dear cousin," she wished "all happiness and success," then she requested that the crown be taken from her head. Apparently, no one had agreed to perform this fearful duty, for no one came forward now. She turned to the young Count Tott, a favorite among her courtiers, and to Baron Steinberg, and commanded them to remove the crown. They would not. She spoke again to them, in earnest tones, until the two consented to fulfil this last command. Together they took the crown from her head, and laid it on the fifth velvet cushion. This awful act accomplished, others now came forward to remove her royal robes, until she stood alone in a simple white dress, "as beautiful as an angel," as old Per Brahe recorded tearfully. Christina made a curtsy to Karl Gustav, and another to the assembly, then retired to a private room, "without the least outward show of reluctancy for what she had done."[6]

Karl Gustav's coronation followed on that same morning. It took place in Uppsala's Cathedral Church, tradition thus reinstated after the ill-fated rupture of Christina's own ceremony in Stockholm. The prince arrived "in his ordinary habit, with a huge troop following him, and the windows and streets crowded with multitudes of people." The archbishop anointed him, and he accepted all the regalia so newly laid down, and with drums and trumpets and loud acclamations he was declared Sweden's new king, Karl X Gustav. He had not sought the crown, and there had been many who had not wanted him to have it. In Stockholm, Whitelocke commented ruefully: "Not many days past they laboured to hinder the doing of it; now they shout for joy that it is done. Thus are the minds and practice of the multitude, whom nothing pleaseth long—nothing more than novelty."[7]

SO ENDED CHRISTINA'S "ten-year rule," and so began the seven-year rule of the "most dear cousin," who had loved her and served her so loyally. Christina had a medallion struck to mark his coronation; it bore the motto *A Deo et Christina*—From God and Christina. The new king was to be in no doubt about the origins of his great good fortune, though some still doubted his entitlement to it. They included the Catholic Poles and their Vasa king, Christina's cousin, Jan Kazimierz, who protested at once, attempting to have himself declared king of Sweden in Karl Gustav's place. The long truce of twenty years between the two countries was soon at an end. By the end of the year, Swedish troops were occupying Warsaw, their brave new soldier-king at their head. His brief reign had begun, as it was to continue, marked by warfare. Poland, and the old enemy of Denmark, would absorb him for the rest of his life.

Chancellor Oxenstierna did his best to serve him, his sense of duty overcoming his dislike of the new king's German blood. But his great old heart did not beat long; within six months he was dead, unable to bear, or so Christina said, the "grievous blow" of her abdication. More prosaic pens recorded that the chancellor had been ill for some time, with influenza, with the ague, with the sad and simple illness of old age. But it may be that, after all the arduous years of war and peace, after a lifetime's strivings for his beloved friend and king and for his royal line, his spirit was broken by Christina's abandonment of it all. She said once that if there had been no Axel Oxenstierna, "this unique remedy to so many misfortunes," the death of her father and her own extreme youth "would have been fatal to Sweden."[8] It is almost certainly true. The chancellor was indispensable to Sweden's age of greatness. It could not have continued, as it could not have begun, without him. But the Vasa crown was at the heart of it all, and it was Oxenstierna's personal tragedy to see the choicest fruit of all his labors cast, unwanted, to the ground.

As for Christina, she could hardly contain her impatience to be gone. She was young, and she was free, and all the world was before her. At twenty-seven years of age, she had cast off the burden of her birth, the stifling constraints of duty, the dullness of her quiet northern homeland. She was not going to Spa or any other Pomeranian town, nor to Denmark, nor to India, nor anywhere she had spoken of as a more or less likely refuge. She would forsake the lovely islands of Stockholm for the lovely hills of Rome; and Rome, to Christina, meant more than the Church, more than freedom, more than truth. Rome was the center of the sunlit southern world, glistening with marble, effervescent with talk in a hundred tongues, the vibrant heart of European culture. There, at last, her fabulous paintings would reflect their wonderful native setting. There she would encounter hearts and minds of her own kind. There she would be as she had longed to be, and begin at last the great adventure of her life.

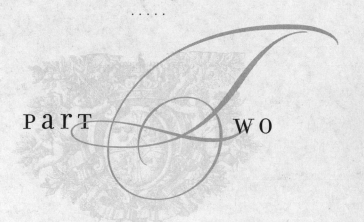

PART TWO

CROSSING THE RUBICON

*I*T CERTAINLY BEGAN adventurously enough. Christina's impatience to leave was made abundantly, not to say embarrassingly, clear within hours of her cousin's coronation. Hardly waiting for the celebratory banquet to be concluded, before the clocks had struck midnight she had set out for Stockholm, the new king riding alongside her, no doubt with very mixed feelings, as far as the first staging-post. In Stockholm she spent a few hurried days in last-minute preparations, finding time nonetheless to attend a public Lutheran service, where she received communion for the last time in the faith of her fathers. It may have been a tactical appearance, designed to dispel suspicions of a possible apostasy, for she needed to confound every rumor that might persuade the tenaciously Lutheran *riksdag* to renege on their financial commitments to her.

Traveling at breakneck speed, she then headed for the Danish border with a harried group of officials in tow. At Halmstad she took her leave of them and also, significantly, of her Lutheran chaplain. It was an unwise move, given her continuing need to conceal her intentions, but Christina was in no mood for caution. With only four gentlemen to accompany her, and none in any official capacity, she rode on to Laholm, effectively as a private person. Here, at this little border town, she paused to make the final changes she had longed for. Casting off her dress, she put on the men's clothing that from now on would be her preferred attire. She sat down only long enough to have her hair cut off, and soon it hung loosely to her chin in the male fashion of the day. To complete the picture, she buckled on a sword. It is said that, as she did so, a rider arrived, dusty from the road, with a letter for her from his new king, Karl Gustav. It contained a last plea for Christina to marry him. She received it, and rejected it, standing, in trousers and boots, a masculine little figure with short hair and a short sword at her side.

But her spirits were high. "Free at last!" she exclaimed, running across to the Danish side of a little stream that marked the border between the two countries. "Out of Sweden, and I hope I never come back!" So at least ran one French report, but the stream was a fiction, and so too, most probably, were Christina's exclamations, embellished now, as often in the future, by a malicious Gallic pen.

To complete her incognito, Christina adopted the name of one of her companions, the young Count Christoph von Dohna, but, although she found the whole subterfuge tremendously exciting, it did not deceive the many spies who followed her out of Sweden. Her disguise was in fact, more than anything else, a revealing of her true inclinations and her real personality. Christina was now twenty-seven years of age. From now on, she would be reluctant to wear women's clothes or a woman's hairstyle. She would appear in public wearing flat men's shoes, often boots, and frequently a sword; princes and popes would greet her with her legs showing and her feathered hat in her hand. Her speech would grow coarser and her habits rougher—even her voice would deepen. Love would come, too, and with it a brief rediscovery of her fragile femininity, but for now, there was nothing but the excitement of escape and the sublime exhilaration of freedom and movement. Formalities and responsibilities lay discarded along with her long hair and her high shoes and her trailing, hindering robes. The already frenetic pace of her journey increased, and her formal itinerary was soon forgotten altogether. Across the Kattegat Strait, in the little town of Kolding, the king and queen of Denmark arrived to prepare a welcome for her after a hundred-mile journey of their own by land and sea from Copenhagen. They found only remnants of her baggage, a few tired servants, and traces of the dust left by Christina's flying horses. The Danes were left to shrug their shoulders while the Swedes stammered apologies, but Christina, irrepressible and unrestrainable, was already five days' ride ahead. It was a thoughtless and even dangerous thing to do, given the delicate diplomatic relations between the two countries; indeed, within two years they would be once again at war. But Christina's eagerness to be gone had overridden every other consideration.

There had been no need at all for her to travel in this way, hasty and unattended. Karl Gustav had placed an entire fleet of ships at her disposal, and she might have traveled in state across to northern Germany, escorted by five thousand soldiers making their own journey onward to Bremen. But Christina had preferred a more dramatic alternative. The spies of sundry European powers reported her progress at every stage, and at every stage the reports became more extraordinary: she was traveling almost alone, she was wearing trousers, she had walked into an inn with a firearm dangling from her neck. By the middle of July, she had reached Hamburg. It was to be the first of several sojourns for her in this busy port, and she lost no time in showing her mettle. The city's magistrates had prepared a house for her, but this she eschewed unceremoniously, opting instead to stay at the home of her new banker, Diego Texeira, whose invitation had been earlier arranged by Don Antonio de Pimentel. As if the snub to the magistrates were not enough, Texeira was a Jew—indeed, he was also known as Abraham. His family had arrived from Portugal some decades before and had prospered in the relatively tolerant environment of this cosmopolitan trading city. But that tolerance did not extend far enough to permit a queen of Europe's foremost Protestant power to take up residence in Texeira's house, and those who had endured her trousers in silence now began to protest. Christina may have felt the need to defend Texeira; certainly she did not often trouble to justify her own behavior, but she now responded with a neat parry to the affronted Christian critics. Jesus Christ himself "had always conversed with Jews," she said. He himself "was come of their seed," and "had preferred their company to the company of all other nations."[1] The riposte was not dictated solely, if indeed at all, by cynical financial considerations. Christina's disdain for religious bigotry was genuine, and the Hamburg pastors could not budge her.

While there, she took it upon herself to visit Duke Friedrich of Holstein-Gottorp, a relative of Christina's on her grandmother's side. The duke lived in the nearby town of Neumünster, and there was talk of a marriage between one of his daughters and the newly crowned Karl Gustav; the king's brother, Adolf, was in Hamburg to further the negotiations at this very time. Christina met both sisters, and advised Karl Gustav to marry the elder one. Seeing portraits of both, however, he chose the pretty blond younger sister, and before the year was out Hedvig Eleonora had taken her place as Sweden's new queen. Whether her efforts had helped or hindered the marriage, Christina did not think much of it, and later insisted that Karl Gustav regretted his choice of a wife. "I shall be miserable all my life," she supposed him to have said, "since Christina has refused me the glory of possessing her. Nothing can console me for it."[2]

Christina consoled herself, at least, with a lively sojourn in Hamburg, courting controversy at every step. Socializing and exploring inside and outside the town, with scarcely any escort, she twice returned so late in the night that the city gates had to be opened especially for her. When an envoy from the emperor came to relay Ferdinand's good wishes, she paid him the honor of donning a short skirt—on top of her trousers. An English spy reported to London the tales he had heard of her "amazonian behaviour." "It is believed," he wrote, "that nature was mistaken in her, and that she was intended for a man, for in her discourse, they say she talks loud and sweareth notably."[3] She received visits and paid visits, accepted lavish gifts and ladled them out in return, and she took the time as well to relive girlhood days with her cousin and old schoolfellow, Karl Gustav's sister Eleonora, now married and a German countess, and resident nearby.

Very early one morning, with no more thought to the local dignitaries than she had given to the king and queen of Denmark, Christina left the city. The magistrates woke up to find that she had gone, without bothering to take her leave of them or of anyone else. All that was left of her was one last, small insult: a copy of Virgil in the pew she had occupied in one of the city's Lutheran churches; she had taken it up to read during the sermon and, inattentively or provocatively, had left it behind.

Christina had set off westward, and if the gentlemen of Hamburg were indignant, one gentleman in London, at least, was pleased. From Ambassador Whitelocke, Cromwell had learned that the queen was headed for Spa, the very town where the executed king's son waited in anxious exile. Charles was now age twenty-four, and as yet he was unmarried. A match between him and the renegade queen of Sweden might revive the Stuart dynasty and persuade its many sympathizers to revolt against England's new protectorate. But Spa was a Pomeranian town, and it lay to the east. Whitelocke had clearly been mistaken. It could not be the queen's destination, after all.

Had Cromwell known it, a quite different marriage was also being considered for the queen. His own good lady wife had not been feeling at her best, it seems, and as she sat palely in bed, glancing without apparent irony through her cherished portrait collection of the many crowned heads of Europe, she came across a picture of Christina. "If I were to die," Mrs. Cromwell mused, "here would be the one to replace me." The thought of the match, centuries on, "still beggars the imagination."[4]

—

Holland, *that scarce deserves the name of* Land,
As but the Off-scouring of the Brittish Sand;
. . . *This undigested vomit of the Sea* . . .[5]

So at least thought Andrew Marvell, writing from an England at war with
its watery neighbor. Christina passed through Holland in July of 1654, and
whether or not she agreed with the poet, her passing through was swift. She
stopped only twice in the space of two hundred miles, and both were scholarly pauses. In the eastern town of Deventer, she met the philologist Johann
Gronovius, then traveled on to the lovely town of Utrecht, where she visited a
famous scholar of her own sex, "the learned virgin" Anna Maria van Schur-
man. A native of Cologne and some twenty years older than Christina, Anna
Maria was a brilliant linguist and a mistress of philosophy, too; she was soon to
publish a celebrated essay "on whether a maid may be a scholar." Descartes
had admired her, and the scholars at Christina's court had spoken of her in
the most glowing terms. What Christina thought, or felt, is not recorded, but
she had never welcomed competition, and the little bud of friendship between the two bluestockings blossomed no further. The queen got into her
carriage and sped her horses southward, away from Marvell's "land of the
drowned," to the Catholic Spanish Netherlands. At the beginning of August,
she arrived in Antwerp, where she passed her first few days as the guest of
Madame Pimentel, before taking up residence on the Rue Longue Neuve at
the home of García de Yllán, Baron of Bornival. Yllán, a Jewish banker, was
known to Christina through Pimentel, his personal friend, and also through
Diego Texeira in Hamburg. She settled in at his magnificent house to await
her invitation from Rome.

It was not so bad a place to have to wait. Once Europe's premier trading
and financial center, Antwerp had by now entered a period of steep commercial decline. The Westphalian Peace had signaled its demise by ending free
navigation on the important Scheldt River, so diverting trade away from the
city and northward to Amsterdam. But its long period of prosperity had ensured it a thriving artistic life, and when Christina arrived its riches were still
everywhere to be seen. It had been the home of Rubens and van Dyck. Jacob
Jordaens, now the leading painter of the Flemish school, was still working in

Antwerp; it was he who had provided the thirty-five paintings for Christina's throne room at Uppsala. Jan Boeckhorst was there, too, an artist known to Christina through one of her own agents, Michel Le Blon, a former spy for Oxenstierna and a superb engraver himself. Advised by Le Blon and her second agent, Johan-Philip Silfvercrona, she began to acquire new paintings and *objets d'art* which she could not afford, including a Rubens ivory of the goddess of love, and she sat for her own portrait—two portraits, in fact, as Minerva and as Diana—by Justus van Egmont.

The city itself held much to interest her. There had been nothing in Stockholm to equal its beautiful churches and its lovely market square. The Storkyrka, where Christina had been crowned, would have been dwarfed by Antwerp's great Gothic Cathedral of Notre Dame, "the most magnificent church in northern Europe," with its spire four hundred feet high.[6] Inside, the master hand of Rubens glowed from the alterpieces, their wonderful red colors a noblest transcendence of the blood of the local pigeons. Christina loved their vibrancy and passion, though their subject matter, the descent of Christ from the cross, was very much less to her taste. Her years of spiritual quest and her imminent conversion had brought her no closer to the person of Jesus. The vital heart of Christian belief remained, for her, unbeating.

This did not prevent her from making frequent visits to the Jesuit fathers, and to their spectacular new church with its paintings by Rubens and the young van Dyck. The church had been built with a view to enhancing the Society's reputation, but instead had invoked only censure. The opulence of the building was said to be out of keeping with the spirit of religious poverty; massive debt was to tarnish it for decades. But debt had never deterred Christina. Lavish spending, she believed, was more or less a duty for all noble hearts, and the church's splendor reassured her that she had cast her newest lot among her own kind. Antwerp could boast as well the spiritual splendor of Christian poverty, and Christina did taste of it, making several visits to a Carmelite convent. Here, the nuns walked barefoot and in silence, in a pure and pious atmosphere conducive to prayer, if not to emulation. Though Christina's plan of conversion was officially still secret, her visits to the Jesuits and Carmelites increased the rumors markedly. From Strasbourg it was reported that she had become a nun herself, and vowed "perpetual chastity."[7] And in London, Oliver Cromwell's spymaster received his own startling report: "Advysed hether from Rom," wrote his agent, "that the queen of Sweden intents thither to imbrace that religion. How lykly, I know not."[8]

In the middle of September she encountered a familiar face from Stockholm days: the Conte Raimondo Montecuccoli, Duca di Melfi, diplomat and generalissimo, hero of the Thirty Years' War, ambassador imperial, diarist of distinction. It was he whom Christina had persuaded to carry her jewels and coins out of Sweden, and to his dazzling martial plumes she had added her own small feather, the Order of the Amaranth.

Though Montecuccoli had traveled from Vienna expressly to see the queen, the two in fact came upon each other by chance as they were driving through the streets of Antwerp. Christina at once invited the conte into her own carriage, where he found two other gentlemen already seated—one a young prince, the other the disgraced former governor of Norway, the two neatly confirming Christina's dual penchant for royalty and roguery. Once in private with the queen, Montecuccoli revealed his mission: he had been sent by the Emperor Ferdinand himself, and indeed carried a personal letter from His Imperial Highness. The letter was gracious but promised nothing, and the conte's mission was of similar ilk. He soon set off for a holiday in England, though not before Christina had alarmed him by declaring that she was about to marry King Felipe of Spain—all of Protestant Holland was convinced of it, she assured him. The king was married already, admittedly, but he had no son; some kind of annulment was expected, and Christina could not forbear teasing the supposedly fearless generalissimo. The rumors were false, but Montecuccoli thought it best to convey them to the emperor, just in case.

Had he waited a week or two, he would have been able to dispel them conclusively, for early in November, Pimentel arrived in Antwerp as envoy from the Spanish king himself. He brought no better news than Montecuccoli had done. Christina wanted to make a grand public profession of faith in the hallowed city of Rome, but this Felipe would not allow. The pope, Innocent X, was seriously ill, and was not expected to survive the winter. Under the influence of his formidable sister-in-law, Donna Olimpia Maidalchini—*la terribile Pimpaccia*—the pontiff had been a firm supporter of Spain; Felipe now feared a successor friendlier to France. He did not want to lose the credit for this latest Catholic coup, and he "commanded" that, for the time being, the queen must remain in the Spanish Netherlands, and keep her proposed conversion secret.

Christina evidently found this hard to swallow, for she pretended loudly that the decision to stay had been her own. In fact she had no alternative but to go along with the king's wishes, for she could not afford to do otherwise. Hardly out of Sweden, she was already short of money. Her jewels had been pawned to raise some ready cash, but within a few months it was gone, and she now wrote urgently to Johan Holm, her steward in Stockholm, that he must sell her gold and silver services there to provide more ready cash, which, in its turn, would melt away. Apparently, her royal plate was more easily sacrificed than any of the valuable manuscripts she had brought with her. The steward, now ennobled as Leijoncrona, though until recently Christina's tailor, opted to pawn the plate instead, and was soon able to buy it back for himself.

Uncowed by the unroyal measures she had had to take, or by the king of Spain's reminder of her dependence, she went on spending, bestowing gifts on every passerby, entertaining lavishly, and in between times playing chess and croquet, driving out in her carriage, "not listening to any sermons," as she wrote to Belle in Sweden, and going to the theater—the last almost every night, and often to see the same play two or three times over.[9] Even this was not enough: she had soon hired the entire theater company for her private entertainment, a luxury that cost her four thousand borrowed francs per month. Incorrigible, she carried on, declaring gaily that the greatest pleasure money could buy was "the pleasure of spending it." Montecuccoli was a constant companion, to the extent that his old comrade-in-arms, the Archduke Leopold Wilhelm, twitted him with the title *Maestà*, insisting that he must be about to marry the queen. The conte laughed it off, but made a bet with "Amarantha" nonetheless that within a year she herself would be "passionately in love."

One or two serious visitors came to interrupt the fun and games. Klaes Tott arrived from Stockholm, with an invitation from Karl Gustav for the queen to return home. Love and loyalty had prompted the new king's invitation, for he had nothing to gain from Christina's return but her own unpredictable company. A newly appointed Dutch envoy came by, too, to pay his respects on his way to Sweden. The Dutch had recently concluded their war with the new English republic, and the two spoke a while of this, the queen asking the envoy whether he did not think it strange "to cut the king of England's head off." The Dutchman replied that he thought it very strange indeed, but Christina disagreed, saying that "they had cut him off a member, wherewith he had served himself very little, or very ill."[10] Treachery, like wealth, it seemed, was all a question of attitude.

One of the Continent's most engaging traitors was in fact now on her doorstep, and eager to pay a visit. Louis de Bourbon, Duc d'Enghien, Prince de

Condé—*Le Grand Condé*—had just arrived in Antwerp. Once the terror of Spanish armies, he had taken to intriguing within his own country. Imprisoned, then released, he had led the "princes' Fronde" against Mazarin and marched on Paris, before fleeing to work for the Spaniards in their war against his own country. He was just five years older than Christina, and his praises had long been sung in her ear by his, and her, onetime physician, Dr. Bourdelot. She idolized the young Mars, and sent a swift invitation to him. It was received with a ready hand. "I want to see for myself this princess who can toss aside a crown," said Condé, "while the rest of us spend our whole lives chasing after one." But two obstacles stood in the prince's way. The first was Pimentel, a frequent visitor to the queen, who distrusted Condé and sought to undermine him. The second, even less tractable, was the prince's own pride, which demanded that royal courtesies be paid him at the Swedish queen's little court. Christina would not agree. Though she had invited him herself, he was not entitled to a proper seat, she declared; he would have to sit on a stool. The prince declined to do so, messengers rushed back and forth, and in the end he came on a private visit, requiring no formalities at all. Christina kept him standing, anyway, and thereafter their mutual enthusiasm was quick to cool.

The nearby city of The Hague sheltered another Frenchman well known to Christina, a hero of a quite different sort, the diplomat Pierre-Hector Chanut. He was now serving in the capital city as France's ambassador to Holland, and Christina invited him to visit her—no easy matter for Chanut, since, because of the war, Antwerp was enemy territory. But he came, and Christina repaid her old friend's efforts by using him for an intrigue of her own. She began to circulate a story that Chanut had come to invite her to act as mediator between France and Spain, and negotiate an end to the war. The story had no basis in fact whatsoever, and it embarrassed Chanut profoundly. He asked her to issue an official denial of it, but this she refused to do. Chanut was eventually obliged to publish a statement of his own, one that made it clear that he had come to Antwerp in a private capacity and at the queen's own invitation. Christina was outraged, and lied loudly that the French king had had a number of proposals to make to her, if she had only stayed in the neutral territory of Holland, instead of going to the Spanish Netherlands. She subsequently did travel to The Hague, but there were no proposals, and she stayed only one day.

It may all have been an attempt to extract money from the Spaniards—or even, perhaps, from the French—or it may simply have been that Christina was missing the feel of the reins between her fingers and wanted to be influential again. Whatever the reason, the concocted story did her no good. Cardinal Mazarin was now convinced that she was in the pay of the Spanish. Furious, he began to finance a series of libelous pamphlets against her: she was a prostitute, a lesbian, an atheist. The pamphlets, irresistibly juicy, made their way to every court in Europe, and tarnished her reputation irretrievably. Her own behavior was, as usual, enough to lend them credence. Even the Archduke Leopold Wilhelm, who knew her personally, suspected they might be true.

Christina was not in the least deterred by the backfiring of her plan. She set out at once to concoct others, some of which revealed that she was already having doubts about the great step she had taken. Her schemes were so wild as to be comical, and yet, in their naïveté, rather sad. King Felipe could oust the emperor's brother, Leopold Wilhelm, and install her as lifetime regent of the Spanish Netherlands. Or she could become Catholic, then return to Sweden, and take back the crown if Karl Gustav should die. Then she could introduce religious liberty, and finally bequeath her throne to a Catholic Habsburg prince. These were not even the most improbable of her ideas. Religious liberty must be introduced in England, she felt. The easiest way of bringing it about would be for Oliver Cromwell to become Catholic and hand England over to the pope. His holiness could then grant the country back to Cromwell, on condition of religious liberty, and Cromwell could then become King Oliver I. She relayed her thoughts to Montecuccoli enthusiastically. He recorded them all dutifully in his diary.

Christina's imagination was the likeliest place for Oliver Cromwell's conversion, but she had her own to dream about, in any case. She wanted it to take place in Rome, and she wanted it to be grand, an exhibition of renunciation to a vast assembly of admirers, a superb display of moral strength to surpass her father's mere vainglory. But Rome was far away, and a public pronouncement was not in the interests of those who were likely to transport her there. "These long journeys cost a lot of money," the archduke remarked to Montecuccoli, and he suggested instead that, for the time being, the queen would do best to make a much shorter one, a journey of just twenty-five miles, say, from Antwerp to Brussels. There a quiet ceremony could be held, a private commitment made under Spanish eyes, which would resound, in good time, to Spanish credit.

So, in late December, Christina's little court packed their bags once again and moved to the capital of the Spanish Netherlands. They arrived in the evening, and fireworks lit the sky as the queen sailed in on the archduke's gilded barge. She was installed in the lovely Palais d'Egmont, where Leopold Wilhelm had in fact vacated his own private apartments to make room for her. The palace also housed his famous collection of paintings; though it was hardly Rome, Christina passed some consoling hours in their company. But a greater matter was at hand for her. The day after her arrival, on Christmas Eve, she was finally received into the Roman Catholic Church. A small chapel adjoined her bedroom, and here, in the evening, attended by the archduke himself, Montecuccoli, Pimentel, and two Spanish diplomats, she professed her new faith:

> I believe in one God, Father omnipotent, creator of Heaven and Earth . . . and in Jesus Christ His only son, Our Lord . . . born of the Virgin Mary . . . and I believe that the holy Eucharist is the true body and blood of Christ. . . . I believe in one holy, Catholic, Apostolic, Roman Church, mother of all Churches. . . . I pledge obedience to the Roman Popes, successors to Saint Peter, Vicars of Jesus Christ. . . . I accept the teachings defined and declared by the Synod of Trent, and I renounce all heresies as the Church has renounced them. . . . I believe in the true Catholic faith, without which none shall be saved. . . . [11]

Father Güemes recited a psalm and absolved her from the heresies of Lutheranism, and Christina rose, Rome's newest and least likely convert.

Well-timed cannon boomed through Brussels, and fireworks burst above the town, startling and puzzling the local people, who had no idea of what was taking place within the archduke's palace. The Jesuit Father Pallavicino later described the event as "one of the most memorable and glorious for our Faith ever recorded."[12] Perhaps it was, but it was not what Christina had imagined. Her grand gesture of apostasy, intended to stun all Europe, had taken place quite secretly, more or less in her own bedroom. In fact, she did not seem to take it seriously at all. At the midnight mass that followed the ritual, her behavior was relaxed, even flippant, and she later spoke laughingly of the Catholic belief in transubstantiation—"that the holy Eucharist is the true body and blood of Christ"—of which she had vowed acceptance. No doubt her French friends in Stockholm, Catholic themselves, by culture, at least, had spoken so, and Christina had simply assumed that this was normal practice for sophisticated people within the Church. If so, it was a wide misjudgment. The devout Spaniards were displeased, and a shocked Leopold Wilhelm declared that he did not believe her conversion was genuine. The queen, he felt, must be anticipating some political gain. He conveyed his doubts to his brother, the emperor, who took no further steps to help her, while to Montecuccoli it was whispered that her majesty would have done better to remain at home in Sweden.

In January of the new year, the ailing pope died, and in April came news of another death, one that touched Christina more closely. Maria Eleonora had died in Stockholm, at the age of fifty-six. She had passed her last months in bleak half-mourning, bewailing her daughter's departure from the land that she herself had taught her to despise. Christina, complaining that "there was nowhere in the world where they mourned the dead as long as they did in Sweden," betook herself to the countryside for just three formal weeks. She spent most of her time hunting, it seems, and at one point returned privately to Brussels for a few less solitary days. It was a modest mourning for a mother whom Christina had loved, in her own words, "well enough," and in any case, there were now more pressing matters to attend to. A new pope had been elected, Alexander VII, the former Cardinal Fabio Chigi, Vatican secretary of state three years before when the question of the queen's conversion had first been raised. He was not very friendly to the Spanish, but he was decidedly unfriendly to the French, and in a time of war, this was as good as outright support as far as the Spaniards were concerned. King Felipe was delighted, and so was his cousin Ferdinand, the Holy Roman Emperor, and so was Christina, who now imagined a speedy departure for Rome. But the new pope had found the papal finances in such a state that he could provide "no temporal help," and her appeal to the emperor for indemnities owed to Sweden was swiftly rebuffed, as the Swedes were demanding the money themselves. She tried to raise a loan from her French banker, Bidal, offering as security some of the lands granted on her abdication. The terms of her agreement, however, had prohibited any such alienation—at the queen's death the lands were to revert to the Swedish crown—and Karl Gustav, hard pressed by the cost of his new war in Poland, intervened for the first time in his life to Christina's disadvantage.

The independence she had longed for, independence from duty, from government, from life in Sweden, was fast melting away to reveal one bare and frightening fact: she had no money. What had come with her from Sweden had been recklessly spent. Karl Gustav had blocked her loan from Bidal. The emperor was distrustful of her. She had alienated the French completely. The pope could send her nothing, and, despite his flowery assurances, there had been nothing—apart from a bed in Brussels—forthcoming from the king of Spain. Christina was in fact more dependent now than she had ever been. She was at last beginning to realize that without the crown she had little real importance, and the realization made her desperate. Montecuccoli must persuade the emperor to send an army against Sweden, she said, to enforce her claims in Pomerania. She would sell all she had, and lead an army against Karl Gustav herself. The emperor must incite the Dutch and the Danes, "who hate Sweden," to attack the country, too. She knew of secret treaties with other powers, allowing Swedish soldiers to pass through their territories; these she would reveal along with other confidential matters of state. Evidently believing that Montecuccoli felt more bound to her than to his emperor, she instructed him to avoid mentioning her name, and to say only that "a well informed person" would provide this information—an act of desperate naïveté to match her desperate treachery.

In the event, her worst fears were realized. In the middle of July 1655, Karl Gustav led the Swedes into Poland, and the lands that were to sustain Christina fell victim to the travel of a mighty army. The demands of war kept Swedish coffers empty, and for years to come Christina would struggle to claim even a fraction of the money due her.

CHRISTINA REMAINED IN Brussels for several months more. She had made up her mind to leave before the winter set in, and at last she resorted to borrowing from her wealthy Antwerp host, Don García de Yllán. Toward the end of September she accepted some hundred and forty thousand riksdaler— two-thirds of her agreed annual income from Sweden—and two-thirds of this were already earmarked for the costs of her journey to Rome. Don García had evidently no hopes of early repayment. He added the handsome credit to his will for the benefit of his heirs.

So Christina left the Spanish Netherlands, traveling via the Lutheran stronghold of Augsburg, and up into the mountains of the Austrian Tyrol, to the imperial city of Innsbruck. It had long been a favored spot for the devout emperor and his family, and it may have been in the hope of eliciting some help from them that Christina had chosen it as the place where she would make her public profession of Catholic faith. Waiting for her at Innsbruck was the Jesuit Father Malines, "Don Lucio Bonanni," one of the earliest guides on her long religious road. He was accompanied by the pope's special emissary, Father Lucas Holstenius, a convert himself and head of the Vatican Library. If Christina was loath to discuss religion, they had at least books to talk of, and talk they did, becoming in the process firm and lasting friends.

In Innsbruck, she was the guest of the Archduke Ferdinand, governor of the Tyrol, a nobleman of modest means who had incurred a hefty debt in order to give her a suitably regal welcome—even his silver candlesticks had been borrowed. He had not been informed of the purpose of her visit, and neither had his cardinal brother, also in residence there. Christina spent a few social days, going to the theater, looking at the fine local art collection, walking in the gardens, and in her honor a new musical drama was performed—*L'Argia*, a work by Antonio Cesti, still a young man, but already celebrated as "the glory and splendor of the secular stage." It is a tale of love, betrayal, incest, and lesbian seduction, with a heroine in trousers and a chorus of pirates, a vast corps de ballet, and plenty of theatrical wizardry, all perfectly calculated to appeal to Christina's adventurous imagination. It lasted six hours, and she watched it—twice—"with great pleasure and attention." Cesti, in fact a priest, and recently rebuked by the pope for his "dishonorable and irregular life," appealed to her, too. Charming and manipulative, he was now treading his path away from Rome just as Christina was treading her path toward it. The real purpose of her visit was now made known to the archduke, and on the third of November, in his own chapel royal, she was publicly received into the Catholic Church.

It had been almost a year since her private conversion in Brussels. She came dressed in a gown of black silk, a diamond cross her only ornament, and pronounced her abjuration in a "clear, loud voice." A solemn mass followed, with the singing of a Te Deum, a Catholic service to honor a newly Catholic queen.

From Innsbruck Christina wrote at last to "Monsieur my brother," Karl Gustav, "to declare myself openly for what I am," and she wrote as well, in rather eccentric Italian, to the newly elected pope. It was the first letter she had ever written in that language, and Father Pallavicino, later publishing it, advised the reader to view it "in a spirit of great piety and generosity." But if Christina's syntax was erratic, her meaning was clear enough:

Finally, I have arrived where I have so long desired to be, folded in the bosom of our holy mother, the Roman Catholic Church, and I humbly beg you to honor me with your benevolent commands. I have shown the world that, in order to obey Your Holiness, I have been ready to relinquish my throne, placed as it was in the midst of irremissible sin. It is a greater glory to obey Your Holiness than to rule from the highest throne. . . . I who have nothing else to offer than my own self, and my blood, and my life, I offer it all to Your Holiness, asking only to kiss your most holy feet.

Your most
obedient daughter,
CHRISTINA[13]

It is a letter of extravagant submission, but Christina's passion for the limelight slips out between prostrations. Here at last was "a greater glory" than even her father had claimed. There is not much penitence in evidence; the "sin" that she had lived in seems to have belonged to her country rather than to herself. But her conversion was now public. Christina had finally "shown the world" that she could submit as she had ruled, with loud panache.

The bells rang out, the cannon boomed, and the day was spent in public celebration. Though Christina enjoyed the festivities well enough, the sincerity of her conversion was widely doubted. Gossips reported that the queen had declared a change of religion "the most diverting thing in the world." Within hours of the solemn mass, she turned up at the theater to watch a far from pious play. "How apt that you should give a comedy for me this evening," she is supposed to have said, "since this morning I gave you a farce myself."

Whether this is true or not, Christina did not stay for any further performances. A few days after her public conversion, she set off on the last stage of her journey to Rome.

CHRISTINA'S RETINUE HAD grown considerably since her midnight ride from Sweden in the company of four young men. In the ensuing eighteen months, she had acquired two clerics, three musicians, eight secretaries, an Italian count and a pair of Spanish dignitaries, and scores of servants, most of them still waiting for wages—in all, about 250 people. They traveled in a caravan of carriages and horses, and numbered among them two unhappy Swedes, puzzled by the recent turn of events, and anxious to reassure their families that they themselves would not turn papist. Pimentel was one of the dignitaries, and with him was his compatriot Don Antonio della Cueva y Silva, now officially the Queen's Master of the Household. Montecuccoli was with them, too; Christina had specifically requested his company, and he took advantage of it to keep the emperor informed as they made their wintry way southward.

The route they took was by no means direct. In fact, on the pope's own orders, it was deliberately circuitous, designed to delay the queen's arrival in Rome until all the different committees had met and the various preparations had been completed. It could not in any case have been an easy journey, since winter was upon them, and all possible paths lay through very high ground, rocky and now also covered with snow. South they rode two hundred miles to Mantova, then, instead of continuing down to Rome; they traveled eastwards along the Po River, the northern frontier of the Papal States. Just before Ferrara, they crossed the river, now in treacherous full flood. A gilded barge had been provided to ferry the queen to the other side, but she elected to cross instead by the plainer but safer bridge of boats constructed for her retinue.

Ferrara was a major stopping-place for a no doubt weary caravan. Here a formal entry was required, and a formal reception, and formal dinners, and formal attendance at religious services. A few such days were more than enough for Christina, and she took to the road again with relief. Now that they were within papal territory, the journey had become easier, at least for the queen herself. She traveled in a new carriage sent especially for her from Pope Alexander, and for her greater comfort along the way, His Holiness had also sent two sumptuous canopied beds with matching armchairs, an elaborate set of table silver, and even a papal chef—the celebrated Luigi Fedele, man of a thousand spices, including, it is said, oil of musk.

To the west they went, to Bologna, with beds and silver and chef in tow, then eastward to Pesaro, across the river Rubicone—Caesar's Rubicon, and Christina's own—zigzagging across the country while a thousand Roman workers hammered and painted and stitched and sewed against the queen's arrival. Christina had been as good as her word: she had embraced her new faith, first privately, then publicly, and she was now to be given a triumphant welcome under the auspices of the new pope. In the humbling aftermath of the Thirty Years' War, Christina's conversion was to be celebrated as a brilliant victory for the Catholic Church.

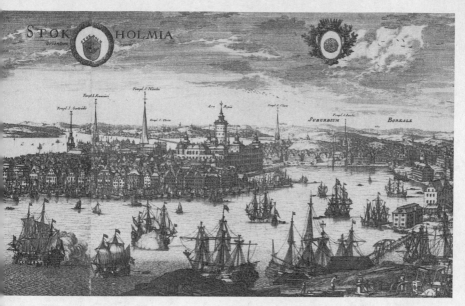

Stockholm at mid-century, built on a series of islands where Lake Mälaren meets the Baltic Sea. At Christina's birth, its population numbered some ten thousand souls. *Engraving by Willem Swidde.*

The interior courtyard of Slottet Tre Kronor (the Castle of the Three Crowns), where Christina was born in 1626, and where she lived until her abdication in 1654. *Engraving by Erik Dahlberg.*

ABOVE Christina in 1632, age 5, wearing the Elizabethan-style collar still fashionable in Sweden. *By an unknown artist.*

TOP LEFT Christina's father, Gustav II Adolf "the Great." *Oil painting by an unknown Dutch artist from the circle of Miereveld.*

MIDDLE LEFT Christina's mother, Maria Eleonora of Brandenburg. *By an unknown artist.*

LEFT Christina, about age 7, wearing a mourning veil following her father's death. *Engraving by Michel Le Blon.*

Christina's aunt, the Princess Katarina, half-sister of Gustav Adolf and mother of the future Karl X Gustav. *Oil painting from the studio of Jakob Elbfas.*

Christina's uncle by marriage, Count Johann Kasimir von Pfalz-Zweibrücken, father of the future Karl X Gustav. *Oil painting by David Beck.*

Christina, about age 8, painted in the Tre Kronor Castle by the court painter and her own drawing master, Jakob Elbfas.

Christina, about age 12, a rather sullen-looking Queen of the Swedes, Goths, and Vandals; Great Princess of Finland; Duchess of Estonia and Karelia; and Lady of Ingria. *Engraving by M. von Lockom.*

Christina, about age 15, at the time of her cousin Karl Gustav's return to Sweden. *Oil painting by Jakob Elbfas.*

Baron (later Count) Axel Oxenstierna, Chancellor for more than forty years, and the effective ruler of Sweden during Christina's minority. *Oil painting by David Beck.*

Johan Matthiae, Christina's beloved tutor, whom she called "papa." A liberal-minded Lutheran pastor, he later became Bishop of Strängnäs. *Oil painting attributed to David Beck.*

Christina's cousin and first love, Count Karl Gustav, later her successor as Karl X Gustav of Sweden. *Oil painting by David Beck, 1648.*

Count Magnus De la Gardie, scion of one of Sweden's first families, and a favorite of Christina's. *Oil painting by Matthias Merrian, 1649.*

Christina, age almost 24.
Oil painting by David Beck.

The beautiful Countess Ebba Sparre, whom Christina called "Belle."
Oil painting by Sébastien Bourdon, ca. 1653.

Christina and Belle visit the French scholar Claude Saumaise, who had feigned illness to escape a dreary academic meeting. Christina roars with laughter when the maidenly Belle is tricked into reading aloud from Saumaise's *risqué* book.
Gouache by Niclas Lafrensen.

Queen Christina at the age of 26, a hunting portrait in very male style. This was her favorite portrait, and it hung in her bedroom to the end of her life.
Oil painting by Sébastien Bourdon.

The great philosopher René Descartes. Christina corresponded with him through the French Ambassador, Chanut, and eventually lured him to her court in Stockholm. He was reluctant to serve "as some kind of exotic elephant in the land of bears," and in fact survived only a few months in the bitter weather. *Oil painting by David Beck.*

Christina and her "academy." The queen sits toward the center of the picture. To her right are the French Ambassador Chanut, the Princess Elisabeth of Bohemia, the philosopher Descartes, and the mathematician Mersenne. The Prince de Condé presides at the table to the left. The scene is a fiction, painted by Louis-Michel Dumesnil a few years after Christina's death.

Christina's abdication in the great hall of the castle at Uppsala on June 6, 1654. Chancellor Oxenstierna refused to remove her crown, and the leader of the Peasants' Estate begged her not to abdicate but to continue "as Sweden's fore-horse." *Engraving by Willem Swidde, after an original by Erik Dahlberg.*

Christina, confident and proud at the age of 26, painted in 1653, the year before her abdication. *Oil painting by Sébastien Bourdon.*

Christina drives through the market square of Antwerp in the late summer of 1654. She spent a restless sixteen months in the Spanish Netherlands, waiting for permission to travel on to Rome. *Oil painting by Erasmus de Bie.*

Christina, painted in 1660, at the age of 33. The reality of life without a crown has begun to bite, and her proud expression has become wary and defensive. *Oil painting by Abraham Wuchters.*

Pope Alexander VII (Fabio Chigi), who welcomed Christina to Rome in December 1655, the Catholic Church's most celebrated convert. Her extravagant behavior soon dispelled his illusions. *Bust by Gian Lorenzo Bernini.*

Cardinal Decio Azzolino, Christina's great love. A man of considerable abilities, he was small, built, with dark hair and strong, though not handsome, features. His personality was markedly warm, with a machiavellian twist. *Bust by Pietro Balestra, ca. 1670.*

Christina and Azzolino talking together in the cardinal's library. Though Azzolino protested their innocence to the pope, his gesture indicates how the relationship was generally viewed. *Engraving by an anonymous artist.*

Cardinal Jules Mazarin, ruler of France during the minority of Louis XIV and beyond. *Oil painting by Pierre Mignard, "le Romain."*

Anne-Marie-Louise-Henriette d'Orléans, Duchesse de Montpensier, "la Grande Mademoiselle." *Oil painting from the school of Pierre Mignard, "le Romain."*

His Most Christian Majesty, Louis XIV of France, "the Sun King." Christina found him "a very handsome, decent fellow," but they were not always on good terms. *Engraving by Robert Nanteuil.*

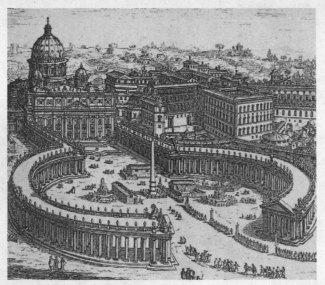

Bernini's oval piazza in front of St. Peter's Basilica. Designed in 1656, it was built during Christina's first years in Rome. *Engraving by Giovanni Battista Falda.*

A lively afternoon at Bernini's Piazza Navona, at the time of Christina's arrival in Rome.
Engraving by Giovanni Battista Falda.

Arrival at the château of Fontainebleau. Here Christina was presented to the court of Louis XIV in September 1656, and it was here, a year later, that she commanded the execution of the Marchese Monaldeschi. *Engraving by Israel Silvestre the Younger.*

Christina's friend, Pope Clement IX (Giulio Rospigliosi). She celebrated his election with a provocative fête in Protestant Hamburg.
Oil painting by Carlo Maratti.

Pope Innocent XI (Benedetto Odescalchi)— "Pope No." Disapproving of Christina's behavior, he closed down her theater and rescinded her papal pension. *Engraving by A. Clouet.*

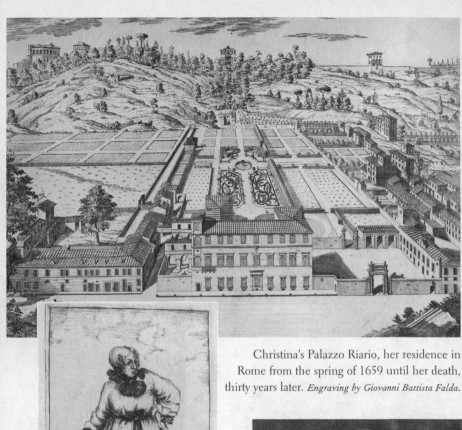

Christina's Palazzo Riario, her residence in Rome from the spring of 1659 until her death, thirty years later. *Engraving by Giovanni Battista Falda.*

Christina in 1662, at the age of 36. "She is low and fat, and a little crooked." *Etching by an unknown artist.*

Christina in 1667, at the age of 40.
Oil painting attributed to Wolfgang Heimbach.

Christina in 1687, at the age of 60, in unusually royal attire.

Oil painting by Michael Dahl.

Christina toward the end of her life, dressed in the simple clothing she always preferred.

Bust by an unknown artist.

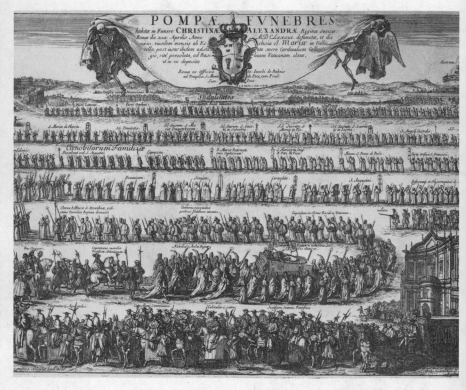

Christina's funeral procession, in April 1689. Following her requiem mass in the Chiesa Nuova, her embalmed body is carried on a bier to St. Peter's Basilica for burial. "The whole Colledge of Cardinals" was in attendance, except for Azzolino, who was too distressed to attend.

Engraving by Robert van Audenaerd.

At Pesaro, the queen was entertained by dancers and acrobats and circus strongmen, by bits and pieces of opera and ballet and pseudo-learned discourses—"rubbish," in the view of one attendee, but Christina enjoyed it all. A star turn in every performance had been taken by two local notables whose acquaintance she now made, the brothers Ludovico and Francesco Maria Santinelli. Both possessed the kind of roguish charm that she had enjoyed in so many other of her favorites, and she decided impulsively that both must join her entourage, to enliven their road to Rome. The Santinellis, never slow to recognize their own advantage, agreed at once, and so they progressed, along the bright seacoast, to Ancona.

Nearby lay the little town of Loreto, a place of Catholic pilgrimage since the end of the thirteenth century. Legend has it that the house of the Virgin Mary, scene of the Annunciation, was plucked from its moorings in the Holy Land during a time of Turkish threat, and transported by a band of angels to the Italian coast. A church had now been built around the little house, and over the years, popes and princes had come to lay their gifts of homage at the feet of the statue of the Virgin, the "black Madonna" supposedly carved by the evangelist apostle Saint Luke. Christina now came to make her own pilgrimage, but if it was genuine, it was half-hearted. Her gift was worthy of any pope or prince: a fabulous jeweled crown and scepter. But, pleading the excuse of the weather—it had been snowing—she did not approach the shrine itself, but sent Father Güemes to make the offering for her, while she waited in her carriage, in residual Lutheran discomfort at all the talk of statues and miracles.

If she felt more at home in the cold, snow-covered Appenine Mountains, further statues awaited on the other side, at Assisi, where a tremendous banquet was held for her near the tomb of Saint Francis, *il poverello*, friend of the hungry. A hundred and fifty miles remained, down the valleys of the great Tiber River, with the cold growing less and the skies bluer, and the landscape more and more the landscape of antiquity. Christina was enchanted, and lost herself in a reverie of Caesar and Virgil. "I dare not speak the name of Jesus," she said, "in case I break the spell!" It was not broken by the Duca di Bracciano, Paolo Giordano Orsini, the last of her hosts on the long journey and a friend by correspondence for many years. It was to the duca that she had written, excitedly cataloguing the treasures looted for her from Prague. The two had written then, as they talked now, of history and art and music, and in the duca's castle, rearing stone among the olive groves, Christina celebrated her twenty-ninth birthday. Two days later she would be in Rome.

HRISTINA'S ARRIVAL IN Rome was almost as lacking in drama as her private conversion had been. She arrived in the dark, with no flares or fireworks to light her way, and made a shortcut through the Vatican gardens into the palace itself, where she was led directly to a private audience with the pope. Whether or not she knelt, as she had wished, to "kiss his most holy feet," is not recorded, but she did make the ritual three obeisances — the first submissive gesture of her life — before taking her seat beside His Holiness.

The seat itself had posed quite a problem. A proper chair, with arms to lean upon, was permitted only to ruling sovereigns in the pope's presence, yet a plain stool had seemed too little recognition of the great sacrifice this former queen had made. The problem had found its way into the hands of a master, the great Gian Lorenzo Bernini, who had sculpted a perfect compromise: a low-backed chair with rounded arms. Official reports described it as "a magnificent royal seat," but the papal accounts recorded a humbler "low stool" for her majesty's use. Christina sat in it for a quarter of an hour, before being led to the apartments where she was to stay. Many curious eyes watched her go, and one observer recorded an impression of her as she made her way to her suite. She was dressed simply, but in women's clothes, a gown of plain gray, and a black scarf around her shoulders. Her bright blue eyes earned a special mention, but she looked very small, and her straight blond hair "appeared brown and curly" — by now she had shaved her head and had taken to wearing a dark wig.

It was a singular honor for Christina to be the guest of His Holiness, as women were not generally permitted to sleep within the walls of the Vatican at all. Eight beautiful rooms had been assigned to her, all decorated especially for the occasion by Bernini's younger brother, Luigi. Every corner was resplendent with brocade and satin, embroideries and lace; vivid frescoes lined every wall. The rooms were at the top of the Torre dei Venti—the Tower of the Winds— and from them Christina could see out over the whole of the city and beyond. She was late to bed on her first night in Rome, but she slept reassured in sheets of finest cambric, with a great fire blazing in the hearth, and a silver bed-warmer at her feet.

She did not sleep long. Morning found her early in the Cortile del Belvedere, a courtyard within the Vatican where a splendid equipage awaited her, a personal gift from the pope. Bernini had come to his aid again, and the design of the gift was his: a beautiful carriage with matching litter and sedan chair, all decorated in blue silk and silver mounts. The pope had included other gifts of more than mortal design: six fast horses to draw the carriage, two mules for the litter, and a small palfrey, nobly draped in blue and silver, too. Whether delighted or insulted by the last gift—it was typically a lady's horse, after all—Christina mounted it at once and put it through its paces, showing off her skill and drawing many admiring comments. She spoke briefly with Bernini himself, then set off with Lucas Holstenius to investigate the Vatican Library.

Christina spent just two days in Rome before departing again, but her journey this time was only a few miles: she had to leave the city in order to enter it again in a formal, triumphal procession. She drove in a papal carriage north to the ancient Ponte Milvio, the oldest bridge in Rome, and there she was welcomed with elaborate ceremony by ranks of nobles and mounted soldiers, headed by Rome's own senator. Together they began their slow procession, along the great old Via Flaminia, toward the center of the town, a vibrant echo of Christina's coronation five years before, though, perversely, the weather was poor, and Rome was as gray now as Stockholm had then been bright. Along the route, they stopped for refreshments, and Christina thereafter declined the carriage in favor of her lively little palfrey. Thus she could easily be seen by the thousands who thronged the way, hoping to catch a glimpse of her, a small figure dressed quietly in gray and black, with a plume in her hat by way of celebration. Her simplicity contrasted favorably with the gaudy riches of the local ladies, and, if the choice was tactical, it was necessary, too, since she had left most of her jewels with the Antwerp pawnbrokers.

She made her formal entry into the city in the time-honored way, through the Porta del Popolo—the People's Gate—and the people were there in force to welcome her. The inside of the great stone arch had been decorated for

Alexander by Bernini, his favorite artist, but the pope had added an inscription of his own to honor the queen's arrival. *Felice Faustoque Ingressui*, it declared, wishing her a "happy and propitious entrance." Looking through to the ranks of crimson cardinals before her, Christina might have been happier just to dismount and slip into the little church to the side of the gate, packed as it was with Caravaggio and Raphael masterpieces, but the formal harangue of greeting could not be forgone.

At length they set off down the Corso toward Saint Peter's, crowds pushing to see through yet more crowds, and the Castel Sant' Angelo cannon booming nearby. Progressing across Bernini's wonderful oval piazza, Christina reached the basilica's entrance, and knelt, on a cushion of golden silk, to kiss the crucifix. With chants ringing out around her, she made her way up the great marble aisle, and at the foot of the high altar, with its huge columns of twisted bronze, she knelt again to pray before the holy Eucharist, "the true body and blood of Christ," according to her new faith. No mass was held; Christina returned instead to the Vatican palace, where Alexander and all his cardinals stood waiting to receive her.

The new day dawned, Christmas Eve, but she had little heart for more festivities. The pope held his traditional banquet for the cardinals, and Christina went along, concealed behind the blue silk of her new sedan chair, to observe the proceedings in peace. She did not attend the midnight vigil, but took herself early to bed in her beautiful sheets, with warm silver at her feet. On Christmas Day she received, from the hand of the pope himself, the sacrament of confirmation, one of the seven "true and holy Sacraments ordained by Jesus Christ for the benefit of all men." New and old friends were in attendance, including Pimentel, determined to receive the dignities due the Spanish king's official representative. Felipe's ambassador in Rome, the Duque de Terranova, was determined to receive the same, and an argument broke out between them; the diplomatic pope resolved it by appointing someone else entirely.

According to Catholic practice, Christina had chosen a new name, a "confirmation name" to be added to the name given her at baptism. She took the name Alexandra, an elegant courtesy to the reigning pope, and perhaps, too, a reminder of the great young general of ancient times whom she had so long admired, and whom, as yet, she still imagined she could equal. At the pope's urging, she added the name Maria, to signal her awakened devotion to the Virgin, a devotion central to Catholic practice. Christina did not want the name, and she was never to use it. Catholic chroniclers would assign it to her, but she herself would sign only Christina Alexandra.

THE ROME THAT Christina now encountered was a city just past a peak of greatness, the highest it had known since the days of the empire. For almost a century, successive popes had poured money into it, developing the central areas, erecting public buildings, patronizing the arts. The repair of a huge aqueduct, still standing from ancient times, had ensured a good supply of water for an expanding population, and the citizens now numbered some hundred thousand souls, more than ten times the people of Stockholm. In years of pilgrimage, half a million visitors would arrive to swell their numbers and their coffers—even the beggars were said to be able to plead in half a dozen languages. And every year, the best of Italy's artists flocked in to take advantage of the Church's endless commissions.

The rebuilding of Rome was an important Catholic stratagem in the years of the Counter-Reformation. The Church had to be seen as strong and progressive, a force for enlightenment against the bleak rigidity of Protestantism. The flourishing Jesuit schools and missions, of which Christina had been a beneficiary, had armed the Church intellectually. Rome was to be its artistic as well as spiritual home, purged of the pagan leavings of empire, Christianized from paving-stone to rooftop. By the middle of the century, a great sea change had taken place: *Roma antiqua*, a small town, peopled by the native-born, a remnant of the days of Caesar and Augustus, had become *Roma moderna*, a cosmopolitan city of beautiful modern buildings and new-paved streets, bristling and bustling with wealth and energy, the thumping heart of the vibrant new culture of the Baroque. So Christina had learned of it from her agents and her diplomats, from Bourdelot, who had lived there, from her books, and, above all, from her willing imagination. The picture was not inaccurate, but, by the time of her own arrival, it had already begun to fade. The vast building projects, decades-long, had sapped papal finances, and armed feuds within the papal states had depleted the coffers to their bare wooden bones. The popes had paid their huge bills by a method well known to Christina—by borrowing, and they borrowed for the long term, relying on their successors to assume the responsibility of repayment. By now, the debts were overwhelming—Christina herself had felt the weight of them, denied as she had been any "temporal" assistance on her journey to the city. Rome's impoverishment had led to a decline in its diplomatic standing and, gradually, in its artistic prestige, too. Christina had plunged her own little daggers into the giant's body: in the heady days of the Westphalian peace treaties, she had lent a willful support, against Sweden's interests, to the rising power of France, undermining Rome's authority even in its own Catholic world. From now on, the sparkling stream of Italian genius would wend its way farther afield, to Paris.

But though the great epoch was now passing, its achievements had been wonderful. There was the Vatican itself, much of it new, with St. Peter's Basilica, only fifty years old, consecrated in the very year of Christina's birth. There were more than a hundred other churches besides, all of them of interest, some of them magnificent. There were countless paintings and sculptures, great monuments, new and old, people to visit, music to hear — including one opera that was being composed especially for her. Scholarly groups acclaimed her with harangues in a dozen tongues, and at the Jesuit's Collegio Romano, she met Father Athanasius Kircher, polymath extraordinaire, Egyptologist, inventor, scientist. In his famed animal museum — designed, in the Renaissance manner, to display every beast and bird known to man — Christina saw for the first time the creatures of the New World. Kircher revealed to her the ingredients of secret medicines, and presented her with a sample of the least inefficacious of them.

On New Year's Eve, she made the closer acquaintance of some of the cardinal princes of the Church. They had seen her already, on the day of her formal entry to the city, and on Christmas Day, at her confirmation, but now they were at leisure to speak with her and to form their own opinions of the new prize convert. The pope had led them to expect a rather pious young woman, and had warned them to be on their best behavior, since "on the other side of the Alps" her majesty may have heard one or two things unfavorable to Rome and her Church. "The Protestants keep a keen look out for any scandal," he had reminded them. "They watch every little thing, and they put it all down in their memoirs."[1] Though none among Their Eminences had disagreed openly, many had been offended at the injunction. A letter of protest had been drawn up against this impugning of their collective reputation, and had been conveyed to His Holiness with dozens of indignant signatures.

The pope himself, at least, had nothing to fear from the gaze of her majesty's recently Protestant eyes. Renowned for his high moral principles,

Alexander VII was a devout and studious man, personally ascetic, but a lover of the arts nonetheless, a native of the beautiful town of Siena. For his coronation, he had permitted no triumphal arches to be erected, none of marble, none of stone, none of cardboard, and on his arrival at the papal palace, all servants deemed unnecessary had been dismissed. His sparsely furnished personal bedchamber could boast just one superfluous item, the characteristically understated memento mori of a wooden coffin. The pope's retiring personality had served him well during his years of diplomacy, but it was perhaps a disadvantage to him now—he was not politically strong, it seems. But he had managed at least to expel his predecessor's greedy puppeteer, Donna Olimpia Maidalchini, and she had soon afterwards died of the plague, humiliated, furious, and sneezing, but very rich—her fortune was said to be two million scudi—enough to have fed Rome's hungry for centuries.

Alexander was relieved to be rid of Olimpia and her jewels and her countless relatives. He wanted a simpler pontificate, unencumbered by grasping Papal Nephews. He wanted men of sincerity about him, and through them, he wanted a more pious flock. During his long years of diplomatic service in the German lands, he had learned to admire the austere devotion of northern Catholicism; by comparison, he found devotional practice in his homeland sadly lacking. "I remember observing, when in Germany," he remarked, "that a great silence prevailed in the churches among lay people and even more so among ecclesiastics, so that if anyone was talking, it was held that he must be either a heretic or an Italian."[2] Christina was neither, though it had already been noted that she liked to talk in church. Her piety, and even her sincerity, were still doubted by some, but for the time being the pope was well pleased with her. On hearing of his arrival at her apartments, she had run "with great strides" to meet him, displaying an appropriate convert's penitence by throwing herself at his feet. Their mutual interest in the arts had given them an easy link, and, not least, her conversion was a diplomatic coup for Alexander in the early days of own pontificate.

Alexander's hopes of good behavior on the part of his cardinal princes was quickly disappointed. Within a few days, one of them had had "the audacity" to fall in love with the convert queen, and the imprudence to tell her so. She laughed it off, saying that she had not come to Rome for scandal. A less adventurous young cardinal was now appointed to serve as liaison between her majesty and the papal court, and to introduce the queen to the ways of Roman society. He was clever and charming, and very polished, in the manner of a practiced courtier, but as yet he did not attract Christina's particular interest. He began to be her daily guide, and directed her to the different shrines with their relics of the saints. Rome's churches concealed myriad items she was now called on to reverence, including, so it was said, the bodies of Saint Peter and Saint Paul, the heads of Saint Luke and Saint Sebastian, and the arm of Joseph of Arimathea, as well as one of the thirty pieces of silver paid to Judas for the betrayal of Jesus, and even a piece of barley loaf, said to be one of the five that Jesus fed to the multitude in the miracle of the loaves and fishes. Christina did not like any of it; Protestant instincts twitched beneath her new Catholic skin.

Happily, relief was at hand—secular, indeed pagan, relief. She had spent only a few days within the Vatican, before moving across the river a mile or so to the famous Palazzo Farnese, near the Campo dei Fiori, where the flower-growers brought their lovely produce to market. The Palazzo Farnese was a beautiful bloom in itself, a jewel of Renaissance architecture and one of the finest palaces in Rome. It belonged to the Dukes of Parma, the Farnese family, but it had been many years since they had lived in it, and until only a few weeks before, the palazzo had been inhabited by Rome's previous trophy convert, in fact a distant relative of Christina's, Cardinal Landgraf of Hesse-Darmstadt. *Il cardinale Langravio* had proved of disappointingly little political use to the Farnese family. For years they had been trying to alleviate the harsh terms imposed on them following the "Wars of Castro" that they had waged against successive popes.[3] Support of a prominent convert had seemed to be a sure route to the papal ear, but Landgraf had been unable to help them. A bigger fish, they felt, might be more likely to draw the pope's attention, and the cardinal was consequently transported to a distant villa to make way for the newest catch.

Christina was pleased with her new abode. On the outside, at least, it was magnificent—Michelangelo himself had had a hand in its design[4]—and, even more importantly, it was free. She was shown the building by the Marchese Giandemaria, the Farnese family's representative in Rome; he was to see that she was made comfortable, and report her activities back to the duca. The first reports of this precise little man were innocent enough: her majesty was enraptured by the wonderful artworks that filled the rooms, sculptures from ancient Greece and Rome, tapestries and paintings, many installed expressly for her pleasure. But only parts of the building were really habitable, and in some rooms the plaster was even crumbling—it was covered over with trompe l'oeil canvases. There had been some renovation, and Giandemaria had managed to provide four principal apartments, each with a bedroom and three other rooms. These had been newly furnished, but they may not have been very comfortable despite that, since Christina apparently slept in a different bedroom almost every night.

One room of the palazzo, at least, was beyond her reproach: the wonderful painted gallery, the work of the great Annibale Carracci.[5] Modeled on Michelangelo's Sistine Chapel ceiling, painted almost a hundred years before, the Farnese gallery had taken Carracci and his assistants eight years to complete. They had painted not only the ceiling, but much of the wall space, too, covering it with frescoes portraying the world of classical legend, a world of violence and of pleasure. When Christina first set eyes on them, the paintings were just fifty years old, their brilliant colors as yet undimmed by the strong Italian sun. They were a riot of beauty and splendor, revealing Carracci at his blithe pagan best. To Christina, they were the essence of all that Rome had promised her—a full-blooded shout of joy, a laughing jettison of the narrow confines of the north.

Exhilarated, she jettisoned more—the plaster fig leaves on the classical nudes, the modest draperies over certain paintings. She hung pictures of her own on the walls, pictures that made Giandemaria draw in his breath, and which elicited an alarmed enquiry from the duca himself. Christina responded stoutly. She was not going to be bound, she said, by rules "made for priests." But there were quieter moments, too. Frowning among the nudes was a bust of her Stoic hero, Seneca, and, in between rhapsodies, Christina had time to review some of the work of "his devoted apostle," Lipsius—a treaty on dogs.

The celebrations in her honor continued, but by now she was tired of all the fuss and was feeling, in fact, rather lonely. Early in January, after only a week at the Palazzo Farnese, she wrote a sad little letter home to Sweden. It was addressed simply, "À la Belle," and it reveals that as yet Christina had found no one to replace her longstanding friend:

> How happy I would be if I could only see you, Belle, but though I will always love you, I can never see you, and so I can never be happy. I am yours as much as ever I was, no matter where I may be in the world. Can it be that you still remember me? And am I as dear to you as I used to be? Do you still love me more than anyone else in the world?

If not, do not undeceive me. Let me believe it is still so. Leave me the comfort of your love, and do not let time or my absence diminish it. Adieu, Belle, adieu. I kiss you a million times.

CHRISTINA ALEXANDRA[6]

Christina's goal of many years had been reached. She had relinquished her throne, she had converted, she was living in Rome in a beautiful Renaissance palace. Now, in a reflective hour, one at least of the costs of her self-sought exile had been brought home to her. Even had she wished it, there was no way back.

SOMETHING OF SENECA'S persevering spirit may have been following Christina through the marble halls of her new home. In any event, she now drew a stoical deep breath, and began to engage herself in Roman life. It was not a difficult time to do so, since it was the beginning of carnival, and it became her carnival, "the queen's carnival," coinciding as it did with the many celebrations held to honor her arrival. Her expectations were high, nurtured as they had been on stories from the golden Barberini years of lavish productions in theater and musical drama. Although the Barberini pope was gone, his vibrant melody lingered in his wealthy and art-loving nephews. They were handsomely installed in a new palazzo designed for them by Pietro da Cortona, and they managed to meet the queen's expectations comfortably. They had begun by commissioning a new opera for her; it was performed in their own private theater, and Christina watched it from a specially built box from which she herself could not be seen—Roman theaters were segregated by sex, but as it was a premiere, and in the queen's own honor, the Barberinis did not feel they could oblige her to wait for a ladies-only performance. The work had a grand double title—*La vita humana, ovvero Il trionfo della pietà* (Human life, or the triumph of piety)—and was in fact the last opera of Marco Marazzoli.[7] It starred Bonaventura, Christina's favorite castrato, and it did not have much to do with piety, including as it did the usual cast of gods and nymphs, though St. Peter's Basilica made a late appearance. The libretto had been written by Giulio Rospigliosi, a talented dramatist and a man of many other gifts besides, as yet a humble *monsignore*, but destined for a greater role as Pope Clement IX.

The Barberini faced competition in their bid to impress the queen from their rival papal family, the Pamphili, whose formidable matriarch, Donna Olimpia Maidalchini, had kept the previous pope in thrall. The Pamphili were hampered by being still in mourning for him, and perhaps, too, for Donna Olimpia, but they did the best they could within the bounds of expensive grief. Their family palazzo was situated on the Corso, and in the weeks before Lent, Christina was a frequent guest, watching from their windows the processions and races of Rome at carnival time. Horses raced, and asses, too, and buffaloes, and old men and little boys, and prostitutes, and Jews—the last against their will, and pelted with every available refuse "from rotten fruit to dead cats"[8]—while anxious Jesuits struggled to contain the flirtatious spirits of the young. Christina loved the races, but felt sympathy for the disdained Jews. She talked of doing something to ease their plight, but the moment passed, and her attention was drawn elsewhere.

To the chagrin of the Pamphili family, the Barberini were now mounting a spectacle to end all spectacles, not in their palazzo but in an arena specially built for the occasion, and seating three thousand people. Designed and stage-managed by the famous Grimaldi, the magnificent mixture of pageant and joust took place after dark on the last day of February.[9] The queen had her own special box, and from it she watched as a fabulous Apollo and other assorted deities processed in glittering train around the arena before coming to rest to serenade her in golden-throated chorus. The singers were followed by an army of mounted warriors—two armies, in fact, of knights and amazons, all brilliantly dressed in red and orange, with huge plumes of ostrich feathers blooming from every headdress. They fell to mock battle, then the victorious amazons turned to face an enormous dragon, with flames and fireworks roaring from its mouth—with the Amazon Queen herself in attendance to urge them on, the dragon was naturally defeated. It was all a far cry from the fainthearted lion and the sad old bear that had amused the Stockholm crowds after Christina's coronation. The bill for the evening's entertainment-*issimo* came to some ten thousand scudi, enough to keep a team of grumbling Roman workmen employed for a lifetime.

If the decorum of mourning prevented the Pamphili from staging an equivalent spectacle, lack of money prevented Christina herself from providing any entertainments at all. She was so short of ready cash that she had not even been able to pay her servants, and they had taken to stealing the silver and pieces of furniture from the Palazzo Farnese—some of the doors had even been broken up for firewood. Christina did nothing about it, and the Marchese Giandemaria, the duca's majordomo, was left writing worried letters to his master in Tuscany. The queen's dream of a new, lavish court in Rome was proving slow to take wing. For Christina, court life meant cultural patronage, but her "royal and bountiful spirit" lay earthbound by the dreary chains of debt—by now amounting to millions[10]—and the money García de Yllán had lent her was already spent. Even the modest evening of opera that she gave by way of thanks to the Barberini family had in the end to be paid for by the Barberini themselves.

Christina's ambitions for the immediate present were consequently shrinking, and she was obliged to contain them in a humble "academy" such as she had had in Stockholm. It was not much more than a regular meeting of local nobles and culturally minded cardinals, though from time to time it did boast some scholarly names. Supposedly a literary society, it featured as much music as anything else. Musicians, even very good musicians, were cheap, and every meeting ended with a concert given by Christina's own court orchestra, led at different times by Marazzoli himself, by Alessandro Cecconi, or by the young Bernardo Pasquini. The academy had been founded to promote classical ideals, but its subjects of discussion were, if anything, more suited to a salon of Parisian *précieuses*, and it is hard to imagine the cardinals in Ciceronian mode on such matters as the teasing cruelty of ladies in love, whether night or day was more suitable to poetry, the *coup de foudre* as a basis for marriage, or, daringly or boringly, the virtues of the pope. The academy fizzled out into a regular social gathering, but it served Christina well, since its likeminded members formed the heart of a new circle of friends of her own. They joined her every Wednesday evening through Lent to listen to an oratorio, the nearest Christina could get, in her own residence, to the opera and drama she so much wanted to stage. Some were old works and a few were new; composition was a modest art to patronize, and a few more little holes of debt seemed to make no difference in the badly torn fabric of her finances.

There was entertainment cheaper still in Rome's many convents, and Christina visited some of them regularly, not with a view to "embracing perpetual chastity," as had once been reported, nor to visit the nun she had fallen in love with, as was reported now, but just to listen to the music. Public performances by women were frowned upon or prohibited outright in Counter-Reformation Rome, and convents had become almost the only places where large numbers of women could gather to make music. Standards were high, despite severe constraints. The nuns were allowed no instrument but the organ or clavier, and any nun wanting to play needed to have spent a studious girlhood, for once inside the convent, she could take no music lessons, even from another nun. Talking about music was forbidden, too, as was singing in harmony—the decadent new "figured" chant, polyphonic music, was prohibited to all nuns, and indeed to all Catholic women. The sisters, in consequence, were restricted to plainchant. Christina enjoyed its austere beauty, and went regularly to hear it, and she may have heard other things besides—it was whispered that, for all the rules and regulations, a note or two of harmony could sometimes be heard escaping through the grilles.

The nuns managed to compose as well, and often with more encouragement than was accorded laywomen living otherwise less restricted lives. The Church itself, a huge employer in most aspects of city life, employed no women musicians at all; those seeking to work professionally were dependent on private patronage, and even this had limitations—there were plenty of women singers in Rome, but very few women were able to work as instrumentalists or even as music teachers. Christina employed some singers, and might have helped to blaze a trail by supporting those who composed as well: Barbara Strozzi, for example, was for many years actively seeking a patron to protect her from the "lightning bolts of slander" that greeted her arias and her innovative cantatas. Christina also knew the compositions of Leonora Baroni, but although, or perhaps because, Leonora was a favorite of the queen's new friend, Giulio Rospigliosi, she did not encourage her.[11] Despite her own unorthodox life, despite her flat shoes and her trousers, Christina was never a champion of unorthodox roles for others of her sex. In music, she felt, as in everything else, there was no point in doing so, since there was no overcoming what was in her eyes "the worst defect of all"—the defect of being a woman.

LOVE AGAIN

❧

$\mathcal{C}$HRISTINA'S MUSICAL EVENINGS and the meetings of her academy were always richly colorful events, mostly red, in fact, thronged as they were with her new friends, the cardinals. Their number was fixed at seventy, of whom about half were resident in Rome, and of these, sooner or later, most found their way to the queen's palazzo. As "princes of the Church," they were entitled to a good deal of ceremony, and Christina addressed them all as "cousin," but she was notably relaxed in their company, and allowed them many extra little liberties to do with hats and seats and wineglasses that she would not allow to secular dignitaries. It was unusual behavior for her, since, always sensitive about her status as a sovereign, she was generally punctilious on matters of diplomatic etiquette. Perhaps she did not take the cardinals very seriously as "princes"; certainly her ideas about their status were somewhat hazy—the engaging Doctor Bourdelot had told her that he had "only to say the word" for the pope to grant him the purple, and she had once promised Montecuccoli that she would make him a cardinal herself if he remained with her until she reached Rome. In any event, she enjoyed having them about her. They were able and cultured men, beneficiaries of the Church's education revival in the years of the Counter-Reformation, and, at least within her own residence, she could move, unchallenged, at their center.

Prominent among the cardinals was the pope's own representative, Decio Azzolino, a native of Fermo, small-built, dark-haired, strong of feature though not handsome, a subtle, witty man, his personality markedly warm, with a machiavellian twist. He was just thirty-two, and of unquestioned ability, but his reputation was somewhat mixed, his name having been tarnished by "certain amorous liaisons less than decent, and some other defects."[1] The "other defects" included a possible cheating of his less capable elder brother over the terms of their father's will, and a reputation for spying, the latter probably responsible for his youthful elevation to the cardinalate. He was a cultured man but not an intellectual; his talents were essentially practical—fortunately for him, since his family was of no great wealth and he had always had to live on his wits. Service to the Church had been the family's financial salvation; of the cardinal's nine sisters, five were nuns, and the eldest of his three or four brothers had also taken Holy Orders. Azzolino had maintained strong ties to his native region in Le Marche, but for now his fate and his fortune lay in Rome.

He had made his swift career within the Vatican secretariat of state, prodded sharply forward in the days of Innocent X by the pope's infamous sister-in-law, Donna Olimpia Maidalchini. Azzolino had obtained his first important position, as head of the *cifra*—the section for secret codes—while still in his twenties. Clever and diligent, he was soon made "Secretary of Letters to Princes" as well, an important step that made him a member of the papal household. Powerful patronage would not have been sufficient to ensure this post for him; a good education and, above all, an elegant literary style were also necessary—the post was alternatively designated "the Secretariat of Compliments." With the *cifra* and the Letters to Princes, the thirty-year-old Azzolino headed the two most important subsections of the secretariat of state, and within a year he had been made a cardinal.

His elevated position had not made Azzolino's fortune. His salary was a modest fourteen scudi per month, with the useful perquisites of paper and books, candles, fuels, and wine, as well as a pair of horses and two pairs of servants. He was thus counted a *cardinale povero*, a "poor cardinal," and this entitled him to extra money from the papal coffers, since every cardinal was expected to maintain himself in a manner befitting a prince of the Church. The required clothing alone would have stalled a less ambitious poor boy from Le Marche: vestments in three different shades of red, hooded cloaks of satin and other stuffs, "festive liturgical dress" (including a miter of white damask decorated with gold), and an equivalent set of country clothes, of fabric rougher but still red, to be packed in "a custom-made cardinal's suitcase"—also red.[2] Azzolino's tastes were elegant, but he husbanded his income astutely. With the extra money allowed him, and the rents from his property near Fermo, he managed to live fairly well; already he was able to keep no fewer than fifteen horses, not counting the ceremonial mule maintained by every cardinal for formal displays of humility.

Christina said that the cardinal reminded her of her old Chancellor Oxenstierna, and in the two men's intelligence and diligence there were, no doubt, many similarities, but Azzolino was "lively and provocative" where Oxenstierna had been "slow and phlegmatic," and besides, the cardinal was Italian, and idealistic, and charming, and young. She soon found plenty of things to talk with him about, besides his liaison work for the pope.

In his early days in the secretariat of state, Azzolino had been set to work summarizing the lengthy reports sent from Cardinal Chigi—now the pope—at the peace negotiations in Westphalia, in which Christina had played her own rather heavy-handed part. His subsequent work at the *cifra* had involved creating codes for confidential Vatican correspondence, as well as intercepting and decoding the secret messages of the pope's enemies. It was just the kind of intriguing thing Christina loved, and before long she was writing in code herself, mainly to the cardinal.[3] They had literature to talk of, too; Azzolino was well read in the Greek and Latin classics and knew several modern languages as well. Like the queen, he was interested in both the new scientific thinking and the rival tradition of Renaissance humanism, and he possessed a number of esoteric philological texts of the kind she most enjoyed.

Presumably they did not discuss the cardinal's love affairs, although they were no secret in Rome, and Christina would certainly have known about them; she does not seem to have minded. She had been horrified to hear of Karl Gustav's escapades during his soldiering years, but jealousy, and an open offer of marriage, may have played their part at that point. Perhaps she was reassured by the nobility of the cardinal's debauches—Karl Gustav had fathered a child with a prostitute, after all, whereas Azzolino had been most famously, or infamously, linked with a daughter of the princely and papal Aldobrandini family. It was an achievement, in its way, since he himself was only a minor noble, though he could trace his ancestors back more than five hundred years—much further than Christina herself, in fact, at least where her Swedish ancestry was concerned. Perhaps, too, she enjoyed the cardinal's notoriety; it gave him a touch of the roguishness to which she was so susceptible. His weakness for women was not shared by his brother cardinals, or if it was, they kept theirs on a tighter leash. This was a time when the priestly vow of celibacy was taken seriously. The Counter-Reformation popes, anxious to avoid Protestant taunts, had renewed the Church's insistence on it, and the priests, on the whole, had seen fit to comply.

Christina and Azzolino swiftly became important to each other. She became politically useful to him, and he, in turn, gave her something to do. Azzolino belonged to a new group of cardinals who wanted to strengthen the papacy and hold it to a politically neutral course in relation to the great Catholic states. For decades, successive popes had depended on either France or Spain for support in their foreign policy. In doing so, they had lost much of their temporal authority, already greatly weakened in the years following the Protestant Reformation. The Westphalian peace treaties had undermined it further—thanks, in part, to Christina herself. Now, with France and Spain at war, neutrality was imperative. The cardinals also wanted to reform the feudal-style administration of the papal states. They wanted to see the kind of modernization that other European countries, including Sweden, had introduced in recent decades, with themselves in the role of senior civil servants. This would put an end to the time-honored practice of papal nepotism—the position of Papal Nephew was formal and powerful—and it was on this platform that they had come to prominence in the recent election of Fabio Chigi, himself an opponent of nepotism, as Pope Alexander VII.

That conclave had markedly increased Azzolino's prestige, but also his reputation for intrigue. It had fallen to him to draw the lottery determining which "cells" the cardinals would live in during the deliberations, and the conclave's official diarist recorded a "curious" circumstance: the Chigi supporters' cells adjoined one another. Though a small group, they managed to hold the balance, repaying French and Spanish vetoes with an intransigence of their own: in round after round, they returned blank sheets, voting for *nemini*—no one— gradually disqualifying all the other candidates before finally ensuring Chigi's election, as the Venetian ambassador remarked, "at the spry age of fifty-six."

Their independence from both French and Spanish influence had earned the new group the dashing name of *lo Squadrone Volante*—the Flying Squadron—swift, energetic, unfixed to any faction. A contemporary described them as "vivacious of spirit, acute in judgement, brave of heart," and all the more inclined to work for the best possible candidate, since as cardinals they had all been too recently elevated to be candidates themselves for the papal throne.[4] At thirty-one, Azzolino was the youngest of them, their undoubted leader nonetheless. The *Squadrone* was a small group, only eleven out of the Sacred College's seventy cardinals. Most came from fairly modest families, and their sudden success at the recent conclave was not enough to maintain them as a force for the longer term. Other groups had looked to a papal nephew, or the pope himself, or some French or Spanish dignitary, to draw them together and give them a collective public identity. The *Squadrone* needed a patron of their own.

Within a matter of weeks, Christina had stepped into the breach. She had known of the *Squadrone*, and even known some of its members, before her arrival in Rome. Three of them had been sent to greet her on her long journey to the city, and they had talked to her of Vatican politics and of their own aims, and no doubt also of their brilliant young leader. The cardinals' rebellious stance struck a chord in her provocative soul. She was delighted to become their royal patron, and took up their cause with all the fervor of a recent convert—all the fervor, indeed, that she had failed to show for her own religious conversion. It was a happy symbiosis. It suited Christina perfectly, allowing her to be politically active without the worry of actual government. She could dabble in intrigues or peddle influence, and any failure could be simply dismissed as the victory of another party—there would be no troublesome financial or political chickens fluttering home to roost. The cardinals were equally delighted. Though their patron had lost most of her power, and more or less all of her money, she was still a queen, and she conferred a certain social validity, a royal cachet, on them all. She could serve them as an informal ambassador with foreign diplomats and visiting royalty. Not least, her royal status could never be revoked. Her influence would continue, while the long train of greedy papal nephews came and went, leaving half the citizens of Rome disgusted in their wake.

Given his talent and his aura of power and his courtly, flattering ways, it is probably not surprising that Christina soon fell in love with Cardinal Azzolino. More surprisingly, perhaps, Azzolino fell in love with her, whether for her intelligence, her emotional intensity, her beautiful blue eyes, or even her political usefulness. This last was probably the most important factor, at least at the beginning of their relationship, and at different times over the years its importance rose and fell as political needs dictated. But it was not the only thing that drew the cardinal to the queen, and kept him prominently by her side for more than thirty years. Neither would a shared interest in the sayings and doings of ancient Greeks or modern Romans have been enough. The perceptive cardinal read beneath the bluster of Christina's public personality to the fearful vulnerability beneath. He saw what underlay the extravagance and boasting and the restless running after glory. He saw her father's shadow on her unheroic soul, and little by little, in her weakness and her compromises, he came to love her.

Azzolino's visits to the queen quickly exceeded the call of duty, and indeed became so frequent and so prolonged that by the end of March he was obliged to write a letter of reassurance to officials at the Vatican. It was preceded by his own reputation, however, and the pope decided to take no chances; he dispatched the cardinal to the country for a few weeks' solitary reflection. Azzolino went, reflected, and returned unchanged. His devotion to the queen, and hers to him, began a long, intensifying climb. Christina was excited by the new sense of political importance that the cardinal's *Squadrone* had given her, but her dormant femininity, too, was awakened. She abandoned her manly clothes and took to wearing décolleté gowns—so deeply décolleté, in fact, that they drew a rebuke from the pope. She kept wearing them, but added a rich pearl necklace to conceal, or to accentuate, the obvious. She may have felt that she could enjoy a flirtation, protected from any real consequence—in theory, at least—by the cardinal's vow of celibacy. If so, it was the first time, and the last, that Christina would play the coquette.

It is impossible to be certain whether or not Christina and Azzolino were physically lovers. Gossip sheets and memoirs declare that they were, and even that the queen bore the cardinal a child. Both were passionate by nature, and Azzolino, at least, had had love affairs before, despite being in Holy Orders. Christina's affairs were more a matter of guesswork and scandal than of any certain knowledge, but she was nonetheless very far from being prudish. "A girl who wants to amuse herself," she wrote, "needs a husband first, especially a girl of my rank." In general terms, her attitude to sex was broad enough for the roughest soldier. She was fond of lewd plays and ribald jokes, and was known to wink an eye at promiscuity and even rape within her own entourage. But for herself, her attitude was different. She regarded the act of sex as an act of submission of woman to man, and this, as she stated many times, she would not endure. "I could never bear to be used by a man the way a peasant uses his fields," she wrote. She recognized her own passionate nature, and for once gave thanks to God for her "worst defect," that of being a woman. It had saved her, she believed, from a life of sexual depravity:

My ambition, my pride, incapable of submitting to anyone, and my disdain, despising everything, have miraculously saved me. And by Your grace, You have added a delicacy so very fine, through which You have protected me from a tendency so perilous to Your glory and my happiness, and no matter how close I have come to the precipice, Your powerful hand has drawn me back. You know, whatever they may say, that I am innocent of all the things they have conjured up to blacken my life. I admit that if I had not been born a girl, my temperament might have led me into terrible disorder. But You, who all my life have made me love glory and honor more than any pleasure, You have saved me from the misfortunes that I would have been plunged into by chance, by the freedom of my rank, and by the ardor of my temperament. I would no doubt have married, if I had not recognized in myself the strength that You have given me to resist the pleasures of love.[5]

There was in Christina a curious squeamishness with regard to sex, "a delicacy so very fine" that a sexual relationship between herself and Azzolino, or any other man, seems unlikely. There had been talk of lesbianism since her girlhood—even those trying to broker her marriage had privately conceded that "the queen will never marry." If she could love women, she could certainly love men, too; there had been Karl Gustav, and Magnus, and now Azzolino, and many lesser flirtations in between times. But physical love was something else. Even in the bloom of womanhood, loving and beloved, Marvell's "virgin queen" stood, in the poet's apt phrase, "shrinking from Venus' captivating toils."[6] Her own native distaste for sex did more than social convention, more even than Azzolino's priesthood, to turn their passion from its natural course. In its place, it seems, they built a "romantic friendship,"[7] intimate and, in the early years at least, heightened by sexuality denied. But their love deepened, and it was to last their lifetimes, in the end not less for having never touched its own physical core.

THE SPANISH DIGNITARIES, Pimentel and others, who still occupied the principal positions at Christina's little court, were not pleased about her new alliance. If the *Squadrone* cardinals were not exactly pro-French, they were not pro-Spanish either. Azzolino in particular was not a man who would ever sit comfortably in Spanish pockets. This fact pleased Christina as much as it discomfited the Spaniards. She began dismissing some of her humbler Spanish servants, replacing them with local people, and before the spring was out she had managed to oust the Spaniards completely.

They had not been helped by their inept ambassador, Diego Tagliavia d'Aragona, the Ducque de Terranova. A dull and clumsy man despite his elegant name, Terranova resented the capture of Spain's prize convert by the smooth Italian cardinals who now surrounded her. He was defensive in their presence, and sorely tried by the Roman practice whereby they took precedence over him—in Spain, an ambassador would have had precedence over any crimson-clad little cardinal. Terranova felt that, in his case, an exception might be made. After all, the queen owed her very presence in Rome to the Spanish, and she had been known to make exceptions before. Waiting for news of a favorable arrangement, he kept away from the court, but instructions from his king did not permit him to absent himself for long. Hoping to avoid a public embarrassment, he requested a private audience with the queen: it was granted him, but he was obliged to seat himself, bareheaded, not in the armchair which he had hoped for, but on a lowly stool. When he heard that the French chargé d'affaires had been permitted to keep his hat on in her majesty's presence, Terranova's humiliation was almost complete. The queen administered the *coup de grâce* by publishing an account of it all in a local news-sheet, whereupon Terranova began a series of anguished letters to his king, demanding remonstrance, retaliation, recall—all to no avail; he was kept in his place.

Christina enjoyed provoking the dim ambassador, and she had no qualms about favoring the Frenchman, Hugues de Lionne. He was a clever, witty, charming man, and she liked him. He liked the *Squadrone* cardinals, too, and between them all they drew her away from Spain's orbit toward its enemy, France. Christina was more than willing to be drawn. France had been her first love, after all, and now that she had reached Rome, the Spaniards had served their turn. She had had enough of them with their elaborate manners and their witless ambassador, and their king's flowery letters with no bank drafts enclosed. She began to flout them openly, offending even old friends like Pimentel, who now left Rome to seek refuge on the battlefields of Flanders, a safer and pleasanter place for a Spaniard, apparently, than Christina's court had become. Pimentel's noble compatriot, Don Antonio della Cueva y Silva, sought to do the same, but he did not escape without tasting a little of the queen's bile.

Della Cueva had been serving as Christina's Master of the Household, and she had recently replaced him with the acrobatic and otherwise very flexible Francesco Maria Santinelli, who had entertained her at Pesaro on her way to Rome. There was no love, nor indeed any respect, lost between the two men, and Christina could not resist a malicious little twist of the knife at della Cueva's expense. One day, while he was absent, she commanded his wife to get into a carriage with Santinelli, and go for a drive with him. The lady was obliged to comply, the insult was received most bitterly, and the della Cuevas removed forthwith from the Palazzo Farnese to the Spanish Embassy. Christina seems to have feared a slanderous reprisal; when della Cueva returned to take his formal leave of her, she warned him that if he said anything against her, she would find him and punish him, wherever he might be, adding injury to insult, or so it was said, by declaring that only her respect for his master the king had prevented her from having him beaten. Della Cueva took his fuming leave, then took his revenge in a complaint to the Farnese family, in which he described her majesty as "the greatest whore in the world." As the French had supplanted them in the queen's favor, so the Spaniards now supplanted the French as chief purveyors of unsavory gossip about her.

EXHILARATED BY HER new independence from the "protection" of the Spaniards, and keen to deliver them a vengeful blow, Christina now became embroiled in a longstanding anti-Spanish plot. The great Habsburg Empire of Spain, though now declining, still had control of many territories beyond its natural boundaries, including the "Kingdom of the two Sicilies," effectively the island of Sicily and most of the Italian mainland south of Rome, ruled from the city of Naples. Taking advantage of Spain's protracted war with France and uprisings elsewhere, the powerful Barberini family, together with Pope Innocent X, had made plans to seize Naples and incorporate the region into the papal states. Through his office at the *cifra*, Azzolino had been drawn into it all; his detection of a spy within the Vatican—in fact, the Papal Nephew himself—had saved the Barberini to fight another day, and probably brought Azzolino his cardinal's hat.

The French, meanwhile, had been making plans of their own to take Naples, at the invitation of an influential group of locals. Some years before, outraged by a heavy new tax imposed by the Spanish viceroy, the working people of Naples had taken up arms against the tax collectors; rioting had followed, then open rebellion, and the viceroy had been tossed unceremoniously into prison. The first phase of independence had not lasted long. The rebels' leader, a booming-voiced young fishmonger's assistant named Tommaso Aniello—*Masaniello*—had at once adopted the haughty and vicious ways of his former rulers, and within a few days, he had been lynched by his own followers, the viceroy emerging from prison just in time to see his body being dragged through the streets by the rebels he had led.

Masaniello's henchmen had then claimed power for themselves, maintaining a rough and ready rule for several months before declaring Naples a republic. Some of their older compatriots, meanwhile, supported by the city's anxious merchants, had decided to approach Cardinal Mazarin with a request for French rule of the kingdom. The king's younger brother, Philippe, the Duc d'Anjou, might be persuaded to accept the throne, they felt. French rule would restore order and add prestige as well to their region, impoverished and backward after 150 years of Spanish voracity. Mazarin had been amenable to the idea, but the prince was as yet too young. The king himself was still only a boy, and until he could marry and produce a son, his immediate heir could not compromise the succession by accepting a throne outside France. Mazarin had suggested that the Prince de Condé take it instead, but le Grand Condé had declined the honor—greater plans were already swirling in his head. The cardinal had decided to capture Naples, anyway. He had appointed the Duc de Guise his generalissimo, and the duc had set off with a will, accompanied by Christina's escapee ambassador to Paris, Duncan de Cérisantes, opportunist extraordinaire, former devout Lutheran, currently Catholic by convenience. They had succeeded, and the duc had proved as bloodthirsty a tyrant as any who had yet ruled the troubled kingdom. His reign had been fortunately brief; the Spanish had intervened, the viceroy was reinstated, and the duc in his turn was thrown into prison—in Madrid, whence he had been eventually rescued by le Grand Condé himself.

Mazarin had not given up. Seven years later, with France's own civil war behind him, he attempted once again to take Naples. A fleet of ships had set out to challenge the Spanish, but, overtaken by storm, they were blown back to port, and the invasion abandoned. He now decided that a third attempt must be made, but before this, a temporary king must be found for Naples. Louis, now age eighteen, was still unmarried; the succession had still to be secured before his brother could be spared. Some other monarch must be found to keep the throne warm for Philippe, some royal person with no great responsibilities of his own, some older person, perhaps, some person without heirs to raise their own claims in years to come. Happily for Mazarin, the very person was waiting, restless for action, in Rome.

There is no clearer indication of Christina's regret for her crown than her eagerness to have the throne of Naples. Had she wished, she might have stayed quietly in Rome among her new friends, enjoying the cultured environment, establishing herself as a patron of the arts. But in Rome, she would never be preeminent. The pope himself must of necessity take first place, and as for patronage, there would always be Barberinis or Pamphilis or some other family of fabulous wealth to steal the limelight from her. In Naples, she would be a real queen again, a queen with a crown and a kingdom, and money, no doubt, from France. And Naples was just fifty miles away, along the sparkling coast—an easy journey for Azzolino to make, and for her to make to him.

Christina now began to turn from her other cardinal friends, toward a new group with more adventurous tastes. They were men she had met on her journey to Rome, and they were now seen at the Palazzo Farnese at all hours of the day and night. They were Italians—Pompeo Colonna, Prince of Gallicano, the alchemist Marchese of Palombara, and the Marchese Gian-Rinaldo Monaldeschi—notorious Neapolitan patriots all, though none actually Neapolitan. Monaldeschi in particular was just the kind of rogue Christina enjoyed most. He came from a family of minor nobles in the little town of Orvieto in the papal states, and, being without great resources, he had been obliged to make his own way in the world. So he had managed to do, by fair means and foul, as occasion had arisen. The rebellion in Naples had provided one opportunity: while the French fleet battled the winds, Monaldeschi himself had been waiting at the head of a militia behind the town, ready to support the invasion. Now living in Rome, still in the pay of the French, he began to ingratiate himself into the queen's favor, encouraging her anti-Spanish sentiments, and at some point, it seems, suggesting the dramatic action that was to replace her angry talk.

In the first months of 1656, Christina began a secret correspondence with Cardinal Mazarin, and gradually, a firmer plan emerged: French forces, under the titular leadership of the queen herself, would secure the throne. Once installed, Christina would rule as she wished for the rest of her life, with France a certain ally, and Philippe of Anjou her agreed heir. So the agreement was concluded, to the anticipated satisfaction of both cardinal and queen. Four thousand soldiers were to capture Naples, and four hundred cavalrymen were to accompany the new sovereign thither. Mazarin envisaged them as an escort for her, a triumphal retinue to announce the new regime to the local people as much as to Spain. But Christina saw it differently. Since her childhood she had dreamed of leading an army into battle. Mazarin's four hundred cavalrymen would allow her to do so at last. They would be her own little army, and she would be more than their titular head. She at once created Monaldeschi her *Grand Écuyer*—her Master of the Horse—an apt title since, for the moment, apart from the pretty palfrey the pope had given her, she had hardly another horse to boast of. Confidently, she placed an expensive order for the new armor and liveries that would be needed when she assumed the throne, a vast wardrobe that included six commander-in-chief outfits for herself. And, in her excitement at the prospect of military action at last, she charged up to the top of the Castel Sant' Angelo, and there fired off a cannon, forgetting to aim, however, so that instead of going into the air, the huge lead ball flew down into the town. It struck the Villa Medici, a Renaissance palazzo with a façade of sculpted Florentine lilies—now one less than before, with Christina's cannonball lodged in its place.

Fair Wind for France

HE QUICKEST WAY from Rome to Naples, or so it seemed to Christina, was via Paris. There she could meet Cardinal Mazarin in person to discuss the plan. Certain things were best not left to chance, or to couriers, even with her new codes in place. It was all as yet too secret, so secret, in fact, that not even Azzolino had been told of it. So she must go herself to France, but no suspicion must be raised; a pretext must be found for her leaving Rome so soon after her arrival.

To conceal her intentions, she let it be understood that she was returning to Sweden to arrange certain financial matters. Malicious Spanish tongues whispered a different reason for her departure — the queen was pregnant, they said, later attributing the missing infant to a convenient miscarriage — but no real scandal was needed to justify her departure. Her lack of ready money was public knowledge, and the need to get hold of some more seemed a perfectly good reason for her to go. An added impetus was quite suddenly provided by a serious outbreak of plague in Rome, which closed off or closed down much of the city. Most who could afford to do so made a swift escape, Azzolino and the pope himself being among the courageous exceptions who remained to organize relief measures. It was a stroke of perversely good fortune for Christina, since the usual way northward was now impossible. The cautious Swiss and Germans had quarantined themselves by closing their borders to all traffic from Rome, and a route across France was the obvious alternative. So she began her preparations for a supposed route to Stockholm, writing to Cardinal Mazarin to tell him of her journey, and receiving from him an encouraging reply.

Christina was so short of money that she could not actually afford to go at all, and she was saved only by a gift of ten thousand scudi from the pope to speed her on her way. Though a handsome sum, it was not enough to cover all the likely expenses, and she fell back on her regular ploys of pawning or selling jewels, and whatever else she retained of any value. These business arrangements she entrusted, as was now her wont, to Santinelli and Monaldeschi, and, as was now their wont, they shortchanged her. Unnoticing or uncaring, she approached Cardinal Barberini for a substantial extra loan; His shrewd and wealthy Eminence agreed to lend it, but insisted on receiving security in the form of Bernini's magnificent carriage.

In addition to his cash gift, the pope had provided almost everything else that would be required, from a small fleet of ships to little parcels of food for refreshment along the way. Christina may have resented this generosity, or perhaps her need of it, for she accepted it lightly, and showed no sign of gratitude. She attended a last mass at the basilica, and kissed the feet of the statue of Saint Peter, after which she had no devotion left, she said, to go and kiss those of the pope. His Holiness had no time to reprimand her, overfilled as his days were with efforts to combat the plague. He could hardly have given her less, in any case. Christina was a famous and recent daughter of the Church. She could not be allowed to traipse about Europe, and especially Protestant Europe, without due splendor, and he had declined to lock her away in a convent, as his Spanish courtiers had recently suggested. He shook his head and turned back to his hospitals, relieved, if anything, to see her go— he was heard to remark that at least one heavy burden had been lifted from his shoulders. "They begin to be weary at Rom of theyr new ghest the quien of Sweden," an English spy reported,[1] and from the Palazzo Farnese, a worn-out Marchese Giandemaria wrote to the duca: "We've been singing the songs of the children of Israel after their escape from Egypt. I can hardly believe she's gone. Every moment I'm afraid of seeing her still in the place."[2]

Christina's Spanish retinue had been dismissed some time before, and now, with a suite of some sixty persons, almost all Italian, and including only three women, she set off for the coast. Santinelli led the entourage as Captain of the Guard and Lord Chamberlain of the Royal Household—though the guard was a band of roughneck adventurers and the household was a motley crew—and Monaldeschi rode along with equal impudence as Master of Horse. The queen's departure from Rome was as muted as her arrival had been blaring. She went without ceremony, without ambassadors, without civic dignitaries. Cardinal Azzolino was the only member of the papal court to accompany her, and this loyalty she repaid by weeping devotedly over his miniature portrait once he had turned back toward the city. Santinelli led the travelers to the little anchorage of Palo, where four gaudily painted papal galleys lay waiting for them.

Palo was only a small harbor, but it was safely distant from the plague-infected areas around the usual Roman port of Civitavecchia. It lay on land owned by the Orsini family, and the scions of this great house now cheered Christina with a lavish welcome. They saw her on board, and on a lovely midsummer's evening the little court sailed for Marseilles, the papal arms prominently displayed on the hull of each of their galleys. In three of them, the *San Pietro*, the *San Domenico*, and the *Santa Caterina di Siena*, Christina's unsaintly retinue was accommodated, and in the fourth, aptly named the *Padrona* (Mistress) Christina installed herself. Her apartments on board were sumptuously furnished in red silk damask, and her every requirement had been foreseen, even to a vast, thronelike armchair covered in rich Neapolitan velvet. Less luxurious were conditions below deck, for these were slave galleys, manned by pitiful crews of criminals and prisoners of war, who pulled the great oars day and night to the accompaniment of drum and lash. Small circles of light reached them through the oarholes, and when the sea was high, so too did rushes of cold seawater, drenching their ragged garments and leaving traces of salt behind to sting their open lash-wounds.

The week-long journey was hot and stormy. According to the custom of the time, they did not take a direct route across the open sea, since the danger of pirates from the "Barbary Coast" of North Africa was felt to be too great. Instead, they "coasted," keeping always in sight of land—a longer route, but safer. Christina claimed to have seen pirates along the way, but she decided they were Turkish, and allowed herself a daydream of life in the harem. Whatever their origins, the other ships did not approach, and the little flotilla met its first confrontation in Genoa, where they were refused permission to land on account of their origin from plague-ridden Rome. Supplies were refreshed, however, and tributes of expensive delicacies sent out to the ships, to show that there should be no hard feelings between the Genoese Republic and those who bore the papal arms.

Toward the end of the month, they reached Marseilles, and here they met fresh objections to their landing: the local people staged a noisy and violent demonstration, claiming that the galleys were plague-infected and demanding that they be turned away. The city authorities acquiesced, and pronounced the quay off-limits. A representative of Cardinal Mazarin intervened, insisting that a royal salute be fired in greeting: the sound of cannon was duly mixed with the hostile shouts of the crowd. The cardinal's stouthearted representative boarded the *Padrona* and pleaded with Christina to make a discreet landing along the way, taking a few companions with her in a small rowboat. Christina insisted that she and her entire suite of sixty would disembark on the quay and nowhere else, and at length they did so, accepting only the compromise that the galleys themselves should be moored at some distance from the shore. Three days of public festivities, by way of eventual official welcome, proved enough to reconcile the locals to their visitors. They acclaimed Christina's royal blood and her many signal virtues, including her male traveling attire, and Christina took it all in good part, remarking that the French had had plenty of good things and bad to say about her, adding provocatively that now they would discover for themselves that there was not so much bad in her as they'd said—nor so much good.

While in Marseilles, Christina took the time to visit the blind mystic François Malaval.[3] Only a few months younger than Christina herself, Malaval was a leader of the new Quietist movement. Quietism taught a kind of Christian passivity, a will-less devotion to God, a quiet contemplation; its followers believed they represented the true tradition of Christian mysticism. Christina had probably first learned about it through Johann Scheffer, "Angelus Silesius," a young German philologist and poet who had come to Sweden at her invitation, and who had since embraced Catholicism. Despite its essential passivity, it had taken the interest of her active soul, and had retained a curious hold on her. Though she could never have kept to its tenets, she needed its restfulness; if so, the noisy splendor of Baroque Catholicism can only have pushed her the faster toward it. She was also drawn to its personal emphasis, its linking of the individual soul with God, without the intervention—or interference—of pastor or priest. This answered her almost instinctive conviction of her own sovereignty, that there was no authority above her but God, that God's will for her was for her alone to interpret. One might pray however one wished to pray, the Quietists held. It was perilously close, for Christina's rebellious power-seeking nature, to believing whatever one wished to believe, and she was soon declaring, without a thought of her conversion, that her only religion was "the religion of the philosophers." She visited Malaval only once, it seems, but they began to correspond. Her interest in Quietism fell dormant, but the seed had fallen on fertile ground.

From Marseilles she set off with her entourage on the first day of August.

They made their way northward through the beautiful region of Provence, past fields shimmering purple with lavender in the midsummer heat. By the standards of the day, their going was not hard, for the roads in France were well maintained, and the main thoroughfares the envy of all Europe. They were "paved with a small square freestone," John Evelyn had noted a year or two before, "so that the country does not much molest the traveller with dirt and ill way, as in England."[4] The queen and her companions encountered their share of potholes, nonetheless, some filled in with little branches of boxwood, some still gaping open. Before them, a detachment of local militia marched complete with trumpets and drums, and past them, now and then, a public stagecoach made its way. Christina saw hardier travelers, too, taking the cheaper way of hired relay horses, their baggage strapped on behind them.

Through the Rhône valley, thick with vineyards, they made their way to the thriving town of Lyon, where they arrived just as night was falling. Here, a century before, Italian architects had built an enchanting Renaissance city of red stone towers and winding staircases, and here, in a thousand silk manufactories, artisans wove their beautiful fabric. Christina's demeanor on arrival did not match the nobility of her surroundings: at the approach of an earnest local dignitary, she was enjoined to step down from her coach to hear his lengthy speech of welcome, but the man was curtly dismissed. "I'm tired," she said. "Let's leave the speeches till tomorrow." And she waved her coachman onward.

She was ready to extend more civility the following day, when she received a gentleman of honor sent from the king—or, more properly, from Cardinal Mazarin. He was none other than the Duc de Guise, former captor and king of Naples, and he was to accompany the queen on her journey to the French court. It was a tactical choice, and a happy one for Christina; she took to the duc at once, and there was nothing she wanted to talk of more than Naples and its crown. The duc himself, a handsome man of forty-two, appears to have viewed his new charge with amusement and irony, and a courtly touch of languid condescension, too. Renowned for his attentions to the fair sex, he paid careful attention to Christina, and after a few days sent a description of her to a friend at Louis's court. In due course it was read aloud to the young king and his mother:

I'm dreadfully bored at the moment, but I would at least like to amuse you by sending you a portrait of the queen I am accompanying. She is not tall, but she is shapely, with a large rump, fine arms, and pretty white hands, but more of a man than a woman, and with one shoulder higher than the other, though she hides this so well with her bizarre clothes and her way of walking that one really could lay odds on whether the defect is there at all. Her face is long but not to a fault, and all her features are long, too, and quite pronounced, her nose aquiline, her mouth rather large but not disagreeably so, her teeth passable, her eyes really beautiful and full of fire, her complexion, despite a few pock marks, quite clear and pretty. Her face is nicely shaped but framed by the most extraordinary *coiffure*. She wears a man's wig, very heavy and piled high in front, hanging thickly at the sides and fair at the ends. The top of her head is a mass of hair; at the back it looks vaguely like a woman's *coiffure*. Sometimes she wears a hat. Her bodice is laced crosswise at the back. It is made almost like a man's vest, with her shirt showing all the way around between it and her skirt. The skirt is very badly fastened and not very straight. She always wears a lot of powder and lots of face cream, and she hardly ever wears gloves. She wears men's shoes, and she sounds and moves like a man as well. She loves to show what a fine horsewoman she is; she really glories in it, and she is at least as proud of it as the great Gustav her father could have been. She is very civil and a great flatterer; she speaks eight languages, and above all French as if she had been born in Paris. She knows more than the whole of our *Académie* at the Sorbonne combined, is admirably well informed about painting as about everything else, and knows more about our court intrigues than I do. In short, she is quite extraordinary. I shall accompany her to court by way of Paris, so you will be able to judge for yourself. I don't think I have forgotten anything, except that sometimes she wears a sword, and a buffalo-hide collar, and her wig is black.[5]

The duc's duties toward Christina required him to spend many hours in her company, and the two were frequently seen engrossed in lively conversation. Rumors of a passionate love affair inevitably followed, but, despite her shapely figure, the duc seems to have remained safely indifferent to the over-powdered little woman with the passable teeth. He himself, on the other hand, was precisely the kind of man whom Christina did admire: tall, handsome, very masculine, a soldier and adventurer, and a clever and cultured man to boot. She enjoyed his attentions to her, and now and then may have allowed herself to misunderstand them.

Juicier rumors soon overtook them, in any case. At a banquet in Lyon, Christina was introduced to the Marquise Elisabeth de Castellane, "*la Belle Provençale*," and within a day the queen's desperate love for her was established fact. The marquise was said to be one of the most beautiful women in the kingdom. She had married at the age of only thirteen years, but her mariner husband had been lost in a shipwreck, and now, at twenty, she was already a widow. Christina found her ravishing; on her account, it is said, she delayed her onward journey—by a day or two. How frequently they met, or what they spoke of, or what the marquise thought, is not recorded, but Christina's captivation at least is revealed in a gallant little billet-doux:

> Ah! if I were a man, I would fall at your feet, submissive and languishing with love; I would spend days, I would spend nights in contemplation of your divine attractions. Your beautiful eyes are the innocent authors of all my woes. I will spend the rest of my life in a state of bittersweet enchantment, while I await some happy reversal that will change my sex. In this sweet hope, I count the days of my life.[6]

Though it goes on to speak of "unsatisfied burning desires" and "never fading voluptuousness," it is not really a love letter. Christina was struck by the girl's beauty, and no doubt very attracted to her. But she did not know the marquise long enough to fall in love with her. A flirtatious, even provocative note, couched in the passionate language of the *précieuses*, anticipating no serious response, was a pleasant game to play; had the marquise responded in earnest, Christina would have been quite disconcerted. She had probably never been a lover of women in the fullest sense, lacking the courage, or the self-knowledge, to engage in sexual relationships with women who were not themselves overtly lesbian, or even, perhaps, with those who were. As with men, it seems, so with women: Christina preferred the unconcluded game. By the end of the same letter she had resigned herself to a "most pure, most firm, most confiding friendship" with the beautiful marquise, and even this does not seem to have outlasted her visit.

Not all Christina's encounters in Lyon were grist for the gossipmongers' mill. She made a point of visiting Claude-François Menestrier, a learned young Jesuit famous for his prodigious memory. Christina, always proud of her own excellent memory, decided to test his powers herself. A meeting was arranged, but Menestrier did not arrive; undaunted, Christina set off to see him at his own house. Without warning or ceremony, she knocked at his door, announced who she was, and sat down. She had prepared for the encounter by making a list of three hundred words, taken at random, and this she began to read. Menestrier duly repeated the list without hesitation, repetition, or deviation. Christina was enraptured, but Menestrier had not finished yet—without waiting for an invitation, he then intoned the list backwards, faultlessly.[7]

From Lyon, Christina took the road north toward Paris, and at the beginning of September she arrived at the magnificent château of Fontainebleau, some twenty-five miles from the city. The château had been built, more than a hundred years before, in the middle of a large forest, "a withdrawn and solitary place." Italian architects, invited by François I, Renaissance prince *par excellence*, had established it as one of the finest palaces in Europe. Later monarchs had added to its splendor; elegant gardens, complete with paths and streams and "a marvellous fountain, with four hundred pipes," had been laid out by Louis's grandfather, and when Christina arrived, the beautiful horseshoe staircase was still being admired as a recent embellishment.[8]

It was probably too early in the season for her to have seen the forest at its most enchanting. Only in the fullness of the autumn would the trees produce their finest aspect in marvelous red and gold. Then, the king would arrive to indulge his passion for hunting, for the richness of game in the forest was legendary. Christina loved hunting, too, but for now she was not tempted. She was tired from her latest journey and was showing signs of a deeper weariness. It had been six weeks since her departure from Rome, and the endless hours of jolting carriages and tiresome formalities of welcome had begun to take their toll. She was irritable and unwell, and relieved to come at last within sight of the beautiful château, rising out of the dense and quiet forest, "like an oasis in the desert."[9]

In the king's absence, the Duc de Guise took charge, and for once Christina had only brief formalities to endure. She was introduced to some of the ladies of the court, each of whom greeted her with a kiss, after the French fashion. Whether amused or annoyed, she could not resist an undiplomatic comment, perhaps with a touch of defensiveness as well. "Why are these ladies all so eager to kiss me?" she exclaimed. "Is it because I look like a man?"

Christina, whether queen or king, was royally installed. One evening's rest was enough to restore her, and so began a lively week of music and dancing and ballets and plays, all to her delight. It was vastly increased by a meeting she had long wished for, with Anne-Marie-Louise-Henriette d'Orléans, the Duchesse de Montpensier, known since her civil war exploits as *"la Grande Mademoiselle."* The two had already exchanged a number of enthusiastic letters. Like Christina, the duchesse was an admirer of the Prince de Condé, though unlike Christina, she was keen to marry him—once his sickly existing wife, "poor Claire-Clémence," had departed the world.

During the second Fronde rebellion in the early 1650s, the duchesse had distinguished herself in the lists against her cousin, the king, and she was now officially in exile at her vast, turreted castle of Saint-Fargeau in Burgundy. For the moment, however, she had lodged herself nearer the court, the better to pursue the possibility of rehabilitation. Hearing that Christina was staying nearby, she had determined to meet her, but first she had sent a strategic message to the king, to solicit his permission for her visit to "a foreign princess." It had duly arrived, and Mademoiselle now sent to establish what kind of reception she might expect from Christina herself. Though currently in disgrace, she was, after all, a princess of the blood. Would she be permitted to sit in the queen's presence, not just on a bench, but in an armchair, with a proper back? Christina was happy to oblige her. Her response arrived at seven in the evening, and Mademoiselle, in public though not private mourning for a sister whom she had never known, changed hastily into a sober gown and set off with a retinue of curious ladies.

She had much in common with Christina. In age, the two were just six months apart. Both loved strenuous physical activity, especially riding and hunting—Mademoiselle had had horses sent to her from Germany, and even a pack of hounds from England. She was a keen billiards player as well, and a veritable fanatic for shuttlecock, at which she spent two hours every morning and two hours every afternoon. Like Christina, too, Mademoiselle was a woman of cultivated tastes, although here Christina had the advantage, for Mademoiselle's formal education had been sporadic and mediocre, and it was only in exile, at the age of twenty-five, that she had begun to read seriously, and also to write. But in her earliest adulthood she had frequented an important Paris literary salon, a point of some envy for Christina, who had spent the same years languishing, as she saw it, in intellectual isolation in the frozen north. And, through the years of her exile, Mademoiselle had developed her cultural interests, and at Saint-Fargeau had installed a theater and even her own printing press. From this emerged portraits and verses, at times pastoral or satirical, but mostly in the style of the *précieuses*, extolling, in paradoxically passionate language, the virtues of celibacy and the life of the mind to which all noble-souled young ladies aspired. They were ideas to which Christina herself responded eagerly and instinctively. They accorded perfectly with her intense admiration for great men and great ideals, and with her personal antipathy to marriage. They answered, and in a way legitimized, her own aspirations and her own unusual nature, and *la Grande Mademoiselle* seemed, to her, a living embodiment of them. Above all, Christina envied the duchesse's amazonian reputation, gained when she had fired a cannon at the king's troops from the top of the Bastille. The act had drawn a ferocious curse from Cardinal Mazarin. "She has killed her husband!" he had declared, and had thenceforth vengefully blocked any plan for her to marry.

For the moment, the suitors' loss was Christina's gain. She was passing the evening at a ballet performed within the sumptuous home of a wealthy magistrate who lived near Fontainebleau. In a beautiful room decorated *à l'italienne*, she sat surrounded by a crowd of the admiring and the curious, most of them perched ignobly on low, backless benches. Mademoiselle's first impression was one of relief, for, as she later wrote, "I had heard so much about her bizarre clothes that I was frightened to death I would burst out laughing when I saw her."[10] On this occasion, at least, Christina was acceptably dressed in a skirt of gray silk and a flame-colored bodice of fine wool, both finished with gold and silver lace. Attached to her skirt was an embroidered kerchief with a flame-colored ribbon and a little braid of gold, silver, and black. Her wig this time was blond, with a womanly bun at the back, and in her hand she carried a black-feathered hat.

Mademoiselle records that the queen's skin was white—no doubt powdered, since days later it would be notably brown. "Her eyes are blue," Mademoiselle continues, "sometimes soft, and sometimes very bold. Her mouth is quite pretty, though large; her teeth are good, and her nose is large and aquiline. She is very short. Her bodice conceals her poor figure. All in all, she reminded me of a pretty little boy."

As for the duchesse herself, Christina was "overjoyed" to see her. She threw her arms around her, saying that she had wanted "passionately" to meet her. She had delayed the start of the ballet expressly so that the duchesse could see it as well. Mademoiselle demurred; she could not stay; she was still in mourning for her sister, who had been a mere fortnight in her grave. Christina insisted; Mademoiselle gave in; the ballet, she records, was "very pretty," and noted too that the queen was abreast of all the latest gossip, and "very eager to let us know it."

The ballet was followed by a play, and now there was general astonishment, not on account of the play itself, but at Christina's behavior. In contrast to the other ladies, and indeed the gentlemen, who in the manner of the day sat bolt upright throughout, she lounged in her chair, swinging her legs over the arms of it, now to the right, now to the left. When a scene took her fancy she praised it aloud, swearing by God, repeating her favorite lines; and when the action wore thin she fell into reveries, emitting deep sighs before springing to attention suddenly to adopt a posture more suitable, as Mademoiselle remarked, to one of the clowns at the commedia dell'arte.

After light refreshments of fruits and preserves, there were fireworks over the lake, after which Christina invited Mademoiselle to speak with her *en tête-à-tête*. She led her into a little private gallery, and shut the door behind them, asking the duchesse to explain all the difficulties that lay between herself and the king. Mademoiselle did so, presumably glossing over her own treasonable activities, and Christina took her part, declaring at once that she was "absolutely right" and the king "absolutely wrong," and insisting that she would tell him as much, and would take it upon herself to bring about a reconciliation between them. Christina would not be deterred; the duchesse was not going to spend her days languishing in the countryside; she was the most beautiful princess in Europe, and the nicest, and the richest, and the grandest; she was born to be a queen and Christina would have it so; she must marry Louis, she must be queen of France; it was a necessary political step, for the good of the whole nation—Christina herself would arrange it with Cardinal Mazarin.

It may have been impulsiveness that underlay this stunning faux pas. It could hardly have been ignorance, for Christina must have been perfectly well aware that the very purpose of the princes' rebellion, only a few years before, had been to rid France of that same Cardinal once and for all. It was to oust Mazarin that Mademoiselle had committed her energies, her reputation, and her money for two years of civil war; for that cause she had raised her own regiment, and for that cause she had now endured almost four years of exile. A glorious promotion at the cardinal's hands was beyond the bounds of any possibility; Mademoiselle was still struggling for permission to show her face at court. As yet, Christina had not even met the cardinal, or the king. Her extraordinary suggestion revealed a grossly inflated sense of her own influence, or perhaps simple naïveté.

An alarmed Mademoiselle interposed a "very humble" objection. She was grateful, she was honored, the queen was so obliging, but there was no need for her to take such trouble—indeed, Mademoiselle begged her not to do so. Christina let the matter drop, and quickly raised a complaint of her own: one of the duchesse's servants had been heard, in his cups at a local hostelry, declaring that the queen of Sweden could go hang—and worse. Mademoiselle was "greatly surprised at his impertinence"; she made "every imaginable excuse" and promised to dismiss the man.

It was well after midnight when the evening meal was served, and Mademoiselle did not stay. She drove back to Petitbourg, where she partook of her own supper, and by the time she was ready for bed it was already broad daylight. Later in the day, she sent her compliments to the queen, and Christina replied that she was in fact on her way to see her. But a second message shortly afterwards relayed that there would be no rendezvous after all; the king's men, who were to drive Christina, had prevented her from visiting the still disgraced Mademoiselle, leaving the queen apparently "very cross." Doubtful of the Prince de Condé, and annoyed by the king's interference, she spent the day in a resentful sulk, fuming with frustration.

THE RISING SUN

N THE AFTERNOON OF the eighth of September, Christina arrived on the outskirts of Paris, a city she had longed to see for half her life, "a place of wonders, the center of taste, wit, and gallantry."[1] She had still a little way to go before she reached the center, and a soberer aspect met her where she stopped now, at the vast fortress of Vincennes, where her hero, *le Grand Condé*, had once been imprisoned by Cardinal Mazarin. The gallantry, however, was already more than apparent in the 22,000 men who stood waiting to greet her, 130 companies of knights and gentlemen, bedecked with swords and feathers, and mounted on gleaming horses. The proud memory of her escort of Roman cardinals on their little gray mules was eclipsed at once in a mighty flash of French armor glinting in the sun. *Here* was a welcome.

The governor of Paris came forward to deliver a grand formal greeting to Her Majesty the Queen of Sweden, and the huge company began its progress along the crowded roads. Some two hundred thousand people had come to watch in the fine, late-summer weather, a crowd so dense that the six-mile journey was slowed to five bedazzled hours.

When they entered the city gates at last, cannon boomed from the Bastille, and the king's own Hundred Swiss Guards stepped forward to form a closer escort for the queen. Slowly, they progressed toward the Place Royale, where the ladies of the court stood waiting—at their head, in the absence of the French Queen Mother, the former queen of England, Henrietta Maria, widow of Charles I, and aunt to the young King Louis. Surrounding her in fabulous finery stood rank upon rank of noble ladies, and behind them, the lesser souls of Church and State and Academe. The satirist Loret described the scene:

There were thirty or forty duchesses
Flirty, haughty too-muchesses
Jostling for rank in their silks and laces
Smiles affixed to their painted faces
And baronnes *a hundred and twenty or so*
(For here we've got plenty of them, you know)
And abbots and bishops, a hundred at least
And each with his own humble priest.
And the scholars and savants were there in droves
Preferring the court to their leafy groves
And authors and academic wits
And critics and similar idiots
And linguists and chemists and other bright sparks
And millions and millions of clerks.[2]

Night fell, and the vast retinue made its way by the light of a thousand torches along the river to the great palace of the Louvre. Like Fontainebleau, it was the work of François I, but unlike its lovely cousin, it stood as yet unfinished, shunned by its king, with Cardinal Mazarin its willing châtelain. There were sumptuous apartments enough in its vast interior, nonetheless, and, to raise hers to a pinnacle, Christina had been accorded a historic bed—draped in white satin embroidered with gold thread, it had been bequeathed to Louis XIII by Cardinal Richelieu. By all accounts, she slept very comfortably in it, while reports of her triumphal arrival drifted through the still September air. They reached the Duchesse de Montpensier at her château of Saint-Fargeau, but Mademoiselle was only partly impressed. It had been a magnificent affair, she admitted, but it was a touch inelegant of the queen not to have had even a single woman in her entourage, and as for that scarlet outfit, that wasn't a good choice, either—she had already been seen in it before.

The day following Christina's arrival had been reserved for the men of Paris's learned societies to pay their homage. They came in their dozens, black-robed from the university, red-robed from the parliament, each group armed with a lengthy speech of welcome. To the hapless dignitary in Lyon, Christina had already made clear that *discours* of this kind were tedious to her, but now, on the whole, she endured them patiently, keeping her countenance even when one theologian, clearly accustomed to less worldly matters, referred to a possible marriage between herself and the young king of France.

He then ceded his place to Olivier Patru, "Seat Nineteen" of the Académie Française, whose own *harangue* to her majesty proved a model of courtly grandiloquence:

> . . . When we consider that a great Queen has deigned to cast her eyes upon us, and to send to us, from the extremity of the North, illustrious marks of her esteem, we cannot do less today than to render Your Majesty that religious devotion which the whole world owes to Her virtue. . . . [3]

The "illustrious marks" of Christina's esteem were a letter and a portrait of herself that she had sent to the Académie from Stockholm. From his *rhapsodie de dévot*, Monsieur Patru proceeded to a *rhapsodie de savant*:

> The knowledge of languages, which consumes our days and nights . . . has for Your Majesty been but a childish game. No flower of *belles lettres* has been left unplucked by Your royal hands. There is nothing in all the sciences that Your mind, so vast a mind, has not encompassed. You have done what very few men have been able to do, and what neither maid nor woman has dared to attempt. . . . And now, we see You at last, we feast our eyes upon You, but alas! our joy is tinged with bitterness, when we consider that in a moment we are to lose—perhaps forever!—Your august presence. . . . However, in our misfortune, Your picture will console us, if anything can.

Monsieur Patru further assured the visitor that he and his *confrères* would find in her picture "the dearest object of our eyes." They would "pay their respects" to it, "render it hommage"—indeed, "offer it sacrifices." Christina thanked the learned gentlemen, got up from her chair, and made her escape to a different kind of theater.

She went now to the Palais-Royal, whose theater had been built at the command of Richelieu himself. The cardinal's enthusiasm had made playgoing respectable, and by the time of Christina's visit, Paris had a large and discriminating theater public from the aristocracy and the bourgeoisie, and the clergy, too. Private theaters existed in the houses of the great, but most theaters were open to all comers, though by the standards of the day it was not cheap entertainment—even a standing place would have cost the better part of a day's wage for a laboring man. Soldiers and liveried servants provided a bois-terous element, and often forced their way in without paying anything at all. Most of the plays were still in the traditional style of the Italian commedia dell'arte, with stock characters acting out familiar themes, more or less ex-temporized, often bawdy or satirical. Christina loved them, and may even have seen Fiorillo in the most famous of all commedia roles, the engaging, in-triguing servant Scaramouche, the archetype of all the more or less lovable rogues she chose to have about her in her own life. She was a little too early, or a little too late, to see the French theater's own rising star, Jean-Baptiste Po-quelin, better known as Molière. A recent stint in prison for debt had per-suaded him to seek a slower fortune touring in the provinces; he did not return to Paris for a year or more.

Regardless of the show, in any case, the audience had an excellent specta-cle in Christina's own eccentric behavior. Rather than seating herself in a no-ble *loge*, as expected, she had chosen instead a small chair in the *gradins*, the shallow stone steps at the far end of the theater. From here, instead of looking out at the fans and jewels of the people on the other side, she had a direct view of the stage, a fair compensation for her uncomfortable position in one of the "cheaper" seats. Whether to make herself more comfortable, or to cause a de-liberate scandal, she propped her legs up on a second chair, adopting a pos-ture "so indecent," according to one shocked Parisian, that "one glimpsed what even the least modest woman should keep hidden."[4]

Christina added to the scandal by similarly unconventional behavior dur-ing a mass at Notre Dame a day or two later. She had not intended to go at all, but the king's own abbé-confessor had suggested that it might be politic to do so; it would be to the edification of all Parisians, mighty and humble, he said, to see so recent a convert as her majesty in an act of public devotion to the faith. Christina agreed, but declined to make her confession to the abbé, re-questing a bishop instead. As to what the queen confessed, His Grace's lips were naturally sealed, but he did let slip that her majesty had not seemed overly penitent, having stared him full in the face throughout the proceed-ings. A similar lack of piety was only too evident during the mass. Christina chattered audibly to the disconcerted bishops on either side of her, and re-mained standing throughout the service, only once or twice kneeling to pray. A little more devotion might have been expected, it was felt. Quite frankly, or so thought *la Grande Mademoiselle* on hearing of it, she should still have been "in the grip of a convert's zeal."[5]

Christina was much more zealous making her social rounds, and she visited every illustrious Parisian who was not likely to bore her with a lengthy hymn of praise. She saw Georges de Scudéry and his sister Madeleine, leading light of the *précieuses*, and the poet Saint-Amant, "big Saint-Amansky," whom she had met in Stockholm, and the melancholic Duc de La Rochefoucauld. The duc's famous *Maximes* were still gestating, but he was on the point of publishing a little "portrait of himself,"[6] and Christina shortly afterwards tried to emulate it—not in French, her preferred language, but, perhaps rather too confidently, in Italian. Like so many of her projects, it did not get very far, perhaps because she lost interest in it, or perhaps simply, as she said herself, "because my Italian wasn't good enough"—for the moment she did not think to try again in a different language. The duc described himself as "very reserved with people I don't know," and by his own admission told few jokes, nobly preferring those subjects that "fortified the soul," but if Christina found him a trifle stodgy, she was certainly interested in what he wrote. He was a fastidious man, "and most particular with the ladies," and he claimed that he had never in his life spoken a doubtful word in a lady's presence. What he made of Christina's colorful language the duc did not record.

Paris delighted and impressed Christina. She liked being fêted, and she enjoyed her grand surroundings, but she was happiest of all when left to make her own smaller, more personal, discoveries. Declining any retinue, she set off for a few hours every day to do some exploring by herself. Paris was burgeoning. Despite being shunned by its sovereign, who preferred to spend his time and his money elsewhere, it was gradually transforming into the brilliant capital of a brilliant state, a city of light for the rising Sun King. The first of the great boulevards and the first fine public squares had already been laid down. The river was alive with commerce, and its banks were alive, too, with people bustling and jostling, tough Parisians shouting out their business, awkward newcomers from the provinces—the disdained "little people"—foraging between boats and boxes for an honest day's work. It was by far the largest city that Christina had seen, three or four times the size of Rome, thirty or forty times that of Stockholm.

Though the building had begun, the medieval city was still very much in evidence. Narrow, crooked streets made the going hard for carts and carriages, and the rich were carried about in practical litter chairs, with a hardy "baptized mule" hoisting the poles at either end. Large signs of painted wood dangled perilously from shops and workmen's quarters, picturing the goods and services available inside. Posters stuck to the walls advertised everything from theatrical productions to remedies for venereal disease. Town criers wandered from corner to corner, and abandoned children cried, too, beneath the eaves of closed doors. The new postboxes drew more attention, and the mail fed to them went three times daily off to the corners of Western Europe and, less successfully, to addresses in Paris itself, with its hundreds of unnamed streets. In the middle of them stood Les Halles, "the stomach of Paris,"[7] a great sprawl of markets, open and covered, selling every kind of food and artifact. Christina saw it all herself from the windows of her own litter chair, and when something took her fancy, she stepped out confidently in her sturdy boots, ready to retreat again if the going became too rugged. "It's true there's a bit of shit in the streets," one famous Parisian had noted, "but we have our chairs."[8]

If Christina found Paris memorable, Paris found her equally so. She was not at all what the city was accustomed to in a *personnage royale*, hopping down from her chair to chat with the local people, popping into a shop to watch cakes being baked or a book being printed, rummaging in the market stalls, teasing the streetchildren. When the mood for familiarity left her, she walked in haughty, almost sculptural mode, head to the right, chin up, gaze fixed, until the next thing took her fancy, and she broke into a laugh, or clapped her hands, or swore.

IN THE MIDDLE OF September, Christina set out for Compiègne, where she was to meet the king at last. She paused in her journey at the château of Chantilly, seat of the Prince de Condé, who, however, was not there to welcome her, being away attending to the business of war on behalf of the king of Spain. Chantilly was about halfway between Paris and Compiègne, and consequently was a good place to spend a night en route. She arrived at about seven in the evening to find Cardinal Mazarin already there. She had anticipated their meeting by a number of letters, sent via Chanut, in which she had urged the cardinal to make more haste with the Naples plan. She was too "zealous in the service of the French," she wrote, to want to wait any longer. If she had hoped for an opportunity of furthering the plan that evening, however, she was to be disappointed, for in her honor Mazarin had arranged what he coolly referred to as "an intimate dinner"—in fact a banquet for forty or fifty people. It allowed him to do her honor, which he did most attentively, providing all sorts of delicate fruit juices for her in place of the wine she disdained, but it made private conversation impossible. Hence he was able to observe her and keep her dangling in expectation, while committing himself to nothing in particular; the agreement they had already made was no more than a piece of paper, after all.

But at least at Chantilly Christina received a flattering token of the king's eagerness to see her. Unwilling to wait until the meeting arranged for the following day, Louis and his brother Philippe had ridden at a gallop the forty miles from Compiègne, and now presented themselves in supposed incognito, dressed as ordinary gentlemen. Having studied their portraits in the Louvre only days before, Christina recognized them at once, but she played along, and when Mazarin introduced them to her as "two of the most accomplished gentlemen in France," she remarked with a smile that one of them at least looked born to wear a crown. She later confided, rather dismissively, to *la Grande Mademoiselle* that she had found Louis "a very handsome, decent fellow," and his brother "very pretty, though rather bashful"—he was in fact a flamboyant transvestite, despite his gentlemanly attire that evening.

It was hot the next morning for the journey on to Compiègne, and Christina, always happiest on horseback, and with the king's example before her, may have ridden some of the way rather than traveling inside the bumpy coach. Approaching the château, she saw her way well lined with four companies of royal guards, a company of Scots guards and Scots gendarmes as well, two hundred cavalrymen, two companies of mounted musketeers, the king's own gendarmes, and detachments of noblemen from four regions of France. At the end of the *garde d'honneur,* half the members of the French court awaited her. Famed throughout Europe for the quality of even their everyday dress, today they were arrayed in splendor. The king himself was wearing a suit that actually glittered, covered with gold and silver and precious stones. His brother stood beside him, scarcely less magnificent, with a line of superbly dressed noble lords stretching away behind him. At a slight distance was the stately figure of the queen of France, Louis's mother, Anne of Austria, surrounded by her ladies-in-waiting.

If Christina had chosen to ride some of the way from Chantilly, she had certainly got into the coach at some point, for she now got out of it, to the absolute astonishment of all those waiting. In contrast to the beautiful, trailing silk gowns of the French ladies, she was wearing a short skirt, revealing her ankles, and a man's shirt fastened with vague decorum at the neck. Her bodice had slipped off one shoulder, and the rest of her *couture,* apart from her determinedly mannish footwear, was her usual uncertain mixture of men's and women's clothing. On her head was perched an ill-fitting dark wig, left uncurled, and now flying in all directions on account of the wind. Her face was covered with dust and sweat, and her hands were filthy. A stunned Madame de Motteville, the Queen Mother's confidante, recorded the scene in her memoirs:

> We saw the arrival of the queen of Sweden, of whom we had heard so many extraordinary things. I was one of those closest to the royal persons, and I must admit I was very surprised at first. She had not taken care of her complexion and she was sunburnt, and she looked like a sort of Egyptian street girl, very strange, and more alarming than attractive. But once I had looked at her for a bit, and got used to her clothes and her odd hairstyle, I saw she had beautiful, lively eyes, and a sweet expression, also rather proud. To my surprise I found that, from one moment to the next, my impressions had completely changed—I realized I liked her. She seemed taller than we had heard, and less hunchbacked; but her hands were not so fine as people had said, though it's true they were nicely shaped, but really so dirty it was impossible to see any beauty in them.[9]

With pomp and fanfare the queen was led to her rooms in the château. The pomp was as borrowed as the fanfare, for she had brought nothing and no one with her, "no ladies-in-waiting, no officers, no retinue, and no money"— the very servants to dress and undress her had to be provided by the king. Madame recorded that she did have "two or three poor-looking fellows" with her, "who were called counts, out of courtesy, but really, she might have been taken for nobody, because apart from them, we saw only two women with her, and they looked more like secondhand shopkeepers than ladies of any quality."

The three "poor-looking fellows" were none other than the Santinelli brothers and the Marchese Monaldeschi. About the two women, Italians whom Christina had brought with her from Rome, Madame was very close to the mark. They were Donna Barbara Rangoni, who was not in fact a *donna*— a countess—at all, and Signora Orsini, who was no relation whatsoever to the illustrious family of the same name. The two had already proved rather good at dealing in secondhand goods, particularly those that had belonged to Christina, and in addition to what they could steal, each of them was paid, or at least promised, a salary of seventy scudi per week. Christina had clearly not chosen the ladies as court adornments. Donna Barbara was approaching fifty; she was short and fat, with hair apparently "more dead than alive," and one conspicuous black tooth. She had the advantage, however, over Signora Orsini, painted up "like one of the old banners of the Landgrave of Hesse," who was even older, and with no teeth at all. Signora Orsini had at least the merit of being devout, attending confession and taking communion every week. Donna Barbara had no such merit; perhaps she was a good shot.

Christina's visit to Compiègne lasted a week, during which she was entertained by a round of dinners, plays, clever conversation, and visits to local churches and other buildings of distinction. As the Duc de Guise had noted, she was well informed about all the current intrigues and scandals, and she managed to tease some of the courtiers about their love affairs, and offend others about theirs. And with respect to the most important of these affairs, that of the king himself, she managed to put her foot in it spectacularly.

Among Cardinal Mazarin's family living at court were his three young Mancini nieces, one of whom, Marie, was Louis's paramour. The cardinal had for some time been trying to discourage this romance, considering a marriage between the two as unlikely to serve his own interests. It was only a few years, after all, since the end of the Fronde, during which he had twice fled into exile with a price on his head. He had still too many enemies who resented his vast power, and who might be prompted into action if he should marry his niece to the king. By all accounts, the king's young love was the least attractive of the Mancini sisters, at least as far as looks were concerned — one ungenerous contemporary described her as "ugly, fat, short, and with the look of an innkeeper's wench," though he did add that she had "the wit of an angel, so that when you listened to her speaking, you could forget how ugly she was."[10] Whatever the reason, the young king loved her passionately, though for the time being in perfect innocence. Christina herself reported to Azzolino in Rome, "I doubt if he's even touched the tip of her finger." The lady was introduced to the queen at a grand dinner held on her first evening at Compiègne. Christina at once declared her so beautiful that she must certainly marry the king, adding to Louis that, if she were in his place, she would certainly marry for love. Amid a flurry of coughs, the cardinal swiftly turned the conversation in a different direction.

Christina had not genuinely admired the girl. She later remarked to *la Grande Mademoiselle* that it was "a shame Louis couldn't be in love with someone more attractive." And to Azzolino she wrote that nothing would come of the romance, since Mazarin would never be such a fool as to allow his niece to marry the king. "Marriage is the best cure for love," she wrote, "and the marriage bed is its tomb"—the cardinal would never risk losing his influence in this way. She had been rather foolish herself in her own remark to the king, however. Whether impulsive or mischievous, it had not endeared her to the cardinal, for, as Mademoiselle noted, "at court, people don't like it when you meddle in things that don't concern you."[11]

She was at least entitled to a private opinion about the great cardinal, and this she conveyed in a letter to Azzolino, describing him as a man possessed by a single passion, and that "the finest of all"—ambition. The war with Spain suited his own interests, she felt, "but he knows how to make the right people believe he wants peace." He was respected by the king, and adored by the Queen Mother, but, despite a great deal of gossip about his relationship with the latter, Christina swore "by all the Saints" that there was nothing illicit between them. "The Queen is the most virtuous woman in the world," she wrote, "and quite incapable of doing anything dishonourable."[12] Of the king himself, she had formed a surprisingly mild impression. She had found him very handsome, and very polite, though shy, and "incapable of any very strong feeling." It is an unlikely picture of the great Sun King, but he was, after all, only eighteen years old, still very much in thrall to Mazarin, and even, to a degree, to his mother. He was aware of the Naples plan, however, and he knew that the queen was to serve as a kind of stand-in sovereign until he had produced an heir. This of course was a secret matter, but the two shared other passions that they might have discussed easily in company at Compiègne: both loved hunting and horsemanship, and all things military, and music and ballet and the theater. Louis may have been shy, but perhaps he simply did not care to reveal much of himself to this curious little woman with her men's shoes and her dirty hands.

Christina's week at Compiègne passed sociably, but before she could leave, *la Grande Mademoiselle*, in the nearby village of Pont, "was seized by an impulse" to see her again. It was the eve of the queen's departure, and Mademoiselle arrived at ten o'clock, only to be told by an Italian servant that her majesty had already gone to bed. "I pretended I couldn't understand Italian," wrote Mademoiselle, and she gave instructions in French, "several times," that her majesty should be informed she was there. She was eventually shown into Christina's room, where she found her sitting up in bed, a candle beside her on the table, and "a kind of towel around her head in place of a bonnet," since she had just had her head shaved. She was wearing a collarless shirt with a big flame-colored bow on it, and her legs were covered by a "villainous" green blanket. In short, "she was not looking very nice at all."[13]

Christina welcomed Mademoiselle warmly, and insisted on seeing her companion, Madame de Thianges, who had not followed Mademoiselle in. She was evidently very taken with this lady, and tried to persuade her, only half in jest, it seems, to leave her husband and return with her to Rome. "Even the best husband isn't worth staying with," she declared, going off into a rail against marriage in general and the "abominable" business of having children, before beginning to talk of the Roman rites "in rather a liberal way." She left Compiègne the following morning, setting off in a carriage with San-tinelli and two other men. "How odd it is," wrote Mademoiselle, "for a queen to be without a single woman of her own." She had not a penny of her own, ei-ther. The king had lent her the carriage, and given her money for the journey, "and off she went, this Swedish amazon, followed only by her pathetic little troupe, without any retinue, without any grandeur, without a bed, without any silver plate, without any mark of royalty."

She left a mixed impression behind her. Her intelligence and learning had made her some admirers, but her sharp wit and her frequent jokes had left many embarrassed and resentful, and they now took their revenge in gossip and pamphlets, describing her as a stocky little hunchback, an ill-bred savage, a lesbian, a clown, a whore. It would not be for the last time, nor would those epithets be the worst. France was to prove Christina's place of nemesis, a nemesis brought on by her own impulsive hand.

FONTAINEBLEAU

CHRISTINA WAS NOW EXPECTED to leave France, not with any show of traveling on to Sweden, for this pretense had been forgotten, but simply to make her way back to Rome, and thence, eventually, to Naples. Her grand diplomatic venture was behind her; the call to arms would be sounded soon. "The treaty we made at Compiègne" would ensure her a sparkling new throne before the new year was old, and she departed the country in the highest spirits. At the border she bestowed an unnecessarily generous gift on the French envoy who had accompanied her. It was a large diamond, and one of her last, but she was confident of reinforcements. Mazarin had already given her money for her present journey, and he had promised to press Karl Gustav, "as a matter of Franco-Swedish friendship," on the subject of her remittances. The throne itself would bring Spanish wealth and French subsidies for the rest of her days. There was no need for parsimony now, and Christina gave in to reverie, picturing herself already installed in the city by the sea, surrounded by her guards in their new violet uniforms. Her court would be a mecca for the good and the great; her patronage would know no bounds; the Barberini star would dim with envy; the new queen of Naples would dazzle the world, bestowing ever grander gifts, ever wider. Overexcited, she wrote to urge Mazarin onward, suggesting that his rewards would be scarcely less than her own:

> Your Eminence should not forget that you are Italian, and a member of the Sacred College, and a cardinal, so that wherever you are and whatever should befall you, you can never be greater than you could be in Italy. Your Eminence understands me, and I salute you.[1]

As she had once promised Montecuccoli to make him a cardinal, so now she was promising Mazarin that she would make him pope.

The cardinal took it all in his stride. He had done a lot of listening, but not a great deal of talking; he had read a good many letters from the queen, but had not replied with many of his own. "The treaty we made at Compiègne" was for him, less conclusively, "the treaty the queen proposed at Compiègne." He would bide his time, uncommitted. If this bird was in the hand, there were others in the bush that might still be caught—Spanish birds, singing of peace, and English birds, screeching of more war. It did not occur to Christina that Mazarin might be playing a double game. Machiavellian tactics, she believed, were her own prerogative.

Confidence got the better of her. Defying the pope's express wishes, on the first day of October she made a jaunty visit to the convent at Lagny, not to pray, nor to listen to the music, but to pay her compliments to Ninon de Lenclos, a notorious courtesan who had been more or less imprisoned for flaunting her charms too openly. Christina was enchanted by Ninon, and reputedly wrote to Louis that only she was lacking at his otherwise perfect court. The pope, still besieged by plague, had time only to roll his eyes to heaven on hearing of it, but a flash of anger escaped them at the next news: the queen had declared herself an unbeliever, a materialist, a follower of Lucretius. She had no religion, she added blithely, but that of the ancient philosophers.

Her ebullient spirits did not last. Crossing into the papal states, she was startled to hear talk of a peace between France and Spain, and she wrote at once to Mazarin, asking him to deny "this news that threatens the world with a great calm. I love the storm," she said, "and fear the calm." A peace would sweep away her chance for Naples, a chance that until now, she had viewed as an absolute certainty. Mazarin did not reply, but the talk of peace subsided. Christina arrived at the little papal town of Pesaro, where the once humble Santinelli brothers had entertained her the previous November—Francesco's rise in particular had been swift, from acrobat to Lord Chamberlain in less than a twelvemonth. They were all now installed at the pope's expense in the Palazzo Della Rovere, and here Christina made the aquaintance of an old friend of Cardinal Azzolino, the cultured and worldly papal vice-legate Monsignor Luigi Gasparo Lascaris. He began to send regular letters to his friend, informing him of all the daily comings and goings, and he did not exclude his own concern at the queen's extravagance and the "band of ruffians" who surrounded her. Lascaris knew about Azzolino's relationship with the queen, and he teased him about it with a rather rude joke that he had made to Christina herself. Her love of bawdy talk was infamous; she is even reputed to have told the devout Queen Mother of France, in precisely so many words, that "fucking is what pretty girls are for"—the Queen Mother had apparently written this off as a Gothic aberration. But despite Christina's promiscuous tongue, Lascaris's playful tone in his letter to Azzolino suggests that he did not believe his friend had been involved in a real love affair with the queen. It is almost as if the cardinal might have been a little embarrassed about her attentions to him:

> Her Majesty is more beautiful than ever [Lascaris wrote], and more devout—and more devoted to you—as well. Last night she was dressed in black velvet and blue ribbons, and a very fine man's collar. It was enough to drive a man to distraction to see her reading from a French play by candlelight. She was playing the lovesick Diana, and she read the part so well that I said to her, Madam, they say I'm a wily old bird, but really this evening you make me feel more like a young *cock*.[2]

French plays at the pope's expense were all very amusing. She was well accommodated, but Christina did not feel comfortable. The plague was abating in Rome at last, but poverty now prevented her return, for Mazarin's money was spent, and she could go no further. There was nothing to do in Pesaro. It was small and dull, and she was bored. She had not come all the way from Stockholm to sit in a crumbling old castle waiting for news from Paris. The only business here was the business of the Church, and no amount of ecclesiastical coming and going was likely to distract her while there were dreams of Naples whirling in her head. She penned a disheartened letter to her friend Holstenius at the Vatican Library, bewailing the lack of intellectual stimulation in the little town. "Platonic works are as rare here as the unicorn," she sighed. Mazarin himself received dozens of letters from her, urging action, asking for money, demanding a response. He sent just two replies, both cool, both telling her expressly not to come back to France.

Christina sat in Pesaro for seven languishing months. She dispatched Monaldeschi to Paris to rouse the cardinal to action, then dispatched Santinelli to rouse Monaldeschi to action, with no result in either case. She had asked the cardinal to send her some money, in fact the very large sum of 300,000 scudi. He sent her 15,000, or rather, Santinelli delivered 15,000 to her. She sent him off to Rome to settle some business matters for her. She had still a few things worth pawning or selling, and she wanted to go back to the Palazzo Farnese; Santinelli was to establish whether or not she might do so. He did—she might not, but he sent her a letter of enthusiastic lies, anyway. Her majesty would be most welcome, he wrote, and he had managed to raise some money for her out of the things she had left there. This at least was true, though he did not say that he had raised much more for himself. In fact, he was living exceedingly well on the proceeds of Christina's remaining diamonds and silver. The plague was in abeyance, and he was in no hurry to return to provincial Pesaro. Monaldeschi returned from Paris toward the end of the winter to find the queen unhappy and restless.

By the middle of June, she had had enough. With no invitation from the French, and without even permission to travel through the country, she left Pesaro and set out for Paris with a small entourage. Though unimpeded by any lengthy formalities, her journey this time was slow. Lack of money and lack of proper arrangements delayed her, and it was the tenth of October when she arrived at last at the château of Fontainebleau, where she was to await her invitation to court. A suite of modest rooms had been set aside for her in the château's *conciergerie*, adjacent to the famous Galerie des Cerfs with its walls lined with stags' heads, the hunting trophies of generations of princes. The magnificent forest surrounding the château was in its richest autumn glory, but, despite the season and the silent promptings of the Galerie, Christina was not in the mood for hunting. She spent most of her time instead writing urgent notes to Mazarin, but the cardinal was suffering from a tactical case of the gout; he could not come to see her, and he could not receive her, either. She paced and fumed, and in her calmer moments thought up wild schemes to raise money from the Swedes and from the emperor, and wilder schemes to spend it on trumpets and drums and helmets for her future Neapolitan guard.

But if too little was happening in Paris, too much appeared to be happening nearer at hand. It was not Christina herself but Monsignor Lascaris who first realized that something was amiss. As early as July he had begun to suspect that his letters to the queen were being tampered with. She had told him that some of their seals had been broken, and he began to take the precaution of binding the packets with fine wire. He urged the queen to be cautious, and he wrote as well to Azzolino, asking him to come to Pesaro for a few days. There was "something serious," he said, that he needed to talk to him about. It seems that Lascaris had got wind of the Naples plan, which Azzolino still knew nothing about.

Lascaris's warning had alerted Christina, but she took no sudden action, biding her time and quietly gathering evidence. Soon after her arrival at Fontainebleau, she wrote to the postmaster at Lyon, asking him to take particular care of the post addressed to her. Ordinarily, a letter would be sent in a sealed box; but at various points along the journey, the box would be opened to admit new letters or remove others for local delivery. Christina now asked that every letter addressed to her or to any member of her household be delivered into her own hands by personal courier. Monaldeschi, meanwhile, had noticed that letters for himself were arriving less and less frequently. He made his own inquiries of the postmaster in Lyon, but nothing had been misdirected, and nothing was waiting. His own post, it seemed, was being intercepted. Had he known it, not only letters addressed to him but also many written by him were now in Christina's hands; indeed, she herself had made copies of them. There is little doubt that some of them, at least, concerned the Naples plan.

From Christina's point of view, the plan had reached a crisis point. It was now a year and a half since her first agreement with Mazarin. She had spent months in France, yet they had scarcely met, and they had never managed to discuss the plan in detail. "The treaty we made at Compiègne" had as yet led nowhere. The attack had been postponed, the troops had been diverted, the cardinal was unwell, and now there were rumors of peace. Christina had spent a lot of money on the strength of Mazarin's promise, and above all, she needed a crown of her own, a real crown, with a court, and lands, and subjects. She felt frustrated and humiliated by the cardinal's dismissive response to her. She wanted to show him, and show the world, that she was a force to be reckoned with. She wanted to be taken seriously as a sovereign, as a queen, and he was not taking her seriously at all. He was keeping her hands tied, while she wanted to strike out, and in the end, she did so, though it was not Mazarin who endured the blow, but her *Grand Écuyer*, the Marchese Monaldeschi.

The prior of the monastery at Fontainebleau was a certain Père Le Bel, and it was he who later recorded the details of the dreadful day of the tenth of November 1657, the blackest day of Christina's life. Her servant, Cesare

Capitone, also gave evidence at an inquiry held later in Monaldeschi's native town of Orvieto. Both testimonies condemn Christina, if not in point of law, in terms of the simpler dictates of humanity.

According to Capitone, the marchese had nursed a bitter hatred for Francesco Santinelli, and had determined to turn the queen against him during his absence. From Rome, his friend Peruzzi had been keeping him informed of Santinelli's roguery, and Monaldeschi had intended to make use of this information in due course. But the queen's evident suspicion had presented him with an apparently quicker way of undermining his rival. Realizing that she was inspecting the post, he had forged several letters prejudicial to her, and signed them with Santinelli's name. They included scurrilous details of the queen's supposed private life. They were intended, says Capitone, to "spark off an explosion that would destroy all Santinelli's plans."

The forged letters, and others, came into the hands of Père Le Bel on the morning of the sixth of November. The priest was standing at the door of the monastery, watching the laborers working in the fields, when he was approached by a servant who said that the Swedish queen wished to speak with him. He followed the servant back to the château, and went with the queen into the Galerie des Cerfs, where, she said, they could speak without interruption. They had met only once before, and very formally, but the queen now told Père Le Bel that she had something confidential to say, and asked him to regard their present meeting as if under the seal of the confessional. The priest acquiesced. Even with ordinary people, he said, with regard to confidential matters he was as if "blind and mute"; with those of royal blood even more so. At this the queen handed him a small packet of papers, three times sealed, but unaddressed. This he was to return to her when she demanded it, no matter who was present at the time, and she asked him to note precisely the date and hour that he had received it. The priest vowed secrecy once more and returned to the monastery. The whole episode had lasted little more than half an hour.

On the tenth of November, in the early afternoon, a servant arrived at the monastery. Once again the queen had asked to see the priest. Taking the packet with him, Père Le Bel followed the servant across the courtyard to the Galerie des Cerfs. Once inside, he found the door shut, and locked, behind him. "I was rather taken aback," he writes.

In the middle of the gallery stood "Her Swedish Majesty." She was speaking to a man whom the priest did not know, addressing him as "Marchese."

Three other men were present, one of whom was Francesco Santinelli's brother, Ludovico. The queen then turned to Père Le Bel, and, "in a rather loud voice," asked him to return the packet of papers to her. She examined the seals carefully, then broke them open, and handed the papers to the marchese. Gravely, she asked him whether he recognized them. Monaldeschi denied it, "but his voice trembled." The packet, it seemed, had contained only copies of documents. The queen then produced the originals and showed them to the marchese, denouncing him as a traitor, calling on him to admit that the handwriting and signatures were his own. He did so, but immediately began to insist that he had meant no harm. Challenged with the forged letters, he replied that Santinelli's originals had been so shocking and so damaging to the queen that he had decided to copy and circulate only the less offensive sections; had he suppressed them completely, Santinelli would have discovered that they had been intercepted. Unpersuaded, the queen put further questions to him, and he protested his innocence, excusing his own actions and casting the blame elsewhere. Seeing the queen unmoved, he then threw himself at her feet, begging her pardon. At this, the three other men drew their swords.

The marchese got up, and drew the queen aside, and for the next two hours walked up and down the gallery with her, putting his case, insisting on his loyalty, imploring her to believe him. She listened to him "very patiently, with no sign of emotion or anger," while the three men stood by, their swords unsheathed. At length, leaning on her little ebony cane, the queen turned to Père Le Bel. "Reverend Father," she began, "you are my witness that I am not acting in haste or without good reason. I have allowed this faithless man more time and more opportunity than he has any right to ask, to justify his actions to me." She pressed the marchese further, and he took from his pocket a few papers and two or three little keys tied together; these he handed to her. At the same time, some silver coins fell out of his pocket. "I do not remember," writes the priest, "which one of us picked them up."

The queen then spoke again, quietly and gravely, though her voice betrayed some tension. "Reverend Father," she said, "I shall now withdraw. I leave this man to you. Prepare him for death, and take his soul under your protection."

The priest and the marchese fell at once on their knees before the queen. "Had I myself been condemned," writes Père Le Bel, "I could not have been more terrified." He begged her to have mercy, but the queen was unrelenting.

She could not allow mercy to prevail, she said. The marchese was a traitor and a criminal. She had given him her trust, in her business affairs and in her own most secret thoughts—more, she had regarded him as a brother, showering him with gifts and favors, which he had repaid with ingratitude and treachery. His conscience alone, she said, must be enough to cause him remorse, and with these words she withdrew to her room.

The marchese threw himself at the feet of Père Le Bel, asking him to go after the queen and plead for mercy. The priest exhorted him instead to plead for the mercy of God, but Ludovico Santinelli went to her, and soon returned in great distress; her majesty had told him to carry out the execution quickly. In tears, he exhorted the marchese to make his confession. "At these words," writes Père Le Bel, "the marchese was beside himself." He threw himself once more at the priest's feet, imploring him to go to the queen. This he did. He found her alone, her expression "perfectly calm, as though nothing was happening." Once again he fell at her feet, weeping and begging her "by the wounds of Jesus Christ" to have mercy on the marchese. The queen remained impassive. She was very sorry, she said, that she could not do so; that after Monaldeschi's perfidy, and the cruelty that he had tried to inflict upon her personally, he could hope for neither pardon nor pity. For far lesser crimes, she added, she had sent men to be broken on the wheel.

The queen then remarked that her decision was not without precedent; French kings, too, had resorted to summary executions. Père Le Bel conceded this, but "took the liberty of reminding her" that the kings of France had at least been in their own country, while Christina herself was a guest in the land. Would not such an act cause great offense to the present king? Would he not feel that his gracious hospitality had been terribly abused? Would Her Swedish Majesty not reconsider?

Her Swedish Majesty would not. She replied that she had not come to Fontainebleau as a prisoner or a refugee, and she owed no allegiance to the king of France. She maintained her right to judge her own subjects, whenever and wherever she chose. She was a ruler in her own right, and she would give account of herself to no one but God. Perhaps to justify the extremity of her response, she then inflated the marchese's crime to the level of abstraction. She was acting, she said, not against the marchese in person, but against his unparalleled disloyalty and treachery, which were matters of universal importance. Such crimes deserved death; she would maintain the same verdict before the altar itself.

Père Le Bel persisted. Such an act would do lasting harm to her majesty's reputation, until now so admirable, for "thought, word, and deed." Her majesty was respected and honored throughout France, but, regardless of her reasons, such an act would undoubtedly be viewed as violent and unthinking. Again the priest pleaded for mercy, "noble mercy." Or could not the marchese be turned over to the law? Could not a formal writ be issued against him? This would give her majesty satisfaction, and it would be acceptable in the eyes of the world.

This suggestion was dismissed indignantly. The queen repeated her right to judge her own "subjects." What need had she to bring a lawsuit against a member of her own household, with the evidence of his disloyalty, in his own handwriting, already brought to light? Père Le Bel objected that, though that was true, her majesty was herself "an interested party." The queen interrupted him, saying that she would inform the king and Cardinal Mazarin about the matter herself. She then ordered Père Le Bel to return to the marchese and attend to the saving of his soul. "On my honor and conscience," she said, "I cannot grant what you ask."

At this point, Père Le Bel writes, the queen's voice faltered. The priest took this to mean that she would now have changed her mind, not about the execution itself, but about the time and place of it. But, he notes, delay might have allowed the marchese to escape, and put the queen's own life at risk. He did not know what to do. "I could not simply go away," he writes, "and in any case I had to help the marchese in the saving of his soul."

The unhappy priest returned to the gallery. In tears, he embraced Monaldeschi, charging him to prepare for death, and to think of his eternal salvation. At this, Monaldeschi shrieked, then sank to the floor, and began his last confession. So distraught was he that his sins emerged in indiscriminate pieces of Latin, French, and Italian, and twice he got up, crying out in desperation. Père Le Bel nonetheless did his duty with care, and had just begun to question the marchese "to clear up a doubtful point" when the queen's own chaplain came in. Seeing him, Monaldeschi got up once again, "without waiting for absolution," and went over to him. Hand in hand, the two withdrew to a corner, where they spoke together at some length. The chaplain then left the room, taking Ludovico Santinelli with him, but Santinelli quickly returned. He raised his sword and said, "Pray for forgiveness, Marchese! You are about to die!"

He pushed Monaldeschi to one end of the gallery, "right beneath a painting of the Château de Saint-Germain." Père Le Bel turned away, but not before he had seen Santinelli thrust his sword into Monaldeschi's stomach. Monaldeschi grabbed at the sword, but Santinelli pulled it back, and in doing so cut off three of the marchese's fingers. Santinelli saw that his sword had been bent, and called to the other two that the marchese must be wearing chain mail under his clothing. He struck at his face, and Monaldeschi cried out to Père Le Bel. Santinelli "considerately withdrew a pace or two," allowing the priest to go to him. The marchese knelt down and asked for absolution. It was granted, and "as penance" Père Le Bel instructed him to endure his death patiently and forgive "all those who had caused it."

Monaldeschi threw himself onto his stomach on the floor, inviting the final blow. The second man came forward and struck him on the head, knocking out a piece of his skull. Monaldeschi pointed to his neck, and the man struck at it "two or three times, but without doing much damage," as the coat of mail had slipped up over it. Père Le Bel exhorted Monaldeschi to remember God, and bear it all patiently, "and other things like that."

Santinelli then asked Père Le Bel whether or not he should continue the execution. The priest replied indignantly that he could not advise him, and Santinelli apologized for having asked such a question. Hearing the door open, the wounded marchese turned to see the chaplain standing at the end of the gallery. The chaplain did not move, and Monaldeschi dragged himself along the paneled wall toward him. Père Le Bel approached, but Monaldeschi seized the chaplain's two hands and began a second confession. The chaplain told him to ask God's forgiveness, then asked Père Le Bel if he might grant the marchese a further absolution. This done, he asked the priest to remain with the marchese, and said that he himself was going to speak to the queen.

The second man at once came forward, and, with his long, narrow sword, ran Monaldeschi through the throat. Monaldeschi fell toward Père Le Bel, and lay in final agony another fifteen minutes, while the priest exclaimed "Jesus! Mary! and other holy words." At a quarter to four, Monaldeschi breathed his last. Père Le Bel began to pray, while Santinelli seized the body and shook the dead man's arms and legs, unbuttoned his breeches and underpants, and felt in all his pockets. He found only a small knife and a prayer book.

The three men, now disarmed, went into the queen's room, where Père Le Bel followed them. Santinelli announced the marchese's death. Her majesty expressed regret, but added that justice had been done. She prayed that God would forgive him, and promised to have many masses said for the repose of his soul. Père Le Bel was charged with disposal of the body. He sent for a bier, and, though the corpse was heavy and the road bad, Monaldeschi was in his grave by a quarter to six.

Two days later, Christina sent a hundred pounds to the monastery to pay for thirty masses to be offered for Monaldeschi. She duly received a receipt.

IF MONALDESCHI HAD INTENDED to betray Christina and Mazarin to the Spaniards, the secret was already out. Even in August, there had been rumors of a possible invasion. Spanish spies had been observing the training maneuvers and other preparations of the French fleet. From Naples itself, the papal nuncio had reported details of the plan to Rome, and from Paris had come the same reports, with the additional information of Christina's involvement. The Viceroy in Naples, sighing at the old news of one more supposedly imminent invasion, had set about strengthening the town's defenses, anyway.

It may be that Monaldeschi was acting not for the Spaniards but for the French, or, more specifically, for one Frenchman. He had certainly opposed Spanish rule in Naples, and was a known local patriot, by convenience if not by birth. He was an adventurer and a conspirator, and penniless to boot. Mazarin would have had no need of him, but he would have appealed to that other noble adventurer and conspirator, the Duc de Guise, who had ambitions of his own for the crown of Naples.

When Monaldeschi arrived in Lyon with Christina in the summer of 1656, he would undoubtedly have renewed his acquaintance with the duc, who had arrived to accompany the queen to Paris. The duc remained with Christina's entourage for some two months. Perhaps, during this time, an alternative Naples plot was hatched to the satisfaction of both men, and to the exclusion of the queen. The discovery of such a plot, with or without a wad of scurrilous letters, would have been a bitter humiliation to Christina. It would have meant more than the treachery of intimates, more than command of an army, more than the loss of the crown. It would have meant that, throughout the summer weeks, through all the compliments and lively conversations and rumors of love, it had been the duc, not she, who had had the upper hand. While she had allowed herself to be flattered by his attentions to her, by his fulsome praise of her intellect and her languages and her beautiful fiery eyes, she had been all the time his dupe. With Monaldeschi, or in the privacy of his room, he had not been admiring her. Instead, he had been laughing at her.

A humiliation of this kind would explain the ferocity of Christina's response. It would have struck at her constant and greatest resource—her sense of personal greatness. The exact nature of Monaldeschi's treachery cannot now be determined, as the intercepted letters, and their copies, have never been found. But it is assumed that Christina herself destroyed them, and indeed, after the marchese's death, servants saw her leaving his room, with burnt papers smoldering behind her in the grate.

AFTERMATH

THE STORY OF the gradual, gruesome killing of Monaldeschi is beyond tragedy, bordering on horror. Not the least horrifying aspect is Christina's own pitiless part in it, her calm determination from start to dreadful end. On strictly legal grounds, she had acted within her rights. By her decree of abdication, she retained the privileges of an absolute ruler over her immediate household, and in a subsequent letter, she referred to herself, significantly, as Monaldeschi's "sovereign." It was of course not the first execution she had ordered; she herself had declared to Père Le Bel that she had had men "broken on the wheel" for lesser offenses. But these had all followed due legal process; there had been trials, and judges, and state executioners. By contrast, Monaldeschi's death seemed brutal and arbitrary.

Even among the cynical courtiers of a violent age, the news fell like a thunderbolt. Already outrageous, it grew ever worse as the gossip spread: Christina had watched the whole thing, laughing at Monaldeschi, mocking his fear, chatting to the killers as the deed was done. At first no motive was sought, for none was felt to be needed. Christina herself had displayed too often her crude and willful nature. Her learning, her culture, all was as nothing. Now, it seemed, she had revealed herself for the savage she was, naturally capable of grotesque cruelty. At the French court, Madame de Motteville, who had earlier "praised the queen" herself, recorded the reaction:

> Our most Christian Queen Mother is scandalized. Everyone at court is horrified by such a hideous revenge, and those who praised the Queen before are now ashamed of it. The King and his brother hold her responsible, and the Cardinal, who is not at all a harsh man, is astounded. Really, we are all horrified. . . .[1]

Madame did add, however, that "everyone at court" was now making fun of the wretched marchese, who had had foresight enough to protect himself with a ten-pound coat of chain mail, but neither the courage to defend himself, nor the wit to run away.

From Rome, the pope denounced Christina as "a barbarian, brought up barbarously and living barbarously," and declared that he would take legal action against the killers—Monaldeschi had, after all, been one of his own subjects. He informed the queen that she should not return to Rome, but instead move to the town of Avignon, within the papal states but at a safe distance from the Holy City. Christina disregarded this warning, but was sufficiently shaken by it to attempt to placate the pope in a subsequent public letter to him.

Soon enough, a motive was felt to be required, and rumors abounded. Monaldeschi had been Christina's lover. He had spurned her and she had responded with the extravagant fury of a woman scorned. Rumors closer to the truth also spread, at least in diplomatic circles: the Neapolitan patriot was in the pay of the Spaniards; the queen had had her own plans for Naples; Monaldeschi had betrayed them. Cardinal Mazarin was concerned at all costs to prevent the truth of the invasion plan from leaking out. He himself concocted the story of a classic Italian *vendetta* between Monaldeschi and the Santinelli brothers. This, perversely, was believed by diplomats from Rome to Stockholm, who viewed the talk of political betrayal as a cover to salvage Christina's reputation. In England it was said that the pope, not the cardinal, had offered her the Naples throne, but that she had been "cheated by His crafty Holiness" and had killed Monaldeschi "for revealing that intrigue."[2] For Protestant Albion, one Catholic was evidently as good, or as bad, as the next.

For her own part, Christina stubbornly refused to allow any pasting over. The background might be hazy, but the scene upstage would be visible to all. Her reputation was to stand or fall on what she had done, in cold blood, with absolute deliberation. To this end, she had herself sent a messenger immediately after the marchese's death to tell Mazarin what had happened. Appalled and anxious, the cardinal had replied at once with a letter expressing his great surprise at "this curious accident," which would certainly prejudice "our current project." The letter was swiftly followed by the cardinal's own emissary, who delivered to Christina the following statement:

The offence that Your Majesty has committed toward the King of France is so serious, and its consequences could be so shameful for you, that the Cardinal has not been willing to inform the King that such an attack has been made in one of his own châteaux. He therefore trusts that Your Majesty will deny any involvement in this distressing affair, leaving all responsibility for it to that unworthy servant who has clearly surpassed his orders, and request him to leave the country.[3]

Christina dismissed this, and instead instructed the emissary to return to the cardinal forthwith, insisting that she herself was wholly and solely responsible for what had happened, and adding:

I cannot believe that the king of France assumes any power over me. That would be incompatible with my birth and my standing, since in that respect I am the equal of any ruler on earth. I recognize no superior save God alone.[4]

Mazarin persisted. He dispatched the queen's old friend Chanut to try to bring her to reason. Chanut appears to have emphasized the scandal that the news had caused, and the harm that would be done to her majesty's reputation if she did not issue some sort of statement denying her involvement in the affair. Christina would have none of it. She sent Chanut on his way, and on the fifteenth of November replied to the cardinal in a letter of staggering defiance:

COUSIN—

Monsieur Chanut, whom I count among my best friends, will tell you that I welcome with respect all that comes from you. Although he has failed to make me panic, it is not owing to any lack of eloquence on his part; he has certainly painted my presumed atrocity in suitably vivid colors. However, we people of the North are rather wild and not very timorous by nature, and you must excuse him if his message from you has not been so successful as you had hoped. Please believe me when I say that I would do anything to accommodate you, except be afraid. Anyone who is past the age of thirty is hardly going to be worried about a little gossip, and as for me I find it much easier to strangle people than to be afraid of them. As to what I did with Monaldeschi, I can tell you that if I had not already done it, I would not go to bed to-night without doing it, and I have no reason to repent of it, but a hundred thousand reasons to feel satisfied. These are my feelings on the subject. If you accept them, I shall be pleased. If not, I shall contine to hold them anyway, and I shall remain all my life,

> *Your affectionate friend,*
> CHRISTINA.[5]

Notwithstanding her "hundred thousand reasons to feel satisfied," Christina appears to have written this with some feeling, for the handwriting is shaky, and the page is covered with ink blots. To Chanut himself, Christina sent a copy of it, with the following cover:

I am sending you the letter that I have written to Monsieur the Cardinal. . . . I know no one who is great enough or powerful enough to persuade me to deny my feelings or to disown what I have done, and I am ready to declare this to the whole world.[6]

Despite her defiance, Christina was sufficiently persuaded to dismiss two of the executioners from her service. But on the same day, she wrote to Francesco Santinelli, who was still in Rome, still living off the unauthorized sale of her possessions, still sending back lying missives of encouragement to her. The tone of the letter is mildly admonitory, but essentially reassuring:

I am sending you news of the death of Monaldeschi [she wrote]. He betrayed me and tried to make me believe that you were the traitor. . . . Now he is dead, after having confessed his guilt and your innocence, and assuring me that he had done his best to incriminate you. . . . Do not try to justify my conduct to anyone. I am responsible only to God, who would have punished me if I had left the crime of treachery unpunished. Let that suffice. My conscience tells me that I acted in accordance with divine and human justice, and that I have only done my duty. That is all I need to say. Try to keep your spirits up. I will do my best to give you the consolation you wanted. Rest assured that I will protect your interests.[7]

The "consolation" that Santinelli had been wanting was no less than a French dukedom. Brazenly, Christina wrote to Mazarin, requesting Santinelli's ennoblement. The incredulous cardinal dismissed the letter indignantly, but the queen continued to support her rogue Lord Chamberlain. She was fully aware of his abuse of her trust. She had herself intercepted Peruzzi's letters to Monaldeschi, and read of her jewels being pawned and her plate being sold. But Santinelli was her *Scaramouche*. He could play the wily servant with impunity, and she would watch, amused, ignoring her own role of foolish aristocrat, gullible and gulled. She could not bring herself to believe that both her favorites had been disloyal to her, and so, from the unreasonable execution of the one, she proceeded to an unreasonable defense of the other.

Her talk was defiant, but she took the precaution of issuing a public defense of her actions as well, and saw that it was widely circulated in the courts and cities of Europe. The story she told was broadly the same as that recorded by Père Le Bel, but in her own tale Christina managed to deflect the blame from herself to Monaldeschi, and to include a tacit appeal to the sympathy of the pope. Her open letter was a clear attempt to save the remnants of her good name. No one believed it.

Monaldeschi's death, and Christina's intransigence after it, placed Cardinal Mazarin in a difficult position. The queen's insistent talk of treachery had led the gossip away from duels and love affairs, and drawn it toward political intrigue, threatening the secrecy of the Naples plan. Paradoxically, Christina was convinced that she was serving her own interests by insisting on her own guilt. Though a throne might be lost, her sense of sovereignty must not be, for this was her very sense of self. Despite her abdication, despite her financial dependence, despite her tiny retinue of motley courtiers, she was convinced that she was born to rule, and rule she would. The queen of Naples she might or might not be in future. She was already a queen in fact, and the world was to take cognizance of it. Let Mazarin play Machiavelli. Christina's role was the Prince.

AFTER THE FIRST RUSH of scandal, Mazarin decided that the queen must leave France. She lacked the means to travel, as he knew very well, and toward the end of November 1657, he wrote to the Swedish ambassador, asking for money for her. It was not the first such letter that he had written, and this time, too, it elicited no material response. He did not wish to insist or to threaten, for the Swedes were France's allies; a treaty signed earlier that year had renewed their mutual friendship. The Swedish king was Christina's cousin, and moreover, he was in arms again. Flush from the conquest of Poland, with the Danes on the brink of defeat, Karl Gustav had now turned his army southward toward Brandenburg. The memory of the queen's great father still shone fiercely. No one wanted to inflame the Swedes to warfare deeper into Europe. The cardinal was therefore obliged to content himself with expressing Louis's disapproval and his own displeasure at the "ill counsel" that had prompted her majesty to wreak such furious vengeance, adding, with more tactic than truth, that his king had nonetheless conveyed to her "every mark of affection."[8] He himself took a quieter revenge by allowing her to languish in the lonely countryside through three long winter months.

At Fontainebleau, she waited restlessly for an invitation to court. It did not come. Just as she had done in Pesaro, so now she spent much of her time writing letter after pleading letter to Mazarin, complaining of her boredom in the isolated château and venting her frustration over the Naples plan. It is a measure of her astonishing naïveté that, after all that had happened, she still believed that the throne might be hers: "I shall not dictate to you the time of my delivery from this present purgatory," she wrote, with unusual humility, "but I beg you to consider that I cannot return to Italy honorably without having spoken to you, nor without being at the head of the army that you promised me."[9]

But the cardinal's gout had become a matter of policy, and it did not improve. She was not asked to visit him, and he took no steps to visit her. So, just as she had done from Pesaro, she decided once again to invite herself to Paris. News of her preparations found their way to the capital before her, and she was swiftly paid a forestalling visit by the king and his brother. They arrived "informally," with a retinue of courtiers, and stayed about an hour. The king chatted affably about this and that, walking through the Galerie des Cerfs to the queen's own apartment, and if he did notice the bloodstains on the wall, he did not mention them.

Two lesser personages also came to pay their curiosity-fueled respects: *la Grande Mademoiselle*, now reconciled with the king, and the ballet master Isaac de Benserade, who had once served at the Swedish court. Mademoiselle found the queen "looking very well" in flame-colored garb with black finishing—a devilish touch, it seems, for Mademoiselle could not suppress "a frisson of fear" on seeing her. She, too, passed through the Galerie des

Cerfs, and noted that "although they had washed it well, the marks were still there."[10] To de Benserade, Christina herself shamelessly relayed the whole story. Seeing him shudder at its conclusion, she smiled and asked him whether he was afraid she might do the same to him. The gentleman's response is not recorded, but he did not come again.

CHRISTINA'S PATIENCE, OR persistence, was finally rewarded. Toward the end of January 1658, the longed-for invitation arrived. She could travel to Paris for the carnival season before Lent, and stay in the cardinal's own apartments at the Louvre. If she interpreted this as acceptance back into the fold, it was not; the apartments had been chosen particularly for her, she was informed, "as she would not be staying long." She did not take the hint. Declaring herself indifferent to creature comforts, she saw no reason for the cardinal to feel discommoded by his own removal to a small "cabinet," and anyway, on her first visit she had not had time to see his art collection properly.

To her surprise, many of her Paris friends found that they themselves had less time now. Partly embarrassed, partly uncomprehending, she did the best she could with those who were willing to see her, among them Saumaise's niece, the Comtesse Charlotte de Brégy, whom she had known in Stockholm, and her own Doctor Bourdelot, now a wealthy—and largely absentee—*abbé*. Despite Christina's now dreadful reputation, Bourdelot seems to have retained his affection for her, for throughout her stay in Paris he accompanied her everywhere, providing a measure of respectability among her otherwise motley companions.

The cardinal himself relented sufficiently to arrange an invitation for her to the court ballet, and in the middle of February she attended a premiere in which the young king, musical and lightfooted, took a modest role. It was no great courtesy to Christina, as it was in fact a public event. "Anyone could go," wrote *la Grande Mademoiselle*. "There were all sorts of people there." The music for the ballet, *Alcidiane et Polexandre*, had been written by a gifted young Florentine composer—and Mademoiselle's former Italian instructor—named Giovanni Battista Lulli, soon to be Gallicized as Jean-Baptiste Lully. It was a historic evening, for this was the first complete ballet that he had ever composed, and here, for the first time, the audience heard his new, slow-quick-slow "French" overture. Though set in motion by an Italian, a native musical style had begun.

The ballet tells a tale of love and adventure: Polexandre, warlike king of the Canary Islands, traverses the world from the arctic to the tropics in search of Alcidiane. After the requisite number of shipwrecks and duels in suitably exotic locations, he duly wins her. En route, the audience is treated to scene after fanciful scene as befitted the ballet of the time—a composite of mime, courtly dance, and tableau vivant, all set to Lully's elegant music. The cast included a full complement of slaves and shepherds, giants and dwarves, the latter "posturing grotesquely." The king took the part of Hatred, alongside Jealousy, Innocence, and other allegorical characters. Other gentlemen danced in women's attire, "diverting the court extremely." For Lully the performance was a personal triumph, and it sealed his destiny as Louis's preferred composer. For the king, it was a significant step toward the splendid, propagandistic *ballets royales* of later years, when, filled with confidence, artistically as otherwise, he would remove his courtiers to the spectators' seats, and apportion to himself every leading role.

But for Christina, this sparkle in the Parisian sky was a shooting star. Though she lobbied hard to be invited elsewhere, the death of Monaldeschi had cast too long a shadow, and her first invitation proved to be almost her last. As was her habit when invitations were wanting, she concocted her own, at one point descending on the Queen Mother's apartments in search of the king and his brother to go out into the town with her. They were both present, but well hidden, and they did not come out until Christina had gone. *La Grande Mademoiselle* herself had not courage enough to be seen with her in public, "for fear everyone would make fun of me."[11] Christina did receive one further invitation, to a grand assembly given by one of the state treasurers. There, pathetically, she made herself ridiculous by her clumsy dancing. The Queen Mother herself, hearing about it the next day, decided she must see this sight for herself. She persuaded Louis to invite the queen to a private ball within the Louvre "so that we could laugh at our leisure." Christina came, but, "just to make trouble," her old Stockholm friend the Comte de Brégy had sent a kindly warning of what was afoot, and she contented herself with a few untheatrical bows.

Uncowed nonetheless, she frenzied herself in a round of public balls and entertainments, arriving masked and costumed, and each time more fantastically. Though she was spurned in society, ordinary people thronged to see her, and she did not disappoint them as she charged about the city, in Turkish or gypsy garb, hailing any public carriage that happened to be passing. Two young Dutchmen, passing through Paris, saw her one evening in this carnival season. They were struck by the proud expression of her "beautiful sparkling blue eyes" and remarked that only the very bold could have endured her gaze for long. They noted, too, her haughty Amazon pose, one foot defiantly planted before the other—an appropriate pose, in this instance, for her head-dress was Amazon, too.

The whirligig continued for four weeks and more, with Christina cavorting about town, showing "very little wisdom, a lot of bad behavior, and a great desire for pleasure."[12] Toward the middle of March, the carnival season began to draw to a close, and with it the round of balls and concerts that had so engaged her until then. With the dancing at an end, it was time for reflection, and Christina's thoughts now turned to the world of letters. A different kind of entertainment was in order. She decided to go to the Académie Française.

During her previous visit to Paris, the *académiciens* had welcomed her with great ceremony, and this time she felt an extra need of their approval. In the aftermath of the Monaldeschi affair, she had been hoping for support from the community of the learned. They at least, she felt, would understand the nobility of her stance, and the justice of her cause. But, although there were some who defended her, on the narrowest legal grounds,[13] the deed itself had been so personally repellent that, among the learned as in society, she remained persona non grata. Consequently, although she had now been in Paris for more than a month, she had received neither visit nor invitation from the Académie. Undaunted, she decided to invite herself, and in the afternoon of the eleventh of March, she duly appeared at the Hôtel Séguier, a grand private house where the Académie held its regular meetings. She noticed at once that her portrait, before which the *académiciens* had sworn to pay eternal homage, had been removed.

The gentlemen were taken by surprise. Despite the queen's letter, they had not expected a visit from her, and they were now in the middle of a debate. It proved somewhat sensitive: it concerned a treatise on pain. The president rose to the occasion, however, remarking that the subject was "singularly appropriate to convey the pain felt by those present . . . at being so soon to lose sight of Her Majesty, owing to Her imminent departure."[14] Her majesty did not depart, however, and the discussion moved on to the definitions to be accepted for the new dictionary of the French language, then still almost forty years away from publication. The word at issue was *jeu*, the French word for "game," and as an example of its usage, a familiar saying was suggested: "The games of princes please no one but themselves." Whether pointed or not, the phrase hit its mark, for Christina was seen to blush, and then, with a strained smile, she rose from her seat and left.

Back at the Louvre, at least, pleasures awaited her in the form of some of her own works of art, forwarded from Antwerp. Most of them she could not keep. The cardinal took them as security for a large loan secured to get Christina out of France and back to Rome. To his ambassador in Stockholm, he wrote, "Tell the King of Sweden that I have exerted myself to prevent this Queen from collapsing into his arms, and it has cost King Louis eighty thousand *écus*."[15] For his own personal collection he chose several valuable tapestries, and these he kept until, little by little, Christina repaid the money, and the cardinal, carpet by carpet, sent them back.

Though the carnival was over and her purse was now refilled, Christina made no plans to leave. In Paris she remained, unforgiven and unreceived, but at least she had a place to stay and plenty of pictures to look at. The pope had told her not to return to Rome, and by now it was clear that she would not be welcome a second time at the Palazzo Farnese. Desperate to be rid of her, Mazarin promised her the use of his own palazzo in Rome, but still she did not move. Finally, the Queen Mother threatened to leave the Louvre herself if the Swedish queen did not go. The cardinal had an old carriage driven up to the gates, and the queen was escorted into it, and so she left Paris, with heads shaking and tongues wagging behind her. She appears to have taken the cardinal's gout along with her, for the very next day His Eminence was seen walking comfortably, and, some said, with a decided spring in his step.

OLD HAUNTS, NEW HAUNTS

CHRISTINA REACHED ROME in the middle of May, and made her way to the Palazzo Mazarini. It was as well she had it to go to, for she was not welcome anywhere else. As she had been in Paris, so now she was shunned in her adopted home. The pope refused to receive her, and pasquinades were pasted up all over the city describing her cruelly as a queen without a realm, a Christian without faith, and a woman without shame. Most of her old friends she found tactically unavailable; the academy evenings she resumed defiantly, once filled with noblemen and prelates, "now seemed a desert." Christina countered by a rejection of her own. She had no intention of staying in Rome, anyway, she declared; she was only waiting for the French to take Naples; then she would be on her way to bigger and better things entirely. No one took her seriously, but her talk was a major embarrassment to the pope, who was trying to broker a peace between France and Spain, at war still after twenty-five years. He sent her a message asking her to leave Rome, but she could not have left even if she had wanted to. The Swedes were at last sending her allowance, not to the queen herself, however, but directly to Mazarin to repay what he had lent her. Her lack of money at last became so pressing that she was obliged to sell the uniforms she had ordered for her future guardsmen; with them she handed over at last her dream of the crown of Naples.

As usual, the business was managed, or mismanaged, by Santinelli. Before long he had got hold of one of Christina's last irreplaceable treasures, her wonderful, twelve-foot coronation robe, with its pearls and its circles of solid gold crowns. She had brought it with her out of Sweden, and for eight years it had lain in storage; now it was to be sold for bread and candles. Some of the proceeds did get back to her, but Santinelli pocketed his customary share.

Loyalty from one quarter, at least, had never been in doubt. Cardinal Azzolino, appalled, embarrassed, exasperated, had stood by Christina despite all through the first shocked days following Monaldeschi's death, through the months of spurning and scorn, through the humiliating lack of money and all the foolish noisy talk of Naples. As the scandal cooled he had done what he could to effect her quiet rehabilitation within the Vatican, and by Christmastime of 1658 the usual good wishes could be sent without disdain from the cardinals to the queen. The pope demanded a final concession from her: she was to vacate the Palazzo Mazarini, which was located, awkwardly enough, directly opposite the pope's own residence, and take with her the crowd of ruffians and ragamuffins that passed for her royal household. She could not have taken this step on her own, as her riotous days at the Palazzo Farnese had made her persona non grata with every noble Roman landlord, but Azzolino found a new home for her, and in the spring of the new year, 1659, she moved to the quiet Renaissance villa on the other side of the river that would be her own for the rest of her life. It was not free, as the Palazzo Mazarini had been, but it was cheap—just over a thousand scudi per year. It needed renovations, but its garden was beautiful. It was part of the lease that the tenant must maintain the garden properly, and it was the one requirement that Christina was delighted to meet. The landlord did not learn this, for the queen's name had not been mentioned—Azzolino himself had signed the documents on behalf of "a person yet to be named."

Christina moved into her new Palazzo Riario[1] in July, and along with her came a new household. Azzolino had taken the opportunity of dismissing most of her servants and installing in their place a capable, reliable, diligent staff of his own choosing. Some were his relatives and many were from his native region of Le Marche, but if this served the cardinal's interests, it served Christina's better. From this point, her financial affairs began to settle somewhat, scandal receded—for a time—from her little court, and she began at last to live, in a modest way, the cultured life that she had dreamed of. While the renovations were being made, she installed herself on the grounds in a *casino*, a kind of grand summerhouse, surrounded by ilex and orange blossoms. It stood at the very top of the Gianicolo, the highest of Rome's seven hills, and from it she looked down over her gardens and her own Riario, and over to the city's stately Renaissance palazzi and the exuberant Baroque buildings of her own day. Opposite her stood the little Palazzo Farnesina, and directly across the river she could see the great Palazzo Farnese, which had been her first home in the Eternal City—Michelangelo himself had planned to link the two by a new Tiber bridge. It was the perfect place for a restless queen to calm down, and Christina turned happily to her books and her music and her garden.

In the same summer of 1659, the French and the Spanish brought their war to an end in the Peace of the Pyrenees, and in the following June, Louis XIV of France married his cousin, Maria Teresa, the infanta of Spain.

CHRISTINA MIGHT HAVE BEEN happy to stay at her new palazzo on the hill above the river, but lack of money, as usual, dictated her plans. For the queen's purposes, Mazarin's pockets were now empty. Most of her payments from the Pomeranian territories were going directly to him, and as Sweden's wars expanded through Denmark and Poland and into Russia, even these began to be sporadic. Karl Gustav did what he could to bridge the gap, allotting her an extra twenty thousand riksdaler—ten percent of her expected revenues—which he could ill afford himself. It was a loyal gesture, but it was not enough to purchase her gratitude. She repaid it by approaching the emperor, who was again at war with the Swedes, asking for money in return for concessions for Swedish Catholics once peace had been concluded. But by now Christina's word was not worth much, and besides, everyone except herself could see that she no longer had any influence in Sweden. The emperor took no notice of her.

Karl Gustav's thoughtfulness to his wayward cousin was his last. In February 1660, news arrived of his sudden death at the age of thirty-eight. If Christina grieved for him, remembering their childhood years together and their timorous, sweet first love, her grief was surmounted by anxiety for herself. Karl Gustav's successor was his little son, just five years old, the very same age that Christina had been when Sweden's throne had fallen to her. The boy was sickly, and he had no brothers. The country was to be ruled by a regency council, its chancellor Christina's longtime favorite, and eventual bête-noire, Magnus De la Gardie. After her departure from Sweden, his rise had been swift, and he was now effectively leading the government. There was no reason to expect that Magnus would be as considerate toward her as Karl Gustav had been. Worse, if the little King Karl should die, some other prince would take his place, someone, perhaps, not well disposed at all toward a renegade Catholic who had proved more than once disloyal to her native land.

Christina's uncertain financial situation might quite suddenly become desperate. This fear dislodged her from her garden and her books, and sent her on an anxious path northward, funded, once again, by the pope. She needed to be in Stockholm before the first Diet was convened in the autumn. Her personal presence there would surely persuade the *riksdag* to agree to continue her apanage. She traveled via Hamburg, and in the middle of October 1660, after more than six years' absence, she set foot once again on the shores of her homeland.

She was by no means welcome. A formal letter, followed by the formal visit of a Swedish envoy to Hamburg and a formal request that she return to Rome without coming to Stockholm, had all been ignored. The regents and senators grit their teeth and gave her a ritual greeting. At the Diet convened in the first week of November, they agreed in principle to continue the apanage, but the war had changed things, and her lands were now yielding only half the sum that they had given before. Her former majesty was now expected to take her leave, but Christina would not go. Not content with flaunting her Catholicism with far greater enthusiasm than she had ever shown in Rome, she wrote to the senators to inform them that, if her ailing little nephew should not survive, she intended to reclaim the Swedish throne for herself. Within hours her statement had been returned to her, without comment, and shortly thereafter she received a new document of the senators' devising that confirmed her abdication and renounced any future claim to the throne. As they held the purse strings, she was obliged to sign it. To add insult to injury, the two priests accompanying her were required to leave the country forthwith. The French ambassador thoughtfully sent his own priest to say mass for her, but his good offices were not sufficient to keep her in Stockholm, and she withdrew, priestless, to her own town of Norrköping.

Here, a hundred miles from Stockholm, she passed the freezing winter months, trying to secure a better return on her local rents. It was effort wasted, for management of any kind was alien to her, and she had none of Azzolino's capable *Marchigiani* on hand to help. There was only the Conte Gualdo Priorato, another plausible figure in the Santinelli mold, and he was with her only when she did not send him slogging off to Paris to beg money from Mazarin or even from the king. When the cardinal died in March of 1661, at the age of fifty-eight, Priorato descended on his relatives, trying to claim those of his bequests to which Christina felt herself entitled. Louis was asked to provide lifelong pensions for several members of her household, including

Landini (her Captain of Guards and one of Monaldeschi's killers), who had escaped Azzolino's purge. His Most Christian Majesty may have been distracted by the charms of his pretty new bride; in any event, he made no response.

In May, Christina left Sweden. Her departure was unregretted, and she left with only one regret of her own: in the six months of her stay, she had not been able to see the Countess Ebba Sparre, her beloved Belle. Magnus had vindictively refused to permit his widowed sister-in-law to meet the queen. Christina believed that Belle tried to follow her to Hamburg, but they do not seem to have met again. The countess was already in poor health, and she did not live another year.

Hamburg was no more pleasant than Sweden had been, but Christina was obliged to remain there for twelve restless months. She took what refuge she could in her books, in the ancient world that she loved, remarking wryly that "the conversation of the dead must console me for that of the living." Her old banker, Diego Texeira, had retired, and her new "Resident in Hamburg" was his son, Manoel. Christina preferred him to his Antwerp associate, Fernando de Yllán, the son of her former host, and decided to transfer the administration of all her affairs from Antwerp to Hamburg. It was no small task, but at length it was arranged, the repayment of loans, the redemptions from pawnshops, the bankers' fees, and all. Texeira was to deal with the queen's administrators in her Swedish lands, and she was to receive from him a monthly income of eight thousand riksdaler—more than enough for her to live on comfortably, provided the Swedes kept paying. Content with this settlement, Christina began to lift her gaze once more beyond "those matters so much beneath her as expenses and accounts." She began to think of politics and she began to think of religion, and for maximum effect she put the two together. The kingdom of Denmark and the city of Hamburg were ready for religious liberty, she decided. There were Catholics enough in both places—though apparently not in Sweden—to warrant a general freedom of observance. This must be demanded, and the great Catholic powers must do the demanding. Accordingly, she sent off letters to France and Spain and Habsburg Vienna. In all places, it seems, memories were longer than her own. The sighing conclusion of the Thirty Years' War—*cuius regio eius religio*—was not about to be called into question now. Louis replied with a pair of elegant, noncommittal letters. The German emperor flatly refused to listen to a word about it. From Felipe, nothing was heard for months, prompting a wry old adage from the lips of a wry old courtier—"If death came from Spain," it ran, "we would all be immortal." By the time the king's reply arrived, Christina was gone, and the project forgotten. But the pope, at least, remembered it, and noted it down with a mark of approval. He was there to welcome the queen when she arrived again in Rome in the bright midsummer of 1662.

HER GREETING WAS affectionate, her old friends gathered about her, and she was ready to settle quietly back into her beautiful home on the hill. So at least it seemed, and so at least it might have been, had not an opportunity presented itself for her to interfere once again in a diplomatic incident, and clumsily make things worse.

Christina's return to Rome coincided with the arrival of His Grace Charles III de Poix, Duc de Créquy, Ambassador Ordinary from His Most Christian Majesty, King Louis XIV of France. The queen and the ambassador arrived on the very same day, and, though Christina's return would have been food enough for Roman society, the duc's arrival was viewed as a welcome surfeit. The great Cardinal Mazarin was dead, and the twenty-two-year-old Louis was now in full command of France. It was the first time in nine acrimonious years that a regular French ambassador had been appointed to the Holy See, and in consequence, the duc provided a focus for renewed discussion of the many disputes that had never been properly settled. The pope had not yet forgiven the French for their extravagant demands at the peace negotiations in Münster, almost fifteen years before; the French had not yet forgiven the pope for his determined stance against the same. The pope had not forgiven the French for opposing his election at the last conclave; the French had not forgiven the pope for opposing their own territorial ambitions. The pope's relatives still had too much influence in papal affairs, or so thought the king; the king only wanted to replace their influence with his own, or so thought the pope. Countless other diplomatic differences, great and small, had kept "the spirits of both sides boiling."[2] The pope's archenemy, Mazarin, had left it all behind him, but there had been no abating of antipapal rhetoric in his adopted country.

France's new ambassador was a dull and heavy man, of dull and heavy mind. He brought no agility to the delicate situation now obtaining between his nation and the Holy See, but instead kept the fires of animosity stoked by countless little tinders of stubbornness and pride. As the Spanish ambassador had once done before him, the duc immediately set about creating difficulties over small matters of diplomatic etiquette, taking every opportunity to offend the pope and his supporters. He began by refusing to call upon the pope's family: they should call upon him first, he declared. He continued by making a fuss at Christina's Palazzo Riario, where he insisted on due recognition of his superiority to the Cardinal Princes of the Church: he was to have a proper armchair to sit down upon, rather than a stool, even when he was her majesty's sole visitor. A formal remonstrance was made to the king, Christina added her own complaint, and in due course a compromise was reached: there was to be no armchair, but her majesty would in future go a little further to welcome the ambassador—specifically, whenever he visited, she would take three extra obliging little paces toward him, significantly stepping off the royal carpet.

The ambassador's residence was none other than the Palazzo Farnese, Christina's former home, long since restored following the unrestrained activities of her servants in their early days in Rome. In front of the palazzo was a little piazza, and every day a band of soldiers, the Roman governor's Corsican Guards, paraded across the piazza toward their quarters. The long chain of command to which the Guards were subject ended in the person of Don Mario Chigi, the pope's own brother, and although Don Mario played no active role in their activities, his nominal responsibility for them provided an excellent opportunity for the ambassador to niggle the pope. The duc therefore demanded that the Corsican Guards should no longer pass by his residence on their daily march home. The demand was viewed as provocatively as it had been intended—there was no other route to their quarters. The matter remained under consideration, and the already bad relations between the two parties grew steadily worse.

One hot August evening, as the Corsican Guards made their usual crossing of the piazza, the duc's men challenged them, flinging out insults and throwing one of them to the ground. Only too ready to respond, the Guards promptly surrounded the Palazzo Farnese, and when the duc himself appeared at a balcony to see what was happening, a shot from an arquebus whizzed narrowly past him and slammed into the wall at his side. His lady wife, meanwhile, trying to return to the palazzo from a visit to a nearby church, found her way barred. The Guards began to batter her carriage, smashing the windows, then they seized one of her pages, set upon the boy, and killed him, and in the fight that followed, dispatched two French soldiers as well.

Within minutes, Christina had heard of the incident. She was enjoying a quiet evening stroll with Azzolino in the Riario gardens when an excited servant burst upon them with the news. The queen was soon no less excited herself, and at once sent word to the Duchesse de Créquy—though not to the duc—that, if the Palazzo Farnese was not yet secured, Her Grace would be welcome to pass the night under the queen's protection at the Riario. Her Grace, however, remained where she was, not wishing, perhaps, to leave her husband alone and unprotected, or perhaps because hostilities had now ceased, or even perhaps because the Roman governor, in whose service the Corsican Guards were employed, was none other than Cardinal Lorenzo Imperiali, a known member of the *Squadrone Volante* and a close friend to Azzolino and to the queen.

Christina had graciously offered her assistance, and her offer had graciously been declined, and that might have been an end to the matter, but the summer was languid, and distractions few, and she decided that there was a role for her to play. She determined to act as mediator between the two parties, and she duly sent off messages to that effect: the pope should send his nephew at once to apologize to the duc; the duc should at once admit him. Her public demeanor was diplomatic, but impartiality was not in Christina's nature. In a private letter to Azzolino, she revealed that, at least at first, her sympathies lay with the French, and she revealed, too, a considerable cynicism, given that three men had been killed, and a sentence of death was the likely reprisal:

> One of the Corsicans will have to be sacrificed [she wrote], and if those responsible cannot be found, the innocent will have to be punished. It must be clear that they are not being protected, and that no tricks have been used to save them.[3]

There was in fact a suggestion that some of her own men had joined in the fight on the Corsicans' side. Certainly they had a reputation for ready brawling; Christina had maintained a remarkably tolerant attitude toward it, shocking Roman society, and particularly the pope, in the process. Whatever the case, her mediating stance did not last long. A haughty reply arrived from the French ambassador, disdaining her majesty's intervention. Christina at once transferred her efforts to a higher authority, and letters were soon flying from her determinedly intervening pen to the French king and his chargé d'affaires,

Lionne, urging them to "paste over" the dispute and put an end to the whole business.

In due course she received their response. Her majesty's arguments had left them unmoved. The king's letter was markedly cool, and it included a piquant reference to the queen's own behavior in the Monaldeschi affair:

> It is all very well to counsel moderation when one does not behave with moderation oneself. If Your Majesty had been ill treated by even the least of Her servants, even if the injury had been of no consideration whatever compared to the injury done to me in the person of my Ambassador, I am sure that Her Majesty would have sufficient spirit and sufficient care for her reputation that She would by no means follow Her own counsel to me to "paste over," as She puts it, this disagreeable picture.[4]

The chargé d'affaires, less succinct than his master, required twelve florid pages to explain to her majesty that she should practice what she preached and mind her own business.

Christina took immediate umbrage. She abandoned the middle ground forthwith and began a noisy support of the Roman party. She sought out the Roman governor, Cardinal Imperiali, official employer of the Corsican Guards. Imperiali had in fact had nothing to do with the incident—on the evening in question, he had been enjoying a good meal with a convivially named brother prince of the Church, Cardinal Aquaviva—but Christina deemed him nonetheless a suitable local hero, and he was soon installed at the Palazzo Riario as an icon of all that was anti-French. Farces were staged at the palazzo mocking King Louis and his ministers, with the audience comprising every noble or cleric who had ever been heard to mumble any remotely anti-French sentiment. Azzolino played his part behind the scenes, urging the pope to make further difficulties, and within a matter of weeks Ambassador de Créquy had left Rome to nurse his wounded pride and await further instruction in the quiet hills of Tuscany. Few of his servants attended him in his indignant exile; the greater part of his household was left behind, fed up in all but the most literal sense, for the Palazzo Farnese had been placed under embargo and no provisions could be delivered there.

Retaliation was not long in coming, and it came in a blaze befitting the ambitions of the rising Sun King. Louis sent his troops to Avignon to seize the papal garrison there, and in order to get it back the pope was obliged to accede to the king's every outrageous demand: Cardinal Imperiali was demoted from his hero's throne at Christina's palazzo and sent into exile; the Corsican Guard was disbanded, and on the site of the incident, outside the Palazzo Farnese, a great pyramid was erected, relaying the whole story in all its ignominious detail—ignominious, at least, to the pope, who was obliged to send his nephew all the way to Paris to make obsequious apologies in person. Christina herself narrowly escaped a serious reprisal: Louis wrote to the regents in Stockholm, making clear his displeasure at the queen's behavior, and suggesting that a reduced allowance might bring her majesty to heel. Her already limited finances were saved by the calming counsel of Louis's Stockholm ambassador, the Chevalier de Terlon, an habitué of Swedish court circles for more than two decades. Terlon knew the queen well, and despite the many provocative traits of her character he remained fond of her. He now wrote to the king, advising him, in effect, to "paste over" this matter, too: "Your Majesty should know," he wrote, "that the Queen is the most blustering princess in the world, and at the same time she is the most timid. She respects only those who speak to her boldly."[5]

Louis relented, instructing only that the goodwill that must be maintained between France and her ally, Sweden, need no longer be extended to Sweden's former queen. The Duc de Créquy, returned to Rome, was to make no further calls upon her majesty, whether she offered him an armchair or not. The duc was only too happy to comply, and appended his own touch by declining to acknowledge the queen or her supporters even when they met by chance: should his carriage come upon hers in the street, he would draw an indignant curtain.

Louis kept Christina in her Roman Coventry for the best part of three years, and finally indicated that he would be prepared to accept a conciliatory letter from her. She sent one through her secretary, but the king replied in his own hand. The clear victor, he could afford to be magnanimous, and besides, as he himself added in a perfectly calculated insult, it was not his custom "to do battle with the ladies."

THE LADY IN QUESTION had by now, in any case, turned her attention elsewhere. She had a home to attend to, a lovely palazzo, and a large garden—almost a park—and many, many pictures to hang. Most had been lying for years in their boxes; Christina had not seen them since they had been smuggled onto the ships in Stockholm Harbor, more than ten years before. She hung them now in a grand first-floor gallery, where visitors to Rome were invited to view them, the queen viewing the visitors themselves, with no less pleasure, through a peephole in the wall. The same floor housed the other public rooms, including a throne room, as well as Christina's private apartments. Though she claimed not to care about material comforts, her suite was decorated sumptuously with Persian carpets and silk tapestries, including the Alexander series that had hung in her rooms in the Tre Kronor Castle. Her bed was made of Indian wood and covered with a canopy of crimson brocade, and there was a luxurious bathroom, with not one but two marble baths, running water, hot and cold, and elegant little niches in which stood statues of Venus and Cupid, and a bust of Christina herself. It was a surprising priority for a woman who had recently greeted the pope "with her unkempt hair tied up with ribbons of various colours, and her face powdered with dust from the roads,"[6] but so it was. It replaced the simple lead-lined wooden bath and the old copper water cans that had served her before. Evidently she had not forgotten Doctor Bourdelot's advice, and was still taking "regular baths," or at least giving the appearance of doing so. The rest of the first floor was decorated with wall-hangings of crimson and green. Christina had revealed the simpler tastes of her northern home by choosing furniture of carved wood rather than the gilt then fashionable in Rome, but she had a number of cabinets inlaid with ivory, and many laquered "Indian" pieces— things from more or less anywhere outside Europe. There were a lot of stools, but very few armchairs; most visitors, it seems, would know their place.

On the top floor, she installed her handsome library, two thousand manuscripts alone,[7] and many more books in many languages, ancient and modern, all beautifully bound, and even one manuscript in Japanese, which, however, only a few local Jesuits could read. Alongside the library was the *sala*, a large hall where her learned friends met to talk, and concerts and plays were performed. The English traveler Edward Browne attended one of these soirées during the carnival season of 1665, and relayed his impressions in a letter home:

> I was the other night at the Queene of Sweden's, she is low and fat, a little crooked; goes commonly with a velvet coat, cravat, and man's perruke; shee is continually merry, hath a free carriage with her, talks

and laughs with all strangers, whom she entertains, once in a weake, with musick, and now this carnivall every other night with comedies.[8]

Though she does not seem to have played herself, Christina possessed some fine musical instruments: numerous strings and six keyboards—two spinets, two organs, one of them in fact Azzolino's, and two harpsichords, one with extra *putti*, fat and golden, perched on top. Musicians were cheap, and she employed them in plenty. They included a flutter of young female song-birds, who lived in nunlike confinement on the little mezzanine above her own bedroom.

There was no separate room for dining, but tables of all sizes lay about, ready to be moved wherever they were needed within the palazzo or out into the garden. Christina herself did not much care for formal dining, "though she was a huge Eater, as the Northern people are"—so one of her servants re-cords. "She lov'd rathr Colliflower," he writes, "or boil'd Chestnuts with her Maids, for which she would slip into their Chambers on purpose, rather than eat the delicate Morsels which were prepar'd for her with royal Magnifi-cence."[9] On her first arrival in Rome she had begun to drink wine, whether out of courtesy to her new hosts or in play of sophistication, but the experi-ment did not last, and now she drank almost none. There was juice to hand, in any case; she had planted some three hundred orange and lemon trees, and now the brilliant little circles and ovals stood out against the lush green, with a quieter chorus of pink and white jasmines—two hundred new plants—behind them.

Gardening was for her a genuine—and, for once, harmless—pleasure. Even as a girl of fifteen, she had summoned the famous André Mollet to lay out gardens for her, but Sweden's climate had worked against them both. Now, in the warmth and light of the south, one of Christina's dreams at last began to blossom, and among the plants and flowers she built a stable to house another passion: her forty-five horses, all named after noble heroes, and her eight mules, all named after Jesuits.

The Palazzo Riario was now a beautiful and impressive place, and full of life. Christina had a large household of some 170 people, higher and lower servants and their families. For formal appearances, she was almost always ac-companied by men, but there were many women living in the palazzo, too, kitchenmaids and laundrymaids and maids of all work, and a few ladies-in-waiting as well. Christina preferred them to be single women, since "she hated, or pretended to hate Marriage. She affected to pass for a Maid," one of her servants wrote, evidently not believing it himself, "and the word Woman offended her horribly." She could not always avoid the wives of her menservants, but she did not like to see them pregnant. When Francesca Landini, the wife of her Captain of Guards, "was big with Child"—not her husband's child, however—she generally refused to admit her at all. "If she had occasion for her," wrote her servant, "she would say, *Bring the Cow hither*, and send her away again as soon as she has done." When another of her menservants came to present his new wife to her, "then (says she to the Woman), *If you come to shew like a Cow, do not come to see me in that condition.*" Christina showed greater kindness to the fruits of their labors; with children she was often tender. She was particularly fond of one little boy of the household, and she would ask for him to be brought to her, "and Caress the child, and sometimes hold it in her Arms; and when the little Child began to go, and came to the Table to embrace her Feet, she would fill his little Apron with Fruits and Comfitures."[10]

Azzolino had done what he could to ensure a stable household, but Christina needed occasional zest in her steady domestic diet. To get it, she opened her doors to a number of unsavory characters who might have been better left outside. Among them was one Vanini, an Italian priest of "some birth, little merit, and much vanity."[11] He was known to have raped one young woman already, but Christina's tolerance was wide, and she allowed him to frequent the Riario with impunity. In due course, his eye fell on one of her servants, and before long, "poor Jovannina found her self with Child, which affrighted her, and made her perfectly desolate." Christina supplied the wherewithal to help her, and Jovannina "had Remedies given her to take it away," to worse than no avail: the girl died. With ghastly irony, she was buried in a nun's habit, a crown of virgin's flowers on her head. "The Queen wept as much as if she had been a Relation," but Vanini was not dismissed. Nor did he reform, being only annoyed that he had not been able to keep the whole affair secret. In this hope he had bribed the queen's new chamberlain, Santinelli's replacement and almost as much of a rogue, the Marchese Orazio del Monte, or as he preferred to be known, the Marchese Orazio *Bourbon* del Monte, an addition that linked him—just possibly—to the French royal family. Del Monte had accepted the money, but had talked anyway, excelling, as he did, at both activities. The queen liked him enormously, and whatever business she could keep from Azzolino's drones, she delivered into his hands. They were large hands, and quick to grasp, and as ready to wield a dagger as to seize a purse of gold. He had married his son, at the point of a blade, to the wealthy but notably unattractive niece of the late Mondaleschi, but the young man had not managed to contain himself for long "within the Duties of Marriage." Christina found it all hugely amusing, and liked to tease him about his "charming" wife. His father, meanwhile, took his own amusements at the queen's expense, by selling his "protection" in the streets around the Riario. Day by day, the district grew livelier, attracting all those who could find no welcome elsewhere, until at last Christina's palazzo stood surrounded, a Renaissance island in a bright Baroque sea of smugglers and thieves and prostitutes.

DEBACLE

AS THE MARCHESE DEL MONTE's pockets grew heavier, so Christina's own grew lighter. Azzolino's reforms kept the wolf from the door, but remittances from Hamburg were dwindling. The agreed eight thousand riksdaler per month had dropped to five thousand, partly through the fault of her banker, Texeira, but mainly through the incompetence, and worse, of her Swedish agents. Of the 200,000 per year that she had expected following her abdication, she was now receiving less than a third. Her household was large, her tastes lavish; more money must somehow be found. From France, Bourdelot wrote of the king's plans to complete the great palace of the Louvre, and suggested that the marble for it might be quarried from Christina's island of Öland. A sample was duly sent, and Louis had a look at it, but he found it "too melancholy" for Paris, and Bourdelot's idea went no further. Azzolino then presented an idea of his own, characteristically practical and risk-free: the queen should send someone to look into things. He proposed Lorenzo Adami, one of his own many capable relatives, and already a member of the queen's staff. Adami set off in the middle of 1665, stopping in Hamburg for a few days to speak with Texeira and to hire an interpreter, and in August he arrived in Stockholm.

He proved to be worth his weight in riksdaler. The queen's governor-general, Seved Bååt, had been taking no notice of her estates, other than to ensure his relatives an income from them. Months' worth of remittances lay uncollected, and in Pomerania, her representative, Peter Appelman, had grown rich on outright theft. Before the end of the winter, Adami had sorted out everything and doubled Christina's income. For his own splendid services he charged a modest price, the largest of his expenses being his interpreter, at two pounds a month.

Greater powers than Adami's, however, were also at work. Since 1664, the English had been once again at war with the Dutch, and were pressing the Swedes to launch an attack of their own. When they declined, the Danes joined the fray on the Dutch side. Sweden now had two good reasons to go to war: the old Danish enemy was in league with the new competitor for the trade routes in the Baltic; defeat of either one would be to the Swedes' advantage. But they held to a precarious neutrality, announcing that the *riksdag* would be summoned shortly to debate the matter. When Christina learned of it, she decided at once that she must attend the session. If Sweden went to war again, Adami's careful work could be undone in a moment. She must ensure neutrality. The French ambassador was working for it already in Stockholm. Christina must lend her influence. And she must start making regular visits to her estates. That would keep the managers honest. But for this she would need the ban on her priests to be lifted. She could not forgo a daily mass on her sorties into Lutheran territory, however desultory her attendance might be in Rome. The question of freedom of worship must be raised at the Diet. She must be there to push the matter through.

She demanded a report from Adami. Who in the *riksdag* would support her? He sent a diffident reply. There were people, he said, some quite important people, who regarded the ban on Catholic worship as unjust. Some of them would, or might, support her, if it came to a vote. He had made the acquaintance of the French Ambassador Extraordinary, Monsieur Arnauld de Pomponne. Monsieur was very friendly, and yes, Louis favored neutrality in the war, and he was paying the Swedes to ensure it. But the drums were beating, and the Danes were a very old enemy; friendship between the two had never lasted for long.

Christina began her preparations, and in May she set off northwards, with money from Azzolino speeding her on, shod in her sturdy little men's boots, girded with imaginary influence. She spent a sleepless first night away from Rome, weeping bitter tears to be apart from the cardinal, but the next day she rallied, and sprang to her stirrup to begin the dusty, month-long ride to Hamburg.

SHE ARRIVED IN BRIGHT midsummer, and took up residence in the house of a friend of Texeira's. She did not plan to stay long. Stockholm was her goal, and she now wrote a mischievous letter to her governor-general,

Seved Bååt, to say that she was thinking of returning to live in Sweden for good. She had no real intention of doing so, but she could not resist setting the cat among the pigeons; Bååt would naturally pass the news on to the regents, and it would give them all a good fright. It did, but the cat rebounded. The alarmed regents postponed the Diet, and sent word to the queen that she would be better not to come to Sweden at all. The ban on her priests would in any case not be lifted. Adami wrote to advise that he had spoken with a number of local grandees about the matter; the ban, it seemed, was largely for public consumption: one priest in her entourage would be accepted, provided he was disguised, and provided the Catholic rites were practiced "with modesty and discretion." It was not quite encouragement enough, and Christina remained in Hamburg, waiting for the Diet to be summoned, longing to be in Stockholm, and longing even more to be in Rome, for the pope had been ill for some time and his death was regarded as imminent.

The excitement of a conclave to elect his successor was already gaining momentum, and Christina was eager to be part of it all. But, to general disappointment, Alexander suddenly rallied. Contenders for the tiara got back on their knees, and the Roman flock submitted once again to the tending of their too-pious shepherd. Christina's feelings were mixed. The whole business had now to be delayed, and delay was never to her taste. On the other hand, had it been sooner, she could not have been there in the thick of it, for here she was, at the end of 1666, still exiled in Hamburg. She shrugged, and sent off a resigned letter to her friend Hugues de Lionne, the French chargé d'affaires in Rome, with the "no longer appropriate" words in code:

> As far as the *conclave* is concerned, I agree with you that it is no longer appropriate to discuss it. The *coffin* we were expecting has been transformed into a *wheelchair*, and I am convinced it will be *in use* for some time yet in this world. It is certainly not a suitable vehicle for the *next world*.[1]

Her letter of the same time to Azzolino is a neat mixture of political satisfaction and girlish piety. The cardinal would have been well able to weigh the merits of both:

> It is the best news in the world [she wrote]. I asked God to allow me the grace of being back in Rome by the time of the next conclave, and

He has granted it. That shows that one's prayers are granted if one doesn't ask too often.[2]

Though Alexander had treated Christina kindly, she was now looking, as no doubt he was himself, to the time beyond his earthly sojourn. Together with Azzolino and the *Squadrone Volante*, she wanted a pope who would do away with the ancient practice of papal nepotism. It had been a heavy charge on the Vatican's coffers, and had stifled administrative reform as well. Bored in Hamburg, Christina allowed herself a relieving rant to Azzolino about "the prejudice and disorder that this plague of nepotism brings to the Church":

Don't you think it's pitiful [she wrote] to see so many millions from the Church treasury used for luxuries for absolute nobodies who turn up to suck the blood and sweat of the poor? Where would we be if that money had been used for the State, or for its defense, or to root out heresy? This private greed gobbling up public money! These popes have ruined Church and State. . . . [3]

Christina's objection was in fact based on one of her most dearly held convictions. The practice of nepotism, she felt, was gradually diluting papal power. It was incompatible with absolute rule, which was quite obviously the best form of government. The cardinal's stand, by contrast, was pragmatic, though overlain by principle. Nepotism was unjust, for it valued birth over merit. The queen was mistaken in thinking that it was diluting the pope's power; on the contrary, it was concentrating power in his hands, when it should have been being diffused to his clever cardinals. Besides, the practice was too firmly entrenched to be ever rooted out. Some form of accommodation was inevitable. He and his *Squadrone* might speak against nepotism, but they were not going to adopt any formal position against it.

Jesuitical ambivalence was not Christina's way. She carried on berating "these popes" and bewailing "poor Rome" for several pages more, and in the end had to beg the cardinal's pardon "if my zeal has carried me away. I have gone on about this for too long," she admitted. She calmed down sufficiently to bewail the fate of another "poor town"—that of London, where "they say twelve thousand houses have been burned up in the terrible conflagration. Some say it was treachery, and others say it was an accident—we'll know the news by next Friday, anyway, and I'll be sure to let you know. It will be the ruin of that kingdom," she added, "but at least it may oblige them to make peace with the Dutch."[4]

The pope's illness had brought Christina one certain benefit, at least: she was persona grata once more with Louis and the French. This she owed to an unwitting Azzolino and his unaligned *Squadrone*. Louis wanted their support for France's candidate in the next papal conclave, and he now invited Christina to help "establish confidence" between himself and them. She proved to be a good servant of the French king, composing regular summaries for him of all the small details of Vatican politics that Azzolino and others relayed to her. She was thrilled to play this new role, imagining herself a broker of political destinies. Louis encouraged the fantasy, praising all her little tableaux of people and problems. They were all "so well depicted," he remarked to her. It was "a singular pleasure" to read them. Christina wrote on, gullibly, assuring Azzolino at the same time that "I know how to dissemble, and how to keep quiet,"[5] and so Louis ensured his supply of inside information, and revealed nothing at all in return.

It was not until the end of April 1667 that Christina began her journey to Stockholm at last. The Diet was scheduled for May, and, though the regents had sent word that she would not be welcome to attend it, nor indeed be welcome in the country while it was in session, she ignored it all and set off for her homeland with a vastly expanded retinue in tow. She had left Rome with a household of twenty or thirty servants, but in Hamburg she had acquired a hundred and twenty more, and of these all but two accompanied her into Sweden.

Christina made her landing in the newly acquired territory of Hälsing-borg,[6] where a large deputation had been awaiting her since February. It was headed by Count Pontus De la Gardie and Count Per Sparre, and with them the two counts had brought the ghosts of Christina's youth, for Pontus was the brother of the chancellor, Magnus, and Per the brother of Ebba, her now dead "Belle." Hälsingborg was a tiny town, with the barest amenities and the bleakest distractions. Here the company had been obliged to wait for almost three months, their only consolation the fine carpets and tables and cups and plates that, like them, had traveled the long, cold road from Stockholm to make the queen more comfortable. On they traveled now to Jönköping, six days' ride in slow procession, armor glinting, banners waving, and Christina waving, too, from the window of her carriage. It was the first royal progress in the region since the days of Gustav Adolf, and the people turned out in force, cheering as the train went slowly by. It was a bright splash of color in their drab and strenuous daily lives, but Christina persuaded herself that it was all and only for her. Their enthusiasm was proof, she said naïvely, of their longing for her personal return.

Little by little, she caught up with all the local news. Some of it pleased her. Magnus was not a very popular chancellor, especially where the country's finances were concerned. On the other hand, he was a good negotiator, and his charm had dispelled a lot of grumbling, and he was proving to be a fine patron of the arts. There was other unwelcome news, too. A new bridge was being built in the capital, and her own coronation arch had been demolished to make room for it. There had been more building as well, and the new queen, Karl Gustav's widow, Hedvig Eleonora, had been behind it. A wonderful new palace had been built on an island near Stockholm, the palace of Drottningholm—the Queen's Island—in Lake Mälaren. It was said to be a marvel in shining white stone, much bigger and much more spectacular than Christina's Riario in Rome. She heard news of two deaths as well, the first that of the pope; this news, however, was false. And from Warsaw came news of the death of the Polish queen, Maria Ludwika. This was most important news, for Maria Ludwika had been the wife of Christina's Vasa cousin, Jan II Kazimierz, and the couple had had no children. In Stockholm, fears had been roused that the widowed king might marry Christina and present a new Catholic claim to Sweden's Protestant crown. Christina herself dismissed the idea outright, saying that while she might well succeed the *king* of Poland, she would certainly not succeed the *queen*.[7] Swedish fears had been allayed, in any case, by Jan Kazimierz's swift decision to abdicate. His wife had been influential and popular, but he himself was weak and widely disliked, and at heart more a monk than a ruler. The Polish monarchy was elective, and Christina would soon be reminding the Poles of her Vasa blood and seeking the crown for herself, but for the moment she was content to acknowledge the connection by entering a period of formal mourning for Maria Ludwika; she had more urgent matters to arrange.

The urgent matters were all financial, and, with patience, Christina might have arranged them easily. But now, as so often, she allowed her pride to obstruct her better interest. She decided to defy the regents' ban and ignore Adami's suggestion that her Catholic rites be practiced "with modesty and discretion." Her priest, Father Santini, had entered the country officially as her secretary, though his true identity was no secret. Christina insisted on a daily mass—a provocation in itself, as she had not been in the habit of attending mass daily in Hamburg, or even in Rome. As long as it remained a private affair, however, no objection was raised, but before long, doors began to be left conveniently open and curious locals found their way inside. Information wended its way back to Stockholm, and in due course, Pontus De la Gardie received a stern letter from the regents. He was to issue a formal reminder to the queen that the presence of a Catholic priest could not be tolerated.

Pontus performed his unenviable task with great care, but not with delicacy enough to prevent an outburst from Christina. She was outraged, she declared. She would leave the country at once. Pontus remonstrated, and she quickly changed her mind, realizing that it would not be in her interest to leave Sweden now. She decided instead to write to the regents to ask them to reconsider the matter, but the letter was anything but conciliatory. She wrote in French rather than Swedish, and, as if to emphasize her allegiance to Rome, she dated the letter according to the new Catholic calendar—her letters to Azzolino, by contrast, are dated in the old Swedish style. Though formally addressed to the twelve-year-old king, Christina's letter was principally intended for the eyes of Magnus De la Gardie. It is technically a petition, but it reveals her fury that Magnus should have dared to try to restrict her behavior:

MY BROTHER AND NEPHEW—

His Lordship Count Pontus De la Gardie has just proposed something to me on the part of Your Majesty. It concerns the person of my priest, and it has greatly surprised me, and I assure you that I was not prepared for it, after all the honors and civilities paid me on Your Majesty's behalf. Although these were no more than my due, nonetheless I was generous enough to be obliged to you for them. But Your Majesty has annulled all this by these latest, unjust steps. . . . I intend to show Sweden and the whole world that there is no advantage on earth for which I would deprive myself for a single moment of the profession and practice of my religion. . . . I would have departed this very evening, had Count Pontus not begged me to wait for your reply, to see whether Your Majesty has enough friendship for me to alter this decision. . . . In any case, to remind you of who you are and of who I am, understand that you were not born to command people of my kind.

> *Despite your peculiar proceedings,*
> *I remain, my brother and nephew,*
> *Your devoted sister and aunt,*
> CHRISTINA ALEXANDRA[8]

There had in fact been no pleading from Pontus De la Gardie. By his own account, he had encouraged the queen to agree to the regents' demand, and had even suggested a face-saving pretense of sickness on Santini's part; the priest could be left behind in Jönköping, whence, after a few days, he could make his way quietly back to Hamburg. Christina would hear none of it. Sure of gaining her point, she announced that she would remain in Jönköping to await the regents' formal reply, then make her way to Stockholm, with Santini still firmly in tow. But, restless as ever, after only three days she felt that she had waited long enough; she would press on to Norrköping, which lay conveniently en route to the capital.

Norrköping was her own place, a busy harbor and textile town whose rents Christina had retained in her settlement of abdication. Here she had intended to host a grand banquet to celebrate her return. She could not have afforded it, but she was spared the expense in any case by the arrival of the regents' reply, addressed, as their previous correspondence had been, not to herself but to Pontus De la Gardie. Christina's letter had failed to placate or to persuade them. It had in fact hardened their attitude, and now they not only repeated their demand that Santini be dismissed from the kingdom, but also informed the queen that she herself would not be permitted to attend mass openly at the home of the French ambassador in Stockholm—a subterfuge would be acceptable, however, and she might attend on the pretext of paying an ordinary visit to Monsieur Pomponne. There may have been mischief in this; Magnus would have known how insulting such a suggestion would be to Christina's enormous pride. If so, it was the perfect riposte to her own disdainful letter, and if it drove her back to Hamburg, so much the better for him.

Pontus conveyed the news to her, no doubt with great care once again, and once again with insufficient delicacy to prevent an outburst. The idea of herself, the queen, going to visit a mere ambassador, was insupportable to her, and she shouted back at the count. "What!" she cried. "I should pay a visit to

Pomponne! If he proposed it to me himself, I would take a stick and beat the man, even if his own king were present!"[9] She declared that she would leave Sweden at once, and vowed to accept no further civilities from the regents. Their representatives, with all their carriages and their courtesies and their keys to cupboard doors, were dismissed immediately. Christina would make her journey back unassisted. Having no means of travel of her own, she was obliged to send for hired carriages—they would be ready on the morrow. With great difficulty, Pontus managed to persuade her to allow him at least to accompany her to the border; without him, he insisted, the garrisons en route would not allow her to pass.

She traveled at a furious pace and in a savage mood, mocking her tired servants for having journeyed so far in Sweden without a glimpse of the capital. She proposed ironic toasts to the king and the regents, then lapsed into a prolonged bad temper. Every delay was attributed to some deliberate act against her, and at times she feared she would be arrested. Pathetically, too, she harried Pontus for news from the regents—had the post arrived yet, would it be waiting for them at the next stage, was she not about to be recalled? There was no recall, and the horses pounded relentlessly onward; Christina's concession alone could have halted them, and this she would not do. Within three days, the company had arrived back at Hälsingborg, and on the fifth of June, at nine o'clock in the morning, Christina and her household embarked to cross the sea to Denmark. Whatever other feelings he may have had, Pontus was relieved to see her safely on board. The journey had been so fast and furious that his eyes had become infected from the dust of the roads; it was two days before he was able to report her departure to the regents. The little queen had worn the herculean count into the ground. "I am so tired, I can hardly stand up," he wrote. "I've never known such hard work in my life."[10]

Christina took the time to pen a misleading note to the French ambassador at Copenhagen, informing him that she was on her way back to Hamburg, and as a postscript to the letter she added, untruthfully, "I forgot to tell you that they were not going to let me attend mass at Pomponne's. After that, I couldn't in all honor remain in Sweden."[11] Five days later, accompanied by her 138 servants, she arrived. Dirty and dusty and burnt by sun and wind, they looked "like gypsies" staggering in from the road. The Marchese del Monte had reached his limit, and exhaustion overtook him. As for the two women of the household, they were apparently "not fit to be seen"; they stumbled directly from carriage to bed, pausing only to wash, perhaps. Only

Christina remained alert, needing neither rest nor refreshment. Rage had kept her going.

As usual, she tried to lay the blame elsewhere. In a letter to Azzolino, she insisted that the regents had never expected her to go to Sweden at all. All the elaborate formalities were no more than a ruse to persuade the people that it was her own decision not to come; the people themselves had been prepared to welcome her warmly. "They tried to put the blame on me," she wrote. "They were simply terrified at the thought of my presense, and the prospect of my staying a long time frightened them more than death. They simply had to get me out of Stockholm and out of Sweden as soon as possible. The whole of Sweden and the whole world holds them responsible, and the consequences could be very bad for them."[12] But there was no one to avenge her, and the only ill consequences were her own. After eleven months of waiting for her chance to visit Stockholm, she was back in "dreary, stinking, barbaric" Hamburg, while Magnus congratulated himself on having served his kingdom honorably, and having settled, at the same time, an old personal score.

WHEN THE AILING Pope Alexander VII went at last to his eternal reward, Rome erupted in a fest of irreverence. Alexander's lofty moral principles and his ascetic, bookish way of life had earned him no popularity with ordinary Romans, and in higher circles his attempts at administrative reform had disrupted many familiar old patterns of influence. Lampoons hung about the necks of Rome's "talking statues" denounced the deceased pope as a miser and a hypocrite and, worst of all, a bore. A swift conclave produced a general favorite to succeed him: Giulio Rospigliosi, Pope Clement IX, an able and easygoing man, a lover of society, and a noted patron of the arts. Azzolino and the *Squadrone* had worked for his election, and so had the French, though Christina did not know it—they had pretended to favor a cardinal of the great Farnese family. Christina was elated, anyway. The new pope was a personal friend, and he loved everything, or almost everything, that she loved. Above all, he loved the theater, and had written for it himself—the libretto for the *Trionfo della Pietà*, performed on her arrival in Rome, had been his own. More, Rospigliosi had pledged to end the tradition of nepotism, and most wonderful of all, he had appointed Azzolino his Vatican Secretary of State — a dual appointment, ironically enough, with his nephew Giacobo.

Christina decided to ignore this inconvenient inconsistency. The position of secretary of state had been Rospigliosi's own before his election, and she interpreted Azzolino's appointment as a sign that he himself might one day wear the papal tiara. Her sulky endurance of dreary Hamburg gave way to a sudden exuberance, and she announced that her period of mourning for the queen of Poland was at an end. Casting off her black garb, she began to lay plans for a celebration in the grandest style, egged on by the canny Marchese del Monte, spotting his chance to turn a quick profit. The preparations for this "publick Testimony of her Joy" were extravagant, and it was soon the talk of the town.[13] Pulpits rocked as one Lutheran pastor after another fulminated against the evils of the queen's Catholicism, while her links to the Texeira family provided the pretext for an outburst of anti-Semitism. Friends grew anxious, and tried to change her mind. Hamburg's city magistrate paid her a visit, warning of the dire consequences that might be provoked by so public a Catholic celebration in so determined a Protestant town. But, "notwithstanding all that could be said, she would follow her own Capricio." Defiantly, or bravely, or foolishly, she pressed on, laying in supplies of arms and ammunition along with the meat and wine, and on July 25, 1667, the festivities began.

They began with a mass, a solemn pontifical mass of elaborate stateliness and spectacle, celebrated in the great hall of Christina's rented palace. The priests donned special vestments, a seventh candle burned upon the altar, and, in a distinctly unorthodox gesture, several cannon were fired off at the elevation of the host. A grand dinner followed for the queen and her guests, while at the front of the building servants prepared the evening's principal entertainment.

It was a highlight in more than one sense, for it consisted of a huge structure that supported no fewer than six hundred torches, as yet unlit, arranged in the shape of the papal tiara and keys, along with the words *Long Live Pope Clement IX*. Together with the pope's heraldic arms, there were "many curious Figures which represented his Vertues." Above them was displayed "a Picture of the Eucharist in a Cloud, ador'd by Angels, and below was an Emblem of the Church in a Pontifical Habit, treading Heresie under Foot."

It was not exactly calculated to placate local tensions, and Christina's martial preparations showed that her imagination had already flown quite far. A crowd of townsfolk, milling about since the morning hours, increased as the evening approached. People wandered over after their day's work, until "a great number of Seamen, both English, Dutch, and Danes, joyn'd to the Populace of Hambourgh, were crowded into the Space which is before the

Palace, being drawn together by the Novelty of the Sight," and no doubt by the two fountains of wine gushing *gratis* in the square.

When night fell, the torches were lit; the pope's name and insignia and all his "Vertues" flamed out above the crowd, and at this, the first stirrings of revolt were felt. But the cannon fired another salute, the wine continued to flow, and the muttering subsided. Christina herself took the opportunity to step outside to see the illumination. No one bothered her, and presently she went back inside, her guests took their leave, and she began to get ready for bed.

A hail of stones and the smashing of glass stopped her in the midst of her preparations. Through the broken windows came the sounds of a restive, drunken crowd. The torches had lasted for three hours, and the fountains of wine for six; both were now exhausted. Stones battered the windows and doors, chipping away the palace façade. Servants rushed to close the gates. The crowd was becoming a mob.

This much is agreed on all sides, but the details of the ensuing riot, from both inside and outside the palace, remain contentious. Christina herself later prepared an official account of it, which she intended for publication; supposedly, it was the work of her secretary-priest, Father Santini, but she is known to have made a number of amendments to it, and there is little doubt that it was in fact her own work. It is entitled "Report on the Insult Offered by the Populace at the Queen's Palace," and it relays an action of considered self-defense, with Christina herself playing an exemplary, not to say heroic, part. The report begins with the closing of the palace gates:

> We closed the gates and defended ourselves against the fury of the populace with their hail of stones and their pistol and rifle shots. We wanted to fire on them, but the Queen forbade anyone to do so without her express order. No one has ever resisted such a temptation, and no one wanted to fire more than she herself. But she judged quite rightly that we should not arrive at such a resolution except in the greatest extremity. . . . Remaining calm throughout, the Queen acted with great prudence and vigour. . . . But seeing the danger increasing rather than receding, she gave the order that the cannon should be prepared. . . . It seemed very likely that she must now prepare to die. She therefore commanded a salvo of muskets to be fired, because there was no hope of any help. . . . The order was no sooner given than

carried out, and so successfully that we killed a number of people on the square. Several others were wounded.[14]

Santini's report, edited or rewritten by Christina, effectively describes the events as she wished them to have been; perhaps she had even managed to convince herself of its truth. A very different tale is told by her chronicler servant. He begins by noting the ill effects of the hours of freely flowing wine, and the ill-considered firing of blank shots into the crowd:

> . . . but this instead of dispersing, incited them the more, and made them redouble their Insults, till they within were forc'd to shut the Palace Gates. The Windows were quickly battered all to pieces. . . . In this Extremity the Queens Servants had recourse to four Falcons [small cannon] that lay in the Hall, loading them with broken pieces of Brass and Iron, and discharged them against the engraged Populace, of whom they kill'd some and wounded others. This indeed made them recoil a little, but the Blood and Cries of their wounded Companions animated them afresh; so they return'd to the Charge with an Intent to break open the Gates and plunder the Palace. . . . A Score of Lusty Fellows brought a huge long Body of a Tree, that lay at a Carpenter's Door in the Street to make a Mast for a Ship, which they moving backwards and forwards like the Battering Rams of the Antients, did many times essay to break open the Gates.
>
> It was then the Queen knew, but too late, that she had done ill; and the fear of falling into the Hands of this Insolent Mob, did so terrifie her, that she knew not what to resolve on. . . . At last two of the Servants took the Queen by the Arms, and led her out at a Back-door, which opened into another Street, and brought her on Foot, in Man's Apparel, to Monsieur le Chevalier de Terlon's Lodgings. . . . When she was come to her self, she began to eat with a good Appetite, and after a while falling into her accustomed Rhodomontrades, she affirmed, That if they would but have let her appear at the Windows, she should have Thunderstruck all this Rabble, which had lost the Respect that was due to her.[15]

The truth may be one or the other, or somewhere in between. But with an angry mob outside, and eight people lying dead in the square, and "a Score of

Lusty Fellows" trying to batter in the doors, it is quite likely that it "did so ter-rifie her, that she knew not what to resolve on," until two sensible servants bundled her out the back way in disguise, leaving others to dispel the chaos.

Christina's public celebration had been unwise in itself, as she had been formally warned, and the huge fiery images of Catholic loyalty, burning three hours above the square, had been provocative in the extreme. She could hardly have created better conditions if it had been her outright intention to provoke an anti-Catholic riot; perhaps it was. The four little cannon and the supplies of lead and brass and gunpowder no doubt provided a whiff of the ex-citement of battle for this heroine *manquée*, but with such small provisions, she cannot have expected any major disturbance. In advance, as in retrospect, she could be brave in perfect safety, "for when the Danger was pass'd," the chronicler observed, "she play'd the Braggadocio, and was infinitely pleas'd when her Sycophants affirmed, That neither Alexander nor Caesar had ever testified so much Bravery in the midst of so many Dangers."[16]

And ten days afterwards, Christina "play'd the Braggadocio" again in the account of the events that she sent to Cardinal Azzolino. It reveals the same tone of exaggerated defiance that is found in the letters that she wrote follow-ing the death of Monaldeschi:

> I am sending you an account of the celebration which I gave here for His Holiness; here you will see the pure and impartial truth, without exaggeration, and you will see that unfortunately I was forced to make blood flow once there was no more wine; but my consolation is that I did what I could to prevent it and that I was forced into it by the most barbarous attack there has ever been. God has preserved us, miracu-lously, for you must know that I defended myself with about a dozen men against more than eight thousand, you could say against the whole town of Hamburg. . . . I flatter myself that I have upheld the Pope's glory, and my own, quite worthily.[17]

Azzolino's response is unknown, but he cannot have been pleased to learn of the eight people killed and many more wounded, effectively in the name of the Catholic Church; Christina's remark that she was "forced to make blood flow once there was no more wine" is, after all, not very amusing. Nor can it have helped to smooth any diplomatic paths for the newly elected pope in an already virulently anti-Catholic city. Though flippant in tone, Christina's letter to Azzolino reveals, between the lines, a contained alarm, a shaky determination to place herself in the right. It is the uneasy defiance of the guilty man who knows he has done wrong but dares not admit it—not for fear of punishment, but because doing wrong is a sign of weakness, and weakness deserves no respect. The Hamburg mob, like Monaldeschi, had lost "the respect that was due to her." Money might be lacking, friends might fail her, love might die, but respect she would have, and must have, at any cost. It is this that lay beneath her frequent "Rhodomontrades," where she bragged of her power and of the force of her own personality—"if they would but have let her appear at the Windows, she should have Thunderstruck all this Rabble." It recalls her wishful story of the Russian ambassadors, awestruck in her six-year-old presence.

An intelligent woman more than forty years old might have been expected to see through it all. But Christina's intelligence and the varied experiences of her forty years were overlain, and too often obscured, by an anxious self-assertion that blinded her to the obvious and led her repeatedly to the extreme. Her unshakable belief in herself was no more than bluster, a tale full of sound and fury, signifying not very much.

She made one last attempt to show what she was made of. Among the territories from which she had been drawing her unsteady income was the large Swedish island of Öland, and to her representative there she now gave orders that all trade goods and all ships from Hamburg were to be seized. As there was in fact no trade between Hamburg and Öland, and consequently no trade goods and no Hamburg ships in the island's harbors, it was largely an academic revenge, but it did inconvenience the very people with whom Christina most needed to be on good terms, namely, the Swedish regents. Angered and embarrassed by this hostile action toward a friendly power, they sent a swift countermand to the island's governor. Christina retaliated by attempting the same tactic with her representatives in other territories; infuriated, the regents revoked her orders again. The authority was theirs, as the lands belonged legally to the Swedish crown; Christina claimed the rents but was effectively no more than a landlord. The regents dealt her a final insult by forbidding her to set foot in Sweden again until the boy-king had come of age, leaving her to stew in her own impotence, with only the cool breath of remorse to remind her of what might have been.

MIRAGES

CHRISTINA SAT STEWING in Hamburg for fifteen further months, while envoys struggled vainly to persuade the regents to reconsider. She had not yet given up. There were still financial affairs to settle, and the *riksdag* would meet again; if the little king were to die, the people of Sweden would surely want her back. Christina's hopes were still alive, but it was left to her agents to act on them, for by now she had neither strength nor spirit to do battle again in person.

She was not well. To her old fevers and fainting fits were added recurrent sore throats and migraine, and "a strange pain in my right side, which makes all my movements painful, even breathing. The doctors say there's no remedy," she wrote to Azzolino, "but they're all idiots. You'll remember I had something similar in Rome, and I cured it with milk. I can't get goat's milk here, so I have to make do with cow's. As for asses, there's only the two-legged variety in this town."[1] By now she was losing three or four days every week to illness and fatigue. The local water being doubtful, she had taken to adding cinnamon to it and also ice, "which leaves the locals absolutely astonished." She was still drinking the small beer that she made herself, though her physician Macchiati regarded this as the cause of half her troubles. He himself could only suggest bleeding her, and in the end he did so regularly—from the foot, apparently. On each occasion it seemed to do her good, and for a few days she would regain her appetite and walk about confidently, and at night fall easily into a sound and restful sleep. But whatever benefit she may have had from the bleedings, she lost again through the intense cold. Christina's years in Hamburg were some of the hardest years of the century's "little ice age"; even in the middle of August, the hardy local people were forced to keep fires blazing in their houses. In the winter months, the cruelty of the weather was "insupportable," and Christina had to keep her ink by the fire as she wrote to keep it liquid at all. "My fingers are so frozen," she said, "that it's all I can do to hold my pen. I think everything's freezing, even my soul, in this godforsaken place."[2] She made things worse herself by refusing to wear a hat or any furs when indoors—"I can't stand wearing furs," she said—and by sleeping, and often reading all day, in a room that was barely heated at all—"because I can't stand stoves, either."

She missed the warmth and vibrancy of her rich life in Rome, missed her lively circle of friends, and her daily involvement in the politics of the papal city. A gray trail of boredom seeps from her letters. At one particularly low point she reported a rumor that the Danes were about to invade Hamburg; this she dismissed with a scoff, but added that she wished they would—an invasion would at least enliven the place. "It's the only possible amusement we could get here," she complained, "and we're not even going to get that."[3] She took little fresh air and, for once in her life, no exercise at all, preferring to pass her time reading, or writing letters, many to do with her business affairs, but most of them to the cardinal.

Everything about her seemed to add to her misery. She loathed the weather, the water, the buildings, and, above all, she loathed the people themselves. "Don't imagine," she wrote to Azzolino, "that there's any difference between wild beasts and Germans; and I can assure you that of all the animals there are in the world, there is none that is less like Man than a German."[4] The German Jesuits, she wrote, were old and lazy and as cold as the climate, and the cardinals were all drunks. As for German surgeons, they were more dangerous than sword wounds. "Curse the place!" she exclaimed, "and the stupid brutes it produces!"[5] Noisily scornful, Christina had overlooked the fact that she was more or less German herself. But this was not the time to be reminded of it. She wanted nothing to do with the place. It was Italy that she was longing for, Italian sun and Italian friends, and one Italian in particular.

For behind Christina's frustration with Hamburg, and her many health troubles, lay the deeper ill of a broken heart. The tears she had wept on leaving Rome had been more than the tears of a temporary separation. Her love affair, whatever it had involved, had come to an end. In an unrecorded conversation, or more probably argument, the cardinal had made clear that, though friendship might continue, *amour,* and every hint of it, must cease. Christina's protestations were to no avail, and her distress is palpable. "The people who told you I tried to sleep are mistaken," she had written to

Azzolino on the road that led from Rome. "My eyes were filled with tears, and not with sleep." He held firm—to a degree, replying that although he did not really want her to stay in Rome, neither did her want her to remain in Sweden. "What!" she replied. "Do you want me to stay away from Rome for good? Can you think that I would ever do so? Your words have wounded me. You have made me really anxious. Tell me what you mean, and believe me, I would rather live on bread and water in Rome with a single servant to attend me than have all the kingdoms and treasures in the world and live anywhere else."[6]

The cardinal's reply has not survived, but it is clear that Christina was now a much less welcome presence in his life than she had once been. It may be that he was attempting to live a more pious life, in keeping with his priesthood, or he may simply have been growing a little tired of her, and wanting to establish some physical distance between them. Certainly their relationship had become something of a political liability to him, particularly in the wake of the Monaldeschi affair. For a time, Christina had been an asset, providing an anchor for his floating *Squadrone*; she had even steadied him personally, drawing him away from the sexual escapades that had dimmed his brilliant reputation. But her own behavior was hardly conventional, and besides, the French and Spanish pamphleteers had done their work well: whatever the truth, the cardinal was widely believed to be, or to have been, the queen's lover. Within the morally cautious Vatican of the Counter-Reformation, this might be enough to stop his progress. He was now secretary of state, the very position the pope himself had held before his election. It may be that Azzolino thought as Christina did, and regarded himself as *papabile*.

Whatever the reason, Christina's Hamburg letters show that there was now a breach between them, a breach of the cardinal's making, and one she regretted bitterly. In page after page, she relayed her "misfortune," her "misery," the "mortal blow" of their parting, and the "most tender passion" that she continued to hold for him. She received his letters with "an excess of joy." They were "life or death" to her, and any slight delay in the post, or any hint of indisposition on Azzolino's part, sent her into transports of anxiety. "I fear for your health more than for my life," she wrote to him, "and I ask God to take from my own allotted days to prolong your own."[7] They wrote to each other every week, Christina always on a Wednesday, sometimes twice in the same day, to catch the weekly post to Rome. Azzolino's letters were always in Italian, and hers always in French, but their most private phrases were concealed in their own numerical code.[8] The cardinal was also kept informed by two members of the queen's household, the Marchese del Monte and Father Santini, the latter's elegant handwriting no doubt a welcome contrast to Christina's impossible scrawl. While Christina continued to protest her love—"Everything is *frozen* in this country *except my heart*, which is more *ardent* than ever"[9]—Azzolino attempted to divert her thoughts elsewhere—principally, it seems, toward her eternal salvation. He had apparently been saying that their relationship offended religious proprieties, perhaps even that it was sinful. Christina responded passionately:

> I have no wish to *offend God*, nor to give *you* any *cause for offense*, but that cannot stop me from *loving you until death*, and since *your piety* frees you from being *my lover*, I free you from being *my servant*, since I want to *live and die your slave*.[10]

In the time-honored way of the sympathetic jilter, Azzolino offered the purer flame of friendship in the place of love. Friendship would not compromise their chances in the next life—Christina must think of that. One bitterly ironic response indicates that she was not deceived:

> Thank you for your expressions of friendship [she wrote]. I only wish I could believe them, but you have already made it perfectly clear that I should not read too much into what you say. I am not likely to fall again into the same error that you have so carefully dragged me out of. . . . It is really most edifying to read your religious reflections on everything that happens. I have no doubt that your thoughts were all of God while you were watching those two young actresses at the French Ambassador's the other night. It must have been mortifying for you to have to look at them. I suppose you went in the hope of converting them. . . . As for me, I think about death all the time, so please stop preaching to me about it. I don't like sermons.[11]

Her grief was real, and she wept many tears, but her active nature saved her from despair. Within a few months she was able to write to Azzolino about returning to Rome "in glory and triumph," and laugh with him about her new admirers in Hamburg, supposedly all captivated by "my beautiful big blue eyes." Besides, there was plenty to distract her. Hamburg was by no means the cultural desert she had pretended, and in any case, there was a good deal of business that still required her attention. Lorenzo Adami, "who has surpassed all my hopes with his hard work and capability," was still looking after things in Sweden, and he had been joined by a new man from Pomerania, Bernhard von Rosenbach, whose impossible task was to persuade the regents to admit the queen to the next Diet. Her hopes were now pinned on a cash settlement for the rents she currently claimed. This, she felt, would free her from the uncertainties of war, a threat brought home to her lately by the destruction of two hundred Dutch ships, laden with the property of Hamburg merchants, by the enemy English navy. The city was resounding with the "terrible uproar" of new-made bankruptcies.

She roused herself sufficiently to pen a halfhearted, perhaps half-guilty, encomium for an old acquaintance, the philosopher René Descartes, and her late "illustrious friend" Pierre Chanut, the dazzled and doting ambassador who had turned from her, bewildered, after the death of Monaldeschi. But she did not want to dwell on them too much, for the thought of them rankled a little, reminding her of her greatest failure of mind, and her greatest failure of humanity.

And in any case, there was life to be lived in the meantime. She was not just going to wilt like a faded rose into the cold Hamburg soil. A new flame had been struck, not love, nor philosophy, nor religion, but an ancient mixture of them all, for Christina had rediscovered alchemy.

She had learned of it as part of the Renaissance education of her girlhood, but, curiously, given her vivid imagination and her interest in other occult learning, it had never particularly interested her. There had been a large number of alchemy manuscripts in the library that had come to her as booty from Prague, but these she had casually given away. Now, however, excited by stories of fantastic alchemical successes, or simply latching on to a new idea in a dull time, she turned her gaze to "the dark world."

It was not a world of seances and black magic, though there was smoke, and there was fire. The occult learning of Christina's day was not much different from the fledgling empirical sciences. There were many charlatans, and some empiricists dismissed the "dark arts" altogether; but many more kept a foot in both camps. It made sense to do so. Occult learning, like empirical science, was a search for understanding of the natural world, and for control of it. Alchemy served as a kind of protochemistry; different substances were observed to interact, to change, to form new substances, all of them, so it was believed, expressions of the same essence. With the right knowledge, any substance, such as a base metal, could be transformed into any other, such as gold. This knowledge, the famed "philosopher's stone," Christina now decided to seek.

"Chemistry is the anatomy of nature," she wrote, "and the true key that opens every door. It brings riches, health, glory, and true wisdom to whoever understands it." There were plenty of alchemists at hand clamoring that they understood it perfectly, and Christina chose the loudest of them, Giuseppe Francesco Borri, a renegade Milanese physician who had recently been making an easier living among the gullible rich. Borri claimed to be in personal communication with the Archangel Michael, and while still in his home city had formed a new sect, its first tenet being total obedience to "the Most High," namely, himself. Summoned to Rome to appear before the Inquisition, he had beat a hasty retreat to Holland, and there had spent several years proselytizing successfully to the surprisingly credulous Dutch. The Inquisition had condemned him *in absentia,* and burned him in effigy in the Campo dei Fiori, along with copies of his heretical tracts. But the smoke signals from Rome did not deter Christina. Borri was just the kind of devil-may-care character she had always most enjoyed, and now, angel-prompted, he came to perch a while beneath her incautious wing.

Azzolino heard the news and wrote to her in alarm. Borri was a charlatan and, much worse, a heretic. The Church's most celebrated convert could not be seen to be dealing with him. For a moment she was swayed, and forbade Borri her palace, but before she had even relayed this to the cardinal, the order had been rescinded. Unwilling to give up her latest protégé, she sent Azzolino a letter of disingenuous defense:

> I have asked all the priests here, and they have all assured me, unanimously, that I should revoke the order and that there was no reason to give it in the first place . . . but anyway I've forbidden him to enter my chapel or to attend mass here. All the priests here told me to revoke the order. If I have done the wrong thing, I ask pardon of His Holiness, and I'm sure he will grant it, seeing my ignorance.[12]

And she added, as if the defense required more weight, that the Marchese del Monte owed his life to Borri. He had been terribly ill, she said, and all the other doctors had been "absolutely at the end of their Latin," and though the marchese had not wanted Borri to attend him, nevertheless he had been "forced" to accept his ministrations, with wonderful results.

The marchese's life was a gain, perhaps, but the "riches, health, glory, and true wisdom" that Christina had hoped to reap had defied all Borri's conjurings. For her two or three thousand crowns, he had been able to produce "only Cinders and Smoak."

CHRISTINA HAD OTHER distractions, in any case. At the cardinal's suggestion, she had begun to write the story of her life. It was not her first attempt to do so. Ten years before, prompted by the Duc de La Rochefoucauld's literary self-portrait, she had begun a memoir of her own, in Italian. Defeated by the language, or by her own unsteady temperament, she had abandoned the project and destroyed her rough-hewn pages. The intervening decade had seen little improvement in her written Italian — she was still relying on Father Santini, and sometimes Azzolino, too, to translate letters for her — and this time she decided to write in French. It was evidently intended to be a substantial work, and its title was certainly grand, though *The Life of Queen Christina, Written by Herself, Dedicated to God* does have a straightforward, Scandinavian ring to it. Azzolino was relieved to have provided some occupation for her at this restless time. She relayed her progress to him cheerily:

> I have been working on the *Life* you asked me to do, but the draft is such a mess that you won't be able to read it. And I can't bring myself to make a fair copy — in my handwriting, it wouldn't be any better than the draft, anyway. And I don't want to give it to anyone else until you've seen it, since I can't speak of myself without speaking of you — though you're not mentioned by name — still, you would be recognizable.[13]

The early chapters, though not long, are mostly about Swedish history; gradually they move on to the lives of Gustav Adolf and his melodramatic widow. Christina acknowledges her father "a great man," and her mother "a woman with all the virtues and weaknesses of her sex," but not until the eighth chapter does she really begin to write about her own life. Despite a number of humble asides to the Lord, her tone is very self-assured, and she is not above telling a few lies to exaggerate her accomplishments. The dedication of the work "to God" was not the result of hubris but was, rather, a reference, pious or ambitious, to Saint Augustine, whose own celebrated *Confessions* were similarly dedicated. Christina liked to compare herself to the great men of the ancient world, but perhaps, too, she began the work in a spirit of humility, intending it to mirror Augustine's passage into the light of true religion. As it stands, her *Life* is by no means a spiritual document; though the first chapter reads almost as a prayer, its tone is ambivalent. Even before God, Christina counts her worth: "Lord, You are everything, and I am nothing, but I am a nothing that You have made capable of adoring You and possessing You. I am, by Your grace, the most favored of all Your creatures."[14] And she goes on to elaborate: her cleverness, her physical strength, her exalted rank, are all evidence of God's favor.

The *Life* is a revealing document, showing a gifted and even noble mind overlain by deviousness and self-deception. It is at times quite defensive, for in writing it Christina was obliged to confront herself. Weighed in the balance, she could not have failed to notice that, on most counts, she was found wanting. Her achievements, at the the age of forty-one, were negligible, and compared with those of her great father, nothing at all. Her power was illusory, and her pride the insistent pride of the weak. Despite her grand conversion, she shows no real sign of spiritual development, and her regret over her abdication is palpable.

> The present King Karl [she writes] has no other claim to the Swedish throne than what I gave to the king his father and to him, so there is no one else in the world who has any rights at all to the Kingdom of Sweden, apart from Karl and myself.[15]

Apart from her ready intelligence, it is above all Christina's immaturity, striking and rather sad in a woman of her age, that starts out from the pages. If she was ever enthusiastic about the project, her enthusiasm did not last, for her *Life* ends abruptly, still in her childhood years. She left it unfinished, and took it up again fifteen years later, only to leave it unfinished once more.[16]

Christina's *Life* was intended to serve a second purpose apart from keeping her occupied. Azzolino was harboring ambitious plans for her, in which, to some extent, she herself acquiesced. Her chance for the crown of Naples had vanished long ago, and her own revived interest in the regency of the Spanish Netherlands had been dashed only months before by an unexpected

French invasion. But since the abdication of her cousin, Jan Kazimierz, his throne had been vacant. It was an elective monarchy, and the race was now on to find a new king for Poland-Lithuania, the largest state in Europe.

The Lithuanians, in the north of the kingdom, favored a Russian successor. This did not please the pope, who was not anxious to see an Orthodox king ascend a longtime Roman Catholic throne. A German prince was suggested, then an Austrian prince, and even Giacobo Rospigliosi, the papal nephew. King Louis wanted the throne for a Frenchman. He had in fact paid Jan Kazimierz to get him out of the way—a "very decent" sum, as Jan Kazimierz himself observed. The Poles disliked all the candidates: they did not want any Frenchmen, nor any Russians, and no Germans, either. But they wanted a man, a man as red-blooded as he was blue-blooded—they had had more than enough of their former-monk, former-cardinal, former king.

Christina was not a man, but this did not deter her. She was at least not French, nor Russian, nor German, at least not officially. But she was Catholic, and, like Jan Kazimierz, she was a Vasa, the last legitimate representative, in fact, of the Vasa line. Her claim to the throne in consequence was quite strong; the Swedes had recognized as much on the death of the Polish queen, fearing a marriage between the cousins that might unite the two crowns and re-Catholicize Sweden by force. Christina was not passionate about the prospect. She liked the idea of it, happily imagining herself a real queen again, with a court, and subjects, and a royal income. But in practice, it would mean leaving Rome for good, leaving the light and the warmth, and, above all, leaving the cardinal. This was more than she was prepared to do, and at one point she went so far as to say that she would "become a Pole" only if Azzolino became one, too.

It was not likely. The cardinal was doing very well in Rome, and had hopes of doing even better. He had much to gain if Christina should be elected, and nothing but her problematic self to lose. Poland was an important crown, and Catholic zeal may have played its part in Azzolino's calculations, but it was also a good idea to keep Christina occupied and out of the way, and he may even have hoped to play a Mazarin-like role, presumably by correspondence, in the Church's great eastern kingdom. Whatever his reasons, the cardinal set to with a will to promote Christina as Poland's future queen, relying for local negotiations on Monsignor Marescotti, the unhappy papal nuncio in Warsaw. At the pope's behest, Marescotti had already two rival candidates to promote—the French Duc de Neubourg and the German

Prince of Lorraine—and he was not at all pleased to find himself with a third. Azzolino battled anyway to put Christina's case: there was no prince more capable of leading an army, he insisted—well, no one apart from the Prince de Condé, and anyway no one more capable of enduring the hardships of army life—the queen hardly needed any sleep, or even anything to eat. She was so full of martial courage—all she needed was an opportunity to prove it to the world. It seems he had been reading a draft of Christina's *Life*, where she had been describing her "indefatigable" self of twenty-five years before. He went as far as promising that she would accept a husband, and almost promised she would produce a number of little Vasa princes as well. As for the question of her sex, he insisted perversely that "everyone regards the queen as a man already, indeed as better than any man."[17]

Marescotti protested. The queen could not be a serious candidate. She was almost forty-two, and she was—well, not inclined to marriage. It was widely known that she was living off the charity of the pope. And her way of life, her dress and her speech, and all the rabble around her palazzo, and inside it, too, not to mention the Monaldeschi affair—it would all stack the odds impossibly against her. In a long series of letters lasting many months, Azzolino did what he could to deflect the accusations, or to deny them, or to ignore them.

He might have saved his breath to cool his *pappa*. As Marescotti knew very well, there was no real chance of Christina's election. Azzolino had promised him a cardinal's hat if she should be successful, but as each of the other candidates had promised him the same, this was no incentive for extra exertion in her cause. The pope had conveyed his own support of "this heroine remarkable for her piety, her wisdom, and her manly courage," but had rather undermined it by insisting that it be kept secret unless and until Christina had won the day. Effectively, he wanted to ensure that whichever of his protégés was elected, the resulting gratitude and diplomatic support would flow back to Rome. Marescotti, shaking his head, took the secret to his bishop, and there confided it under the seal of the confessional. The bishop, it is said, on hearing Christina's name, gave a sharp intake of breath, then crossed himself swiftly and raised his eyes to heaven.

Already besieged, Marescotti now had the queen's own correspondence to deflect as well. She asked him to remember that she "surpassed all the other candidates in birth, and perhaps in some merit," saying that she would never have left the throne of Sweden had that been a Catholic country. The Poles should not do her the "injustice," she wrote, of choosing "some foreigner less worthy than she to occupy the throne of her ancestors." Finally, making a virtue of necessity, she pointed out that, at almost forty-two, she was not "of an age to marry," nor had she any "inclination" to do so. This was supposedly an advantage for the Poles, since they would have the chance of electing another king when her own life should have run its course.[18] Though this was directly counter to Azzolino's insistence that she would certainly marry, Marescotti was not concerned. The point, he felt, was academic.

Christina cared enough to follow the proceedings as the months progressed. "The French are only playing with the Duc de Neubourg," she wrote to the cardinal. "They really want Condé to succeed, but their game is so subtle that the thick-headed Germans can't see it. If His Holiness supports me, and if my sex is not an insurmountable obstacle, I think I have a good chance of succeeding. By the way, I have recommended the Duc de Neubourg myself—a good joke, isn't it?" But in a sense, she herself was only playing, perhaps on the off-chance of success, more probably to please Azzolino. She did not much care whom the Poles chose for "their master, or rather their slave. If God calls me to this throne," she wrote, "within two years or even less I will take on the Turks, and make my name resound with glory," but her enthusiasm lasted no longer than a sentence: "But if God wills otherwise," she added, "I shall be just as happy."[19]

Years later, when the need for propaganda was far in the past, Christina returned to the composition of her *Life*. She added some new thoughts about ruling, thoughts that she had not had, or not wanted to admit to, in the days of her efforts to claim the Polish throne. They are curious thoughts for a woman who believed that she carried her right to sovereignty personally, within herself, and who had sought no fewer than four thrones, including the one she had herself relinquished. Perhaps they were a belated justification of sorts for the abdication that she had so often regretted, or they may have been an excuse for her general failure to accomplish anything remarkable. In a way, they are a recognition of her own weaknesses, all of them attributable, or so she had decided, to the single great fault of being female.

> Women should never be rulers [she wrote], and I am so convinced of this that I would have barred my daughters from the succession, if I had married. I would have loved my kingdom more than my children, and it would have been a betrayal of my kingdom to leave it to girls.

And I should be believed all the more since I am speaking against my own interest—but, then, I have always made a point of speaking the truth, whatever it has cost me. It is almost impossible for a woman to be a good monarch or a good regent. Women are too ignorant, too weak in body and soul and mind. Everything that I have seen or read confirms that women who rule, or who try to rule, only make themselves ridiculous one way or the other. I myself am no exception, even though I was groomed from my cradle to be a Queen. . . . The defect of being female is the greatest defect of all.[20]

The Poles, at least, were ready to agree. Though Maria Ludwika had been strong and clever and popular, they wanted a man to replace their king, and in due course they found one. Azzolino may have been disappointed, but Christina was not. When Jan Kasimierz died, in a Parisian abbey, she remarked, "I am glad to hear he has at least died among men. If he had died in his own country, he would have died among beasts."[21] She contented herself, or discontented herself, with contesting his will for years to come. Jan Kasimierz had owned lands in Italy, and Christina felt that she could do with the rents from them.

CHRISTINA HAD BEEN two and a half years in Hamburg and she was ready to go home, not to Sweden—never again to Sweden—but home to her Palazzo Riario, among the ilex and the sweet-smelling jasmine. Late in the summer of 1668, she began to make the preparations. She had nothing to show for her time of exile. There had been no progress on the question of her income since Adami had worked his miracles, without her assistance, almost three years before. In time of war, it would not be guaranteed, and her own impolitic behavior had made the Swedish regents less inclined to help her now than they had ever been. True to form, Christina persuaded herself, in the teeth of the evidence, that it had all been a resounding success. "Things are going so well for me," she wrote to the cardinal. "It's all going just the way I want." By the end of the autumn she would return, she said, "in glory and triumph" to Rome.

Azzolino knew enough to read between the lady's loud protesting lines. He shrugged his shoulders, but his cousin Adami, less wise or less experienced, decided to disabuse the queen of some of her more costly delusions.

He drew her attention to the doubtful activities of the Marchese del Monte, in which, apparently, even Father Santini was now involved. Refusing to accept that she had been duped yet again by someone whom she had trusted absolutely, Christina sent a furious flurry of letters to the cardinal, berating the cousin he had recommended to her. Once a marvel whom she "could not admire enough," the loyal *Marchegiano* had metamorphosed into "that infamous Adami." "If you knew how badly he has served me," she wrote, "you would hardly be able to keep yourself from stabbing him with your own hand."[22] But she stayed her own hand long enough to hear the story out, and at length was obliged to believe him. Her fury was now transferred to the "treacherous, criminal, thieving" marchese, and he was swiftly dispatched to Rome for judgment.

The marchese set off unconcerned, carrying in his own pocket the queen's letter of condemnation. It was effectively a list of his misdeeds, loudly entitled "Principal Accusations Against the Marchese del Monte." The list, in numbered paragraphs, was as follows:

1. *That he is the Cardinal's deadly enemy.*
2. *That he and Don Matteo Santini have been selling my interests and secrets to anyone curious enough to buy them, and that they are the Cardinal's deadly enemies.*
3. *That they betray me in every way that servants can betray their mistress.*
4. *That he sent goods worth five thousand écus to his wife.*
5. *That he gave a thousand ungari to the Chevalier Castiglione, when he was here, to take to his wife.*

And she concluded, "On all these counts, the said marchese must exonerate himself, or, if he is guilty, he must die to expiate such enormous crimes. That is the reason for his journey to Rome. It is up to the Cardinal to condemn him or to find him innocent, and up to me to carry out his sentence."[23]

Despite the pseudo-legalism, the list is not very specific, and it might be almost comical if Christina had not offered to put the marchese to death. It is also a list of surprisingly mild failings, considering the other crimes imputed to him at about this time—embezzlement, extortion, seduction, kidnaping, and even attempted murder. Here, as elsewhere, Christina does not seem to have much cared what her servants did, as long as they were loyal to her. Del

Monte himself cannot have taken the whole thing too seriously; not only did he carry the letter himself to Rome, but he also handed it over, in person, to the cardinal.

The letter was not really any kind of "case for the prosecution." More than anything else, it was a latter-day defense of Christina's own actions in the Monaldeschi affair, and it circuitously reveals her continuing need to justify herself on that count. The unspecified betrayals mirror those of that other, less fortunate marchese, which were never clarified. They are equally leveled at Father Santini, who seems to have escaped scot-free. And the letter, addressed personally to Azzolino, refers more than once to "the Cardinal," as if it had been intended for other eyes, possibly even for publication. Christina must have known, as the marchese clearly did himself, that Azzolino would never condemn him to death; she was even sending him with business instructions to relay at the same time. But he had proved disloyal and untrustworthy, outwitting her again and again. She did not know how to deal with him, so she sent him, as she now referred all her difficulties, to Azzolino.

So del Monte set off for Rome, "to exonerate himself or to die at your feet." He did neither. Azzolino no doubt reprimanded him, but he continued in the queen's service, defrauding her cheerfully for another twenty years, in fact for the rest of his life. Adami resigned, but Santini carried on, blessing Christina when in priestly mood, and otherwise writing letters for her in his beautiful copperplate handwriting.

GLOry Days

OWARD THE END OF November 1668, Christina arrived back in Rome. "I don't expect you to come to meet me," she had written to Azzolino. "It will be enough if you don't leave Rome yourself when I arrive." But when her entourage reached the little castle in the Sabine hills where she was to pause before entering the city, Azzolino was there, with twenty-three of his brother cardinals, and Christina's friends in force, and thousands of flowers sent from the pope himself, and a train of servants ready to prepare her a great, grand banquet. After a long and unhappy absence, she was welcomed home at last, and the following day she rode into the grounds of her own Palazzo Riario.

The garden was taking on the aspect of a soft southern autumn, with the last bright flowers fading into the green and gold. Inside, fragrant woodfires warmed every room, and Christina wandered, smiling, from one to the next, stopping to look again at her pictures, or to stroke a cool marble arm or head. In her private room hung her favorite work of all, the equestrian portrait Sébastien Bourdon had painted of her as a young queen in Stockholm, when strength and confidence had pulsed in her veins and everything had lain before her. There were few enough of those dreams left to her now, but she was home, safely ensconced in the middle of what she loved, and of those who loved her. The money from her rents was now as well assured as it could be, thanks to her new administrator, Johan Olivecrantz, who had replaced Azzolino's cousin, Adami. Olivecrantz was a man in the same capable, honest mold, and Christina was now receiving a reliable five thousand riksdaler from him every month, a solid if not sumptuous amount. To this was added a small pension from the pope. As a kind remembrance of her friendship, he had awarded her the sum of twelve thousand scudi per annum; it added an extra quarter or so to her income from Sweden.

The pontificate of Clement IX was a golden time for Christina and Azzolino. The secretariat of state was one of the highest Vatican offices, and the cardinal was an able occupant, balancing his prestigious diplomatic work with the quieter steps of administrative reform. Christina received her own honors: private visits from the pope, invitations to every grand event, a prominent place at official Vatican ceremonies. The pope was her friend, and as eager as she was to see a flourishing artistic life in the city. There was not the money that there had been in the great early days of the century, but his outlook was liberal and his modest purse always open. They were generous and generous-spirited days.

With Clement's permission and encouragement, Christina now proceeded to promote a new theater. It was intended for public performances of opera, and it was the first such theater in the city. It filled a decided gap, since, where opera was concerned, Rome had been lagging behind other Italian cities. Lacking a secular court, and periodically hampered by popes unsympathetic to artistic extravagance, Rome had developed instead a tradition of private performances in the houses of the rich. The new public theater had been established by a member of Christina's own household, the Comte Jacques d'Alibert, officially her French secretary, but in fact a talented and active impresario. He hoped to fund his venture by public subscription, in the style of the Venetian theaters, and intended to employ independent musicians rather than borrowing those attached to private houses. Summoning patronage and investment from every available source, D'Alibert had built the theater on the site of the old Tor di Nona prison, and the name was retained for the new building, which became the Teatro Tordinona. Christina was not rich enough to be its only patron, but she ensured its financing through her wealthy friends in the Vatican, and half the operas commissioned during its short life were dedicated to her—like most other Roman entertainments, the theater met its end when a villainously puritan pope arrived on the scene.

The prelates came as well to her private soirées at the Riario, evenings of music or drama, often interspersed with an hour or two of pseudo-academic debate. Pasquini, a brilliant harpsichordist as well as a composer, was the most fashionable of the many musicians she employed; Christina called him her "Prince of Musicians" and would often stop her carriage in the street to chat with him if she saw him passing by. But among the discerning it was the young "foreigner" Alessandro Scarlatti, a frequent visitor from Naples, who claimed the highest laurels. Though he lived with the Bernini family, for several years he was attached to Christina's court as her *maestro di cappella*, directing her orchestra, playing, and composing. He dedicated his second opera to her—as well he might; after a prim papal banning of his first, she had stormed a Jesuit College with a band of Swiss Guards and commanded him to begin playing, anyway. She wrote one opera libretto herself, in collaboration with a more experienced friend, and asked Scarlatti to provide the music for it, but it seems that he declined the honor.[1] His years in her employ overlapped with those of another gifted young "foreigner," Arcangelo Corelli from Ravenna. Corelli was a violinist, indeed the premier violinist of his day, and a leading composer for that instrument. He was living at the palazzo of Christina's friend Cardinal Ottoboni, whence he was regularly borrowed to lead the queen's own orchestra. In Scarlatti's absence he conducted a legendary performance at the Riario one evening, in honor of the English ambassador: with 150 musicians, most of them violinists, the ensemble was the largest Rome had ever seen.

"There be excellent Musicians at Rome," one of Christina's servants recorded, "who are most them Castrated to preserve their Voices."[2] The queen made use of them all, poaching them whenever she could from their bread-and-butter duties at the Sistine Chapel. But her favorite castrato, Antonio Rivani—*Cicciolino*—she kept in her own employ, paying him so well that he was quite soon able to buy a large estate in the country, which, however, she did not permit him to enjoy often. Lent for a time to the court in Savoy, Cicciolino made the mistake of overstaying his leave of absence, and even considered not returning to Rome at all. Christina's response was true to form:

> I want it to be known [she wrote to the agent charged with his recapture] that Cicciolino is in this world for me and only for me, and that if he doesn't sing for me he won't be singing for long for anyone else, no matter who they are. Get him back whatever it costs. They say he's lost his voice. I don't care. Whatever has happened, he'll live and die in my service, or he'd better watch out!'[3]

Cicciolino returned to Rome, where, with occasional reprieves at his country house, he continued to sing for Christina to the end of his days. From the heavenly choir, even she could not reclaim him.

And when the music stopped, there were plays, Italian and Spanish, twice a week, always popular, always well staged, and, according to contemporaries, generally *sporchissime*—very dirty. For one whole year she managed to engage the famous Scaramouche, on leave from Paris, whose own productions did nothing to raise the tone. There were sometimes ladies present—aside from the queen herself—but on the whole they concealed themselves within small curtained boxes, with the visible audience comprising mostly cardinals.

But the most illustrious of all Christina's artistic acquaintances in these years was not an actor, nor a playwright, nor a musician, but the great Gian Lorenzo Bernini, Neapolitan born, long resident in Rome, most lavishly gifted, and as pious as any cardinal to boot. A legendary figure even in his youth, he had been the virtual creator of the Baroque style, and he was now its uncontested master. John Evelyn has left a picture of his capacities:

> A little before my Comming to the Citty, Cavaliero Bernini, Sculptor, Architect, Painter & Poet . . . gave a Publique Opera (for so they call those Shews of that kind) where in he painted the Seanes, cut the Statues, invented the Engines, composed the Musique, writ the Comedy and built the Theater all himself.[4]

Scene painting was by now behind him, but Bernini did not disdain the lesser arts of design and decoration. Christina herself possessed a beautiful gilt-framed mirror made by him, and she had financed her journey to France on the strength of the magnificent carriage he had made for her at the pope's request. The very face of Rome was largely of his making; his churches and fountains stood everywhere, and at the heart of the city, before the great Basilica of St. Peter's, his wonderful oval piazza curved around Rome's oldest obelisk, seized from Egypt in the days of the emperors. As for his sculptures, in the words of the overwhelmed Evelyn, they were "plainely stupendious."

Christina thought so, too, and made no bones about her reverence for Bernini's genius. She visited his workshop frequently, and was once caught bestowing a kiss upon a discarded smock which the master had worn. She loved his fiery temperament and his quick imagination, and he was good company, too, "a very acute conversationalist, with a very special gift of expressing things in words, with his face, and by gesture."[5] An artist's status at the time was generally humble—not so long before, the young Andrea Sacchi had been listed in his master's books along with "three slaves, a gardener, a dwarf, and an old nurse"—and although Bernini was famous and rich, Christina's deference to him is nonetheless remarkable. In social or diplomatic matters, she was extraordinarily, almost morbidly, insistent on her royal status, but where artists were concerned, she gave place without demur. It was not a question of *noblesse oblige*—in any other respect, she could never have condescended so far. Rather, it was a genuine recognition of abilities that she valued, and did not possess—and, no doubt significantly, abilities to which she herself did not aspire.

Though she had no creative gift, there is no doubt of her own artistic sensibility. Bernini himself paid her the highest compliment he could have paid, and it would have pleased her, perhaps, above any other. While visiting Paris, where he had come to complete the design of the Louvre, he was asked about his friend the queen of Sweden, whose reputation in France had never risen beyond her savage deed at Fontainebleau. Bernini made no comment about Monaldeschi's death, but replied simply to his inquirer, "She knows more about sculpture than I do."

Admiring the man, and admiring his work, Christina was nevertheless unable to become his patron. Sculpture, let alone architecture, was too expensive for a northern Pallas with a few modest rents from Norrköping and Pomerania. After twenty years, she possessed only a few of his drawings and paintings, none important, none especially valuable. Curiously, she declined Bernini's own gift of his very last sculpture, a marble bust, larger than lifesize, of the *Salvator Mundi*. By way of explanation, she told him that she was "not worthy" to receive it—not worthy of the artist, perhaps, rather than of the subject, for she was still, as she had always been, utterly without devotion to the person of Christ.

There is a bronze bust of Christina herself that is attributed to Bernini's hand. She stares alarmingly out from it, and this, together with her tangled Baroque wig, makes her rather Medusa-like. The two young Dutchmen who had once seen her in Paris had been struck by the proud expression of her bright blue eyes, and had remarked that only the very bold could have endured her gaze for long. Perhaps, in her happy Riario days, it was, at least sometimes, still true.

Unable to commission sculptures, and unable to carve them herself, Christina decided to dig some up. Inspired perhaps by her archaeologist librarian, Benedetto Mellini, or perhaps by Athanasius Kircher, whose unearthed Roman obelisk was now the centerpiece of Bernini's great fountain in the Piazza Navona, she took to archaeology. The pope granted her permission to excavate the ruins of the palace of the Emperor Decius, infamous for his persecution of Christians, on the Viminal Hill. As yet, archaeology was less an investigation of a vanished way of life than a sort of treasure hunt for valuable flotsam, and Christina found some—no martyrs' bones, but mosaics from the Roman period, and several statues, including a beautiful Venus with a dolphin. In the custom of the day, missing legs and noses were swiftly replaced, though Christina stopped short of adding the fig leaves and draperies normally supplied for the palazzi of Counter-Reformation Rome. Her sculptures were seen as they had been intended to be, green and crimson wall-hangings notwithstanding. They filled three whole rooms, and must have dominated many others, for the Riario was not vast, and there were some hundred and sixty of them, not counting vases and urns and even columns.[6] The collection became as famous as her paintings, and is a surer sign of her own individual taste. None of the sculptures had come with her from Stockholm; they were not tributes or war loot, but had all been chosen by herself. Most were early Roman copies, but she did have several Greek originals, and her favorite of all, which stood opposite Bernini's beautiful mirror, was a bronze head of a young Greek athlete, which she believed to represent Alexander the Great. Not for her the meek and suffering *Salvator Mundi*; heroic Alexander was the model for her life, the image of her own most cherished illusions.

"GOD SAVE OUR Pope Clement, for your good fortune will last as long as his life."[7] So Christina had written to Azzolino from Hamburg, and so it was to prove. Clement's life did not last long. He had been pope not much more than two years when news came, on a bright autumn day in 1669, of the fall of Crete to the besieging Turks. The same night, he suffered a stroke, and though he rallied long enough to make a pious visit to each of Rome's seven pilgrim churches, in the early days of the winter, at the age of sixty-nine, he died. Christina was present, along with others of his friends. He made a tender farewell to her as she wept beside him.

She had lost more than a friend. Clement had been a fellow traveler in all her best artistic endeavors, and his own love of theater and opera had allowed them to flourish in an illiberal time. His personal attachment to Christina had ensured her social standing through much flouting of convention and no small number of scandals; he had even given her a pension. Azzolino had still more cause to lament. He owed his powerful position to the favor of the Rospigliosi pope, and there was no guarantee now that he would keep it.

The months following Clement's death were, in consequence, an anxious time for them both. Azzolino, being a cardinal, was involved on a daily basis in the business of electing the new pope. The conclave lasted more than four months, an unusually long time, and for its duration he remained in the cold Vatican "cells" reserved for the members of the Sacred College. The Vatican became a closed city; the cardinals could not leave, nor could anyone be admitted without a Vatican passport. Christina did not have one, but she was not prepared to wait for a new pope before she could see the cardinal again, and besides, she wanted a hand in the election. Her own Riario was in Trastevere, outside the Vatican boundaries, so she shrewdly rented another palazzo, in Azzolino's name, on the Borgo Nuovo, near St. Peter's, taking a three-year renewable lease at the very reasonable price of five hundred scudi per annum. It was the Palazzo d'Inghilterra,[8] so called because it had once belonged to the English king Henry VII, and it was perfect, being within the Vatican, yet a private residence. She equipped a little study there, and moved in. She was now able to exchange daily, passportless messages with the cardinal. It was a definite bending of the rules, and it did not go uncriticized, but the notes went back and forth, insouciant.

The conclave, as always, was effectively a contest between Spain and France, each seeking to dominate the proceedings and ensure the election of a pope favorable to itself. With unwarranted belief in her own powers of dissembling, Christina offered to work for both sides. Both accepted, though neither relied on her, and neither told her any secrets. The French assumed, correctly, that she would be receiving information from Azzolino, as she had done during the conclave for Clement IX; they felt they might as well hear whatever she heard. The Spaniards sent a spy of their own to wait upon her, one Monsignore Zetina, who relayed to her what he wanted Azzolino to know. In her own mind, Christina inflated her role of go-between to that of mediator between the two great nations. She reveled in the excitement of it all, and did not notice that her own views were not persuading anyone. Azzolino was not succeeding, either. His *Squadrone Volante*, once a vital fulcrum in the political balance, had now divided into smaller, inevitably less influential, groups. With their power had gone their popularity. While the conclave was in progress, a satirical play entitled *Il Colloquio delle Volpi*—The Foxes' Conversation—was staged in Rome, making fun of them all and accusing them of enriching their families at the public expense:

CARDINAL AZZOLINO: It has not been difficult, since we are almost the only people in the Vatican who can read and write. We have more or less had the field to ourselves.

CARDINAL OTTOBONI: The people loathe us. The other cardinals are calling down curses on our heads, and the ambassadors won't tolerate us any longer.

CARDINAL AZZOLINO: It's true. I'm afraid we won't be able to block the election of a pro-Spanish pope. If that happens, I'll betake myself to my land in Le Marche and wait there until the danger and disgrace have blown over—provided my lady the Queen will give me leave.[9]

Azzolino's character was correct. They were not able to block the election of a pro-Spanish pope. At the end of April 1670, after much to-ing and fro-ing on behalf of other *papabili*, the cardinals quite suddenly lighted on Emilio Altieri, and declared him the new pontiff. He had not been by any means an outstanding candidate, and the sudden swing in his favor suggests a compromise rather than any belated recognition of genius. Altieri himself objected that, at eighty, he was too old, but he was installed despite his protestations, and took, or was given, the name of Clement X. An anonymous pamphlet of the day provided a succinct biography: "Altieri is a Roman, of decrepit age, an intelligent and zealous man. He has no relatives and is highly esteemed by the cardinals. Unfortunately his memory is gone."[10]

"The popes they elect these days are too old, and too far behind the times." So Christina declared, but, despite his lack of memory, and despite having no male relatives, the new Clement managed to appoint a papal nephew in due course—the uncle of the husband of one of his nieces.

It was the end of Azzolino's glory days. All his cleverness and conniving were insufficient to persuade the new pope that he had voted for him. He lost his position as secretary of state, though he continued to work within the Secretariat, where he had served assiduously for more than thirty years. He retained most of his lesser offices, too, and his wealth did not decline, but from now on, his influence was waning. In later years, he had the satisfaction of seeing other men bring to fruition some of the ideas that he had so long promoted—a more professional administration and, gradually, less nepotism—but he was never to regain the position he had enjoyed in the brief, bright days of the Rospigliosi pope.

CHRISTINA GAVE NO banquet to celebrate the new pope's election, and there was no rioting recorded in Hamburg. On hearing the news, she simply swore, then hurried off to make her obeisances. She remained on civil terms with him, but no favors were forthcoming, for her or for Azzolino, from Clement X. The six years of his reign were quiet years for the queen and the cardinal, steady, domestic, unspectacular. Azzolino went daily to his work at the Secretariat; Christina cultivated her garden.

The pope survived to celebrate the Holy Year of 1675, but shortly afterwards he was assailed by dropsy, then fever, and within a few months, these, or his "decrepit age," had carried him off. His successor, Benedetto Odescalchi, Pope Innocent XI, was elected virtually by diktat of the now mighty Louis XIV of France. His pontificate might once have brought a second spring for Azzolino and Christina, for Odescalchi had been one of the original members of the *Squadrone*. He had left them, however, to join an emerging group of *Zelanti*—Zealots—and his new, puritanical ways did not bode well for Rome's many seekers of the good life. He began by closing down most of the theaters, including Christina's Tordinona, which was converted into a granary. Women were barred from appearing on stage altogether, and Guido Reni's celebrated *Madonna and Child* was ordered to be painted over—a breast, apparently, was exposed. The pope did not stop there. Many Romans remained to be saved, especially Roman women. What had once applied to nuns only now applied to the whole female population; unearthing a prohibition from the previous century, Innocent forbade them to take music lessons from any male teacher, and further discouraged them from learning music at all. "Music is completely injurious to the modesty that is proper for the female sex," he declared piously, "because they become distracted from the matters and occupations most proper for them."[11] Innocent himself, within the luxury of the Vatican, lived in the barest asceticism, forbidding all amusements so assiduously that the people swiftly dubbed him *Papa Minga*—Pope No. One of his first acts of *deluxurization* was to withdraw Christina's pension. With a neat twist of the knife, he instructed Azzolino to give her the news. She responded by describing it, most unconvincingly, as "a great favour from God. The pension was a blot on my life," she wrote to the cardinal, "the greatest humiliation I have known. The grace of its removal is worth a thousand realms."[12] Pressing his advantage, the pope then tried to put a stop to her soirées at the Riario.

It was no fun. Rome was beginning to look like Stockholm. Disheartened, Christina even thought of leaving, and considered an estate in Prussia, where she might at least do as she pleased and stage whatever plays she liked. She wrote to Olivecrantz in Sweden that the thought of leaving Rome was "like a dagger to my heart," but she owed it, she felt, "to God and to my glory." But she did not go. In spite of all, life at the Palazzo Riario was good. Prussia was far to the north, and very cold, and a long way from the cardinal. A hostile and infuriating pope notwithstanding, Rome was where she belonged.

With Azzolino demoted and the *Squadrone* effectively dissolved, there was less chance now for Christina to play at the games of intrigue that she had so enjoyed. Though she did not acknowledge it, the conclave of Clement X had been her last real foray into politics, papal or otherwise. She continued to dabble, and might even have made a real difference had she been steadier of temperament or purpose. At one point, she made plans to become the formal protector of Rome's Jewish population. She had spoken in defense of the Jews before, during her first, much criticized, sojourn with the Texeira family in Hamburg. Now, apparently in response to a violent incident, she spoke out again, identifying herself, for the first time in her life, as the friend of all lowly folk as well:

> Moved by great compassion, Queen Christina, defender of the poor, the oppressed, and the downtrodden, declares that she has taken under her royal protection the Jewish ghetto and all its inhabitants. Let it be known to all who read this declaration that anyone who dares to insult or ill use these inhabitants in any way in future will be severely chastized.[13]

But no one read it, for the declaration was never published. Perhaps Christina realized the impossibility of punishing every insult cast at Rome's Jews, or perhaps, her moment of outrage passed, she simply forgot about it. "The queen frequently undertakes things," one of her servants remarked, "and then forgets them in the middle."

Christina's frequent readiness for a fight now revealed itself, too, in a potentially bloody confrontation with the pious and unfriendly pope. It concerned her royal franchise over the streets around her palazzo, itself a papal courtesy rather than a right. In Rome, the residences of foreign ambassadors—and royal persons—had long been exempt from papal authority or any of the usual laws of the land. This courtesy had gradually been extended to the streets around each residence, so that distinct quarters had developed, within which the ambassador—or royal person—was sovereign. In practice, this provided large areas of the city that were effectively free from the law, and over the years the ambassadorial quarters had attracted all kinds of half-legitimate business folk seeking to avoid taxes and trading regulations, as well as the usual appendage of thieves and prostitutes. The pope now decided that enough was more than enough. Rome was to be washed clean of these shamefully colorful characters, and the whole city, to the very gates of the ambassadors' residences, was to be taken back under papal jurisdiction. He managed it by a simple expedient: when an ambassador left the city, his replacement was not received at the Vatican until he had relinquished control over his quarter. In due course, he had regained them all except—by virtue of Gallic subterfuge—the French ambassador's, and, of course—since she was never replaced—Christina's. As the French king was particular about the tone of his ambassador's neighborhood, a sturdy guard was placed around it to keep the undesirable element at bay. This last alternative avenue being now closed, the undesirables found their way, sooner or later, to the streets around the Riario.

Christina's quarter soon became a refuge for all the "Thieves, Assassins, and Debauch'd Women" of Rome.[14] She was not concerned about the scandal, nor about the people themselves, many of whom she happily took into her service. But she allowed herself to be flattered by the franchise that remained to her; it presented her with an opportunity, she felt, to make a patronizing gesture toward the detested pope. If it was a wise and gracious gesture, too, that was incidental. She wanted to show that she was in a position to grant favors, so she wrote to Innocent, offering "to resign for ever" the franchise of her quarter, "reserving nevertheless the Respects due both to my Habitation and Domesticks," and adding, "As for my self, I neither pretend to, nor desire, any thing of your Holiness"[15]—except, possibly, the reinstatement of her pension.

It was not long before she regretted the letter, since it was not long before one of her own servants was in trouble with the pope's *sbirri*—his guardians of the peace. It seems that the valet of one of her guardsmen had cheated a banker of some barrels of brandy, and the *sbirri* seized him as he was making his way to church. They carted him off to a nearby tavern, and there they were surprised by a trio of the queen's guardsmen, who set upon them, retrieved the valet, and returned him in triumph to the Riario, while a chorus of excited onlookers stood by, "Hooting at the *Sbirri*."

The pope was not pleased at this flouting of his authority. His own men had been attacked in the streets—or, rather, in a tavern, but still, they were his own men. Unable to wrest the valet from his refuge in the Riario, the pope had him tried *in absentia*, along with the guardsmen who had freed him. With a heavy-handed twist, all were found guilty, not of theft, nor of brawling, but of sedition, and all were condemned to death. Notices were placed upon the walls of the queen's own palazzo, with a bounty offered for each of them. Seeing them, Christina went "stark mad." She sent off a letter at once to the condemning magistrate, threatening wildly that if her men "do not die a natural death, they will not be the only ones to die." Invited to a banquet at the Jesuit residence, at which the pope was to be present, she turned up defiantly with a dozen men in full armor, among them three of the condemned. Innocent pretended not to notice, and courteously offered her "some Basons of Fruit," including a bunch of green raisins, a rarity, apparently, for the time of year.

Her first provocation having been defeated by politesse, Christina tried a second. She gave orders that any of the *sbirri* passing the Riario should be taken prisoner, and in charge of this order she placed one "Captain" Merula, in fact a former bandit from Naples and a person of no glorious reputation, for it seems "he would Kill a Man upon the least occasion, or for Money, if you pleas'd." Shortly after Christina's order was given, in fact, he did, and his victim was one of the *sbirri*. The pope, enraged, began to shout about excommunicating the queen, until it was pointed out to him that "Crown'd Heads" must be approached with caution, and that, anyway, her involvement could not be proved. Instead, he considered sending a troop to her palazzo to take the condemned men by force. Christina, hearing of this, decided to mount an armed defense. She called all her servants together after Sunday mass, and gave them the choice to fight or to flee. "I will be at your Head," she told them, "and expos'd to the same Perils with all of you. He may be Pope, but I will show him that I am Queen"—queen, at least, of one small palazzo, and a ragged quarter of the wanted, and the unwanted. She was met with cheers, but not very loud ones. Her majesty was keen for the thrill of battle, but her servants did not want to fight. They were hewers of wood and drawers of water, after all, and the pope's soldiers were, after all, soldiers.

When Innocent heard of this, it is said that he "fell a Laughing," and, in his lightened mood, he reconsidered his own plan. Violent confrontation was never a good idea, and there was no knowing how far the thing might go. Fighting could so easily spread, there could be a riot, there could even be a real uprising—it was true, he must admit, that his efforts to keep the people on the straight and narrow path had not made him very popular among them. In the end, he let the matter pass with a shrug of his shoulders and a comment that Christina would not have appreciated: "*È donna!*" he sighed—"Women!" There was no march on the Riario, the seditious quartet escaped scot-free, and the servants went back, relieved, to their peaceable hatchets and kitchen knives.

CHRISTINA RETREATED INTO private pastimes. Much of her time she spent reading and writing, and now, for the first time in her life, she took to attending daily mass, perhaps in an effort to reform her ways, but more likely to placate the pope. Though political life was behind her, it was not in her nature to be inactive. "Doing nothing is what makes one old," she said, and though her health was not robust, and she had given up hunting long before, now and then she still attempted to defy the passage of the years. She had a calèche, a little low-wheeled carriage she liked to drive in, and she would sometimes take the reins herself to drive about the Riario gardens, displaying her skill to the men of the household. One day she set off at a great speed, running "like a Fool up and down the Field," and quickly lost control of the horses. Carriage and queen overturned together, and for some minutes Christina lay on the ground, with her skirts up around her waist, calling for help, while "no Man durst come near her in this condition." At last getting up unassisted, she laughed at their embarrassment, declaring that at least they would know now that she was "neither Male nor Hermaphrodite, as some People in the World have pass'd me for."[16]

The cardinal, at least, had not. In the years since her return to Rome, he had reconciled himself to his own love for her. It was chaste, almost certainly, but it was strong, and his political hopes, now vanished, no longer stood in its way. He seems to have spent most of his evenings with her, and during the day they wrote to each other frequently, sometimes every few hours. One tender note from Azzolino, from December of 1679, shows the warmth of his love for Christina, a quarter of a century after its beginning:

Your sweet little letter arrived just as I was thinking particularly of you. Dearest, I did not enjoy the play so much yesterday evening because it prevented me from being with you as usual. . . . Dearest, I thank you a thousand times for the comfort you have given me, and I embrace you with all my soul.[17]

But even after so long, she remained a demanding mistress. Only weeks later, with the cardinal enduring severe gastric trouble, she accused him of manufacturing the illness to avoid coming to see her. Abandoning his "Dearest" and retreating to "Your Majesty," he wrote snappily to her: "Perhaps, Madam, Your Majesty may doubt my condition. I can assure you that no one who had seen me with a beard of five days' growth, and after an evacuation, could say that I was well." He was aware that the queen herself was no stranger to the indignities of "evacuation." His letter continues, in the rather exasperated tone of a faithful, but doubted, lover:

I say to Your Majesty, as a man of honor, that I am now well, and I will be there tomorrow, if I live. And if Your Majesty does not believe this, and does not put Her soul at rest and assure me that She has done so, then this very evening, once it is dark, after the Ave Maria, I will pay Her a visit, and She will see that it is absolutely true. For the love of God, Your Majesty, let me know that You believe me, otherwise I am not even going to wait until it's dark.[18]

But, man of honor or not, the cardinal was not well, and he did not see her the next day, nor indeed for more than a week. A further evacuation he endured with Christian patience, and the following day, "having felt the effects of it and slept very well," he wrote a rather more timid note to the queen, expressing his "infinite, infinite, infinite thanks" for her constant concern for him.[19]

AZZOLINO HAD COMMISSIONED a history of his family, and Christina now turned back to her own *Life*, which she had begun so long before in Hamburg. She had evidently not acquired any modesty in the intervening years, except in the most roundabout way—recognizing her "many faults," she blamed herself for having failed to correct them, "since among the many talents that the liberal hand of God has bestowed on me, I have such absolute and admirable power over myself that I can make of myself whatever I want."[20] She did not mean it ironically, and it is the more amusing, or the sadder, for that; of her many undoubted talents, self-delusion was not the least.

Her *Life* has a quality of apology, nonetheless, though it is far from apologetic in tone. She seems to have realized how little she had achieved from a position of so much privilege, and her words are by turns defensive and defiant. She thought again about her great father—great in everything, she admitted, but all the same, "too fat, and too quick to anger, and too fond of women, and he drank . . ."—all of it untrue. She revisited her heroes of the ancient world, Caesar and Alexander and Cyrus; they did not pass muster, either. "It's true they had all the heroic qualities," she wrote, "but the world was very different then. I'm convinced that if they had lived in our own time, they would never have done anything."[21] In short, there was an excellent excuse for her own lack of achievement, and if she had grown a little portly of late, her father had provided a good historical precedent.

Attempting to redress the balance while life and hope remained, she began a new *Accademia* for learned debate, and all the cardinals turned up dutifully to discuss such topics as "True love lasts till death"; "Love exists for its own sake"; "Only the stupid are wicked"—and all the usual suspects, which had not progressed since the days of her first academy in Rome, or indeed, of her academy in Stockholm, thirty years before. The Accademia did not last, but the list of topics survived in the form of reflections in the style of La Rochefoucauld, whose own work she had read admiringly, and annotated, too.[22] Christina labeled hers at first *Heroic Sentiments*. From these, she sifted out a set of *Moral Reflections*, renaming them *Reasoned Reflections*, then *The Fruit of Leisure*—the changing titles revealing, perhaps, her own odyssey from tough northern stoicism to a luxuriant garden in sunny Rome. She began the *Sentiments* with pious declarations about the Catholic faith, but within three pages she was talking of Caesar and Alexander, and of pride and hypocrisy and pleasure—though now and then a Jesuitical strain does sound: "Everything we like is permissible," she states, "but we must not like anything that is not reasonable." Given her intelligence and her very ready wit, the little maxims are surprisingly lacking in pungency and often trite, though there are traces of unconscious irony: "A minor prince can do much harm, but little good." "Modesty is the finest virtue of all." "People are only ever fooled by themselves." "One must save money, but nobly, not sordidly."[23] Christina did not often take advice, and particularly not from herself.

She worked on her maxims and *Sentiments* on and off for several years, and they went through no fewer than eighteen drafts. Santini did most of the copying, aided by Anders Galdenblad, a new Swedish secretary who found it hard to penetrate the mysteries of Christina's atrocious handwriting. He made a good many bad drafts. "The King must always be the minister," read one. "The *master*," the queen retorted. "The King must always be the *master*." "Wrongdoers must be secretly punished," read another. In the margin came her box on the ears: "You've written *secretly*, you idiot. I put *severely*." One way or other, the drafts were produced, though they were never really finished. In the end, the maxims, each mostly a single line or two, numbered more than a thousand, and the *Sentiments* 444, but Christina was not satisfied with them—or with herself, perhaps. There is one pause among the pages, oddly muted after the stentorian tone of the rest, that suggests a moment's humility on her part. She is speaking of the Delphic oracle:

> Know thyself, the oracle says. And everyone wants to make of this the source of all human wisdom. But it's not. It's the source of all human misery. We can't help knowing ourselves, and we can't help being miserable because of it.[24]

And she followed it with a sad reflection, which would certainly not have escaped the Vatican censors had she sought to publish it in Rome:

> Man is a nothing covered in a scrap of life. Knowing this makes him miserable, not wise. And he can't change it, not for all the philosophy in the world.[25]

She did not seek to publish it, in Rome or anywhere else. Her secretaries knew the text, as did the cardinal, no doubt, but otherwise it remained a private testimony to Christina's illusions, and her disillusion.

Azzolino was writing, too, not maxims or memoirs, but a huge history of the papacy, beginning at the beginning, with the reign of Saint Peter. He had begun it many years before and had worked on it intermittently, but his appointment as secretary of state had left him no time to complete it. Now, relieved or bereft of his post, he returned to his book. The pope was the beneficiary, and he may have permitted himself a moment's ambivalence when the cardinal presented him with the completed work—two giant volumes, some fourteen hundred pages altogether.[26] He appreciated the effort, at least, and was also pleased with the cardinal's improved behavior—the ladies were long gone, and as for the queen, well, they were just good friends. He approved Azzolino's steady progress in the ranks of the workaday cardinals, and eventually made him Cardinal Bishop of the Church of Santa Maria in Trastevere, very near to the queen's Palazzo Riario—but of course, they were just good friends. Azzolino tried to step a little higher by pretending that he had never really been inimical to Spanish interests. It did not help him, but his administrative capacities kept him in demand, and he continued to live well, with a good salary and, importantly, his rents from Le Marche. He had now no fewer than twelve carriages, most of them large six-seaters, including one expensive "French" vehicle. He must have been personally well decked out, too, as he owned—apart from his trousers and shoes and his cardinal's regalia—thirty-seven shirts of Dutch cloth, thirty-two pairs of light summer socks, forty-six handkerchiefs, eleven nightcaps, and nineteen pairs of underpants. No matter if his palazzo was rented. He certainly had enough to make it worthwhile for someone in Geneva—some scandalized Calvinist, or perhaps just a rogue with an eye for an opportunity—to attempt to blackmail him over his "public concubinage" with the queen.[27] But the blackmailer had missed his chance. If it had ever been true, it was true no longer.

Christina was almost sixty, and Azzolino older. Their passions were now the quieter passions of later life, gardening and reading and charitable works, and they shared, as much as anything else, their gout and gastric problems and troubles with their teeth. Like a long married couple, they made frequent visits together to the cardinal's family in Le Marche; Azzolino had endowed libraries and religious institutions there, and a home for poor unmarried women—thinking, perhaps, of his five sisters who had become nuns, apparently for lack of an adequate dowry. He and Christina read each other's writings, and inscribed their own with mutual dedications; even their letters were sometimes written together. They had grown to depend on each other. "Fidelity in love," Christina wrote at this time, "is a necessity." "Fidelity alone can distinguish the true from the false."[28]

This much the cardinal had given her. With her garden and her works of art and her musicians and her friends, this love was now her life. All in all, it was a fair consolation for the phantom glory she had pursued for so many fruitless years.

JOURNEY'S END

A FRENCHMAN BY THE NAME OF Misson, visiting Rome in the spring of 1688, sent home to his family a description of Christina, which, as it happened, was the last:

> She is more than sixty years old, very small of stature, exceedingly fat and corpulent. Her complexion and voice and face are those of a man. She has a big nose, large blue eyes, blonde eyebrows, and a double chin, from which sprout several tufts of beard. Her upper lip protrudes a little. Her hair is a light chestnut colour, and only a palmsbreadth in length; she wears it powdered and standing on end, uncombed. She is very smiling and obliging. You will hardly believe her clothes: a man's jacket, in black satin, reaching to her knees, and buttoned all the way down; a very short black skirt, and men's shoes; a very large bow of black ribbons instead of a cravat; and a belt drawn tightly under her stomach, revealing its rotundity all too well.[1]

She was clearly no longer standing on ceremony, at least within the walls of her own residence. Misson had come to the Riario to view her collections as part of his grand tour. Christina welcomed him, remarking that these days she herself was more or less one of the ancient monuments of Rome.

She was beginning to feel her age. Sad news had come to mar the comfortable duties and pleasures of her later years, news of illness, and news of death. Heinsius and Vossius had died, scholars from her far-off days as the Pallas of the North; Montecuccoli had died, and *le Grand Condé*, and her old friend Doctor Bourdelot. "How does it feel to be eighty?" she had written to him teasingly, but Bourdelot had not lived to reply. Gone, too, was Magnus

De la Gardie, and his widow, Christina's schoolfellow cousin, Maria Euphrosyne; on her tombstone was carved a history that might have been Christina's own—eleven children, eight already buried; there would be no grandchildren at all. In Rome, Bernini had died; in pious hope, he had bequeathed his sculpture of the *Salvator Mundi* to Christina. She still did not want it, but honored the memory of her great friend by commissioning his biography.[2] And the Marchese del Monte had made his exit, too, leaving nothing but a whiff of brimstone behind him. At the age of seventy he had begun a new affair with a married lady of the town; bereft of his erstwhile stamina, he had "endeavour'd to support his Vigour by Art," and the concoction had been the end of him—though it was said in Rome that the devil had smothered him at last.

It was time to think of the next life. In her daily writing, Christina was reflecting on how little she had accomplished, but she was reminded, too, of certain things that she had done that she ought not to have done at all. Monaldeschi's death was the worst of them. Though she had never publicly excused or regretted it, now, with her own last judgment perhaps not far away, it was on her mind again. A note of understanding, even remorse, had crept into the lines, very different from the stamping and shouting of previous years. "Men would never be traitors or liars if they weren't weak and foolish," she wrote. "The Emperor Theodosius's law was just and wise: he said that no one should be executed within thirty days of the sentence of death. It is a safeguard for the prince's conscience. People can always be put to death, but you cannot bring them back."[3]

It was in this reflective and perhaps vulnerable frame of mind that Christina encountered the last and greatest of her persuasive, plausible rogues, the "golden-mouthed" Spaniard, Miguel de Molinos. Suitably enough, he was a priest, a former confessor to a community of nuns, and he preached the newly fashionable doctrine of Quietism, the perfect balm, or so it seemed, for Christina's ruffled soul. Quietism encouraged a passive attitude to life, abandonment of the will, religious contemplation, and, conveniently, the denial of conscience. Sin belonged to the lower, sensual part of man, said Molinos. It was instigated by the devil and was not subject to man's free will, hence man could not be blamed for it. In short, "he would abuse the finest Women and Maids to whom he had access, by perswading them that Whoredom was no Sin."[4]

Christina, at least, was seduced, and Azzolino, too, and thousands of others in Rome and elsewhere, including the puritan pope, "who believed him a Saint," at least for a while. Christina had never liked ostentatious piety or any of the extravagant, Baroque Catholic rituals, whose images and relics and smoky ceremonies offended the Lutheran leavings in her soul. She had been drawn to Quietism since her meeting with François Malaval on her first visit to France, thirty years before. Its simplicity appealed to her, and besides, she did not like being told what to do; a Quietist needed no intermediary between himself and God; he could choose whatever form of devotion he pleased. It all suited Christina very well. "To love God and one's neighbor is real piety," she now wrote. "All the rest is just farce."[5] Molinos was soon taken into her service, not as her confessor but as her personal theologian, and the two of them would pass three hours in pious discussion together every Monday morning. She considered him a genuinely holy man, though not a saint—"I can't believe in saints who eat," she remarked, for the priest's appetite was legendary. But it was a small price to pay for the comfort of his words, and the Riario servants did not mind the extra cooking. Molinos had a way of cooling the queen's hot temper; his visits always calmed her, and saved them many a beating. In consequence, the apostle of Quietism was always made to feel at home at the Riario, which remained, nonetheless, as the French ambassador wryly noted, "the most unquiet house in the town."

Azzolino went along with it all to a certain degree, but his native Catholicism tempered his enthusiasm. He was not ready to jettison all the time-honored ways of his faith. Though his writings became markedly pietistic, he clung to his Roman instincts, and went so far as to purchase the bodies of four martyred saints for the Church of Saint Filippo Neri, a Saint Francis–like figure to whom he was particularly devoted. He warned the queen to tread warily, but in the end it was the French who exposed Molinos and denounced him to the Inquisition. Christina refused to believe the charges, which were, she was informed, "worse than your majesty could possibly imagine." She suspected a plot masterminded by the "cursed tribe" of the Jesuits and exerted herself to free Molinos from his cell in the Castel Sant'Angelo. Her petitions vanished into the void; she sent food after them; it warmed the prisoner's stomach and, it is hoped, his heart, but he was not freed.

Molinos was tried "in the presence of an immense concourse of people," curious, no doubt, but also encouraged by the undeceived pope's promise of indulgences for all who attended the proceedings—they exceeded their brief with a great roar of "Throw him in the Tiber!" Molinos was duly found guilty and his teachings declared "heretical, suspect, erroneous, scandalous, et cetera." Among his private papers were found more than a hundred enthusiastic letters signed by the queen. It is said that Azzolino had managed to destroy a hundred others.

Molinos was not burned but was sentenced to life imprisonment "in penitential garb," with daily recitations of the creed and the rosary, and confession four times a year. For so devout a gourmand, his bread-and-water rations must have been the hardest penance of all, but Christina did what she could to sweeten them by sending him regular supplies of jam.

CHRISTINA'S LAST EXPERIMENT with religion was at an end. She turned back to her tried and truer pastimes, to her paintings, and her garden, and, above all, to music, the least expensive, and so the most indulged, of them all. Cicciolino was no more, but his place had not been vacant long. It was now occupied by the young Angelica Quadrelli, "a Virgin, incomparable both for Beauty and Wit," and a fine instrumentalist at harpsichord, lute, and oboe. Angelica was a tall and elegant blonde of ruby lip and sparkling eye, so beautiful, indeed, that Innocent XI, "a very severe and angry Pope," had felt that Rome's worldly temptations would be the fewer if she were kept out of sight; he decided to put her into a convent. Angelica, alarmed, had at once sought refuge with Christina, who had long admired her voice and who, besides, was very fond of her. She took her at once under her protection, promising the pope at the same time to see the girl safely into the convent. Forgetting this, or ignoring it, she had soon installed Angelica at the Palazzo Riario, together with her mother and her sister, "who was also a most Beautiful Virgin, but had not so great a faculty of Singing."[6]

Angelica's musicianship was a decided asset at Christina's frequent musical parties. Her sweet temper and her flaxen beauty recommended her likewise, reminding the queen, perhaps, of her long lost Belle. She came to love Angelica, and kept her in daily attendance.

One afternoon Christina was holding court at a sort of musical picnic in the lovely gardens at the Riario, in the company of "a great many Ladies, Knights and Gentlemen." It was a hot summer's day, and they had all seated themselves in the shade of some leafy branches, where they nibbled fruits and sipped wine as the various numbers were played. One of the servants described the scene:

It was about the time that the Song of *Flon, Flon* was in request; and the Trumpets . . . being posted upon a little Hill in the same garden, sounded the same Tune. The Eccho's repeated the last Words. All the World Sung it, and the Queen her self sung *Flon, Flon.* . . . A thousand fits of Laughter accompanied the Musick, insomuch, that others of the Queen's Maids walking in the Garden, were drawn thither by the Noise.[7]

Angelica took a star turn, extemporizing on her "Guitarre," and afterwards, when all the guests were gone, Christina sent for her so that they might sit together quietly for an hour or two. It is a happy glimpse of "a happy Season," the queen, round and contented, sitting in a great armchair, reading in the candlelight, looking up now and then to smile, like a fond grandmother, upon the gentle young beauty at her side.

It did not last. Christina, perhaps feeling unwell, began to have presentiments of her own death. In the autumn of the same year (1688), with no trust left in priests, she sent for her tailor and had him make for her a white satin gown of her own design, with gold buttons and gold lace, and embroidered all over with flowers. On Christmas Eve she tried it on, walking about in her dressing room, with a tall looking-glass in front of her and another one behind. For some time she said nothing, but walked, and looked, in silence. The tailor was present, and some of the servants, among them the curious Giulia, a young girl once believed to have been a boy, now serving as assistant to the queen's alchemist, but better known for her foretelling of the future; for this reason, Christina called her Sybil. She turned to her now. "This gown makes me very thoughtful," she said. "I believe I shall soon have call to wear it at one of the greatest events of my life. Can you guess, Sybil, what that might be?" Sybil looked uncomfortable, but the queen pressed her, and at length she answered. "I can guess," she said, "but I do not wish to say the words." Christina insisted. "Then," said Sybil, "Your Majesty thinks she will soon be buried in this gown." The others remonstrated, but Christina confirmed that Sybil had guessed what she had been thinking. "We are all mortal," she said, "and I as well as another." The tailor, looking for levity, then declared he would make a cover to match the gown, "since if Your Majesty is correct," he said, "you will have to take care the worms do not eat it." Christina "fell a Laughing, and was well pleas'd with the Repartee."[8]

She remained defiantly stouthearted. A letter arrived from a German astrologer, advising her to take care; the first three months of the coming year would be, he wrote, a time of serious illness for her. The prediction did not trouble her, and in fact, a second warning left her roundly amused: when a soothsayer advised her to dispose of her "indecent" paintings before she died, her only response was a loud burst of laughter. She left them hanging on the walls of the palazzo, and set off for a short winter tour of the warmer southern provinces.

For once, it seems, she found the traveling strenuous; in any event, it did not do her good. In February, shortly after her return, she began to have fainting fits; she became feverish, and a dropsy swelling that had often troubled her now returned to add its familiar nuisance to her other ills.[9] Her physician Spezioli diagnosed an infection; it was severe, and it was not soon to pass. Though Christina had dismissed it, the astrologer's prediction had been correct: for many weeks she lay gravely ill, feeling close enough to death to dictate a new will, confirming Azzolino as her principal heir. Toward the end of March, she lapsed into unconsciousness. The cardinal kept a prayerful watch beside her, but revival seemed impossible.

Quite suddenly, with the first days of spring, she awoke. The fever had abated; the swelling was reduced; the danger, it seemed, had passed. Azzolino commanded three masses of thanksgiving, and the Comte d'Alibert, summoning all his impresario skills, arranged the performance of his life, with celebrations across the city in church and street: trumpets sounded, fireworks blazed, cannon fired off. The rejoicing was general, for Christina had many dependents and, moreover, was a notoriously generous benefactress—notorious, that is, among her creditors, who had long stood the cost of her easy charity. She leaped back into life with a bound and was soon penning a characteristically ebullient letter to her administrator Olivecrantz in Sweden:

> I had given up hope [she wrote] and had resigned myself to making the last journey of all, but God has snatched me from the embrace of death. I am now full of life again, thanks to the combined miracle of God's grace, my own nature, and the physicians' art. Together they have given me back my health. The strength of my own temperament has pulled me through a sickness strong enough to kill twenty Hercules.[10]

Olivecrantz was soon to visit Rome, and she wrote that he would find her in her usual high spirits, "plump and happy." And so indeed she remained, for a fortnight or so.

Christina was now sixty-two, a fullness of years but no great age, even for her day. The spring had arrived, the spring of 1689. Christina was in good heart; with care and time, she might have recuperated fully, or so at least her servants thought. But a violent incident, occurring beneath her own roof at this very time, undermined her sanguine mood and dealt a fatal blow to her recovery. While she had been convalescing, a vicious plan had been hatching within the palazzo against her lovely protégée, Angelica Quadrelli. The girl's own mother had effectively sold her, for a thousand silver crowns, to the Abbé Vanini, seducer extraordinaire who years before had caused the death of another young woman of Christina's household. The abbé had long laid siege to Angelica's virtue, but to no avail: "All his Presents were rejected," it seems, "and his Sighs despis'd."[11]

One day, as the queen lay resting, Angelica's mother led her into a small chamber on one of the upper floors of the Riario. Angelica was "extreamly surpriz'd" to find the abbé waiting there for them, but her surprise turned to horror when her mother quit the room, leaving her alone with the abbé, who was quick to take his advantage. The girl did her best to defend herself, crying out and knocking over tables and chairs in her struggle. The noise brought help, though too late to save her. The abbé was surrounded and a hasty command given that he should be "hewed in Pieces," but, on consideration of his many high connections, the guards' hands were stayed and he was permitted to escape. The ordeal left Angelica in despair, and she kept to her room for many weeks, stricken and grieving.

Christina, meanwhile, had begun to recuperate. She was accustomed to seeing Angelica often throughout the day, and she asked for her repeatedly, only to be told that the girl was unwell and could not attend her. Concerned that the shock of the news would hinder the queen's recovery, Azzolino had forbidden the servants to tell her what had happened, but at length, "either by Accident, or out of Malice," the alchemist's assistant, Giulia, hinted to her that Angelica was not ill after all. The girl was immediately sent for. She came weeping to the queen, but could not be persuaded to say what was the matter. Christina questioned Azzolino, and then each of the servants, and gradually, by threat and inveiglement, she discovered the truth.

She fell into "an horrible Passion" and sent for "Captain Merula her Bravo"—recent killer of one of the pope's men—and commanded him in raging tones to bring her the abbé's head. Merula's scruples were not dainty, but this time, it seems, he preferred a less bloody transaction. He accepted payment from some of the abbé's friends to do nothing while his supposed quarry decamped to Naples. When he returned to the queen with only his own head to show, she fell into a fury. She "Scratch'd him in the Face, and gave him twenty blows with her Fist, reproaching him with Perfidiousness, and had certainly Strangled him, if she had not wanted Strength." Too weak to kill her faithless captain, Christina was also too weak to endure the shock of Angelica's assault and the girl's continuing distress. As the cardinal had feared, she quickly relapsed into fever.

Sure now that her death was near, Christina made her confession and a final act of contrition. She received communion, and sent to the pope to ask his blessing and his forgiveness for all the differences they had had. A letter of absolution was duly brought, and Christina passed her last few days telling stories of bygone years, praising her old rival, Chancellor Oxenstierna, resurrecting him for an hour or two from his long oblivion in the grave. She spoke, too, of the "incomparable" cardinal, of his great merit, and of his devotion to her. She wanted him to know of her affection for him, and her esteem, and her gratitude. Toward the end, she became unconscious, and on the nineteenth of April she breathed her last. She lay as if sleeping, leaning on one side, her hand resting at her throat.

It was six o'clock in the morning. Azzolino had been at her bedside through the night, and now he turned away, the many small, sad duties of bereavement before him. The first of them was to inform the Church authorities, and to this effect he took up his pen. His mind was elsewhere, it seems, for he began this first, formal letter with a date awry by many days:

> The Queen is dead [he wrote], and she has died in all holiness, a true and faithful child of God and daughter of the Catholic Church. She has done me the honour of making me her heir. She wished to be buried privately, without any undue ceremony, but the glory of God and His Church and the honour of His Holiness demand otherwise. A simple funeral would be a triumph for the heretics, and a scandal and disgrace to Rome.[12]

He went on to propose instead a grand public ceremony, with a formal procession, in the presence of the whole Sacred College of Cardinals. In a second letter, he wrote that the queen had bequeathed to the pope her Bernini sculpture of the *Salvator Mundi*. "I beg His Holiness to accept it," he wrote.

"It is among the very finest of all the master's pieces." It was a fine irony, too, for Christina had disliked the sculpture almost as much as she had disliked the pope. She had also left instructions that three chaplaincies be founded at St. Peter's. This, the cardinal felt, was "a very strong motive" for laying the queen's body to rest in the basilica itself. All the priests at St. Peter's wanted it to be so, he wrote, and all the townspeople, too; many of them had approached him about it before, and there would be "a good deal of discontent, and a good deal of bitterness" if it did not come to pass. There was little precedent for it, or indeed for any woman to be buried within the confines of the Vatican, but, by suprising the cardinals and by pushing them to make a quick decision, Azzolino gained his point. Within a day, the pope had given his permission, though without having learned of Christina's final, characteristically extravagant demand: the three new chaplains at St. Peter's were to say no fewer than twenty thousand masses for the repose of her soul.

Though written only hours after Christina's death, Azzolino's two letters contain many details of the proposed funeral. It seems that he had considered it quite carefully, perhaps during her last days, when she had lain unconscious. He knew very well that she herself had wanted "no pomp or exhibiting of the body or other vanity," but if he had hesitated at first to override her wishes, no doubts remained by the time he sat down to relay the news of her death. Christina had been too public a figure, too great a convert, too noisy a sinner, to disappear without the final grand embrace of the Church.

Thus, Christina's departure from Rome was as splendid as her arrival had been more than thirty years before. For four days, her embalmed body lay exposed at the Palazzo Riario. The Carmelite brothers kept a silent watch, while half of Rome and folk from the nearby country towns filed past, sorrowful, awed, and curious. Christina had been dressed in her new white satin gown, "stitch'd with Flowers," and a purple cloak embroidered with gold crowns and trimmed with ermine, like the mantle she had worn on the bright, distant day of her coronation. A crown she had on this day, too, a small crown of silver, and a silver scepter in her cold hand, and a silver mask covering her face.

On the fourth day, her body was carried from the palazzo to the Church of Santa Maria in Vallicella, the Oratorians' Chiesa Nuova that Christina had long admired. Its entrance was decorated with sheaves of wheat in homage to the Vasa queen, and inside, its beauty was made glorious by the light of three hundred torches, flaming from silver mounts. Through the night, she lay surrounded by wreaths and candles, and the next day her requiem was said, with hundreds of members of the sixteen brotherhoods and seventeen ecclesiastical orders in attendance, and "the whole Colledge of Cardinals," save only Azzolino, who had not the strength, nor perhaps the heart, to endure this farewell to his most beloved friend.

As night fell, the queen's body, "with her Face discover'd," was carried across the river to St. Peter's, with the cardinals and other clergy and the Roman nobility and diplomats and scholars and artists and "all her Domesticks" processing before and after in a vast train of mourning. There, in the great basilica, it was placed in a coffin of cyprus wood, then inside another made of lead, engraved with her name and her coat of arms, and finally in a third, wooden coffin, bearing her effigy in large medallions. So, this vital, vibrant, brilliant figure, the northern star who had compared herself to the very sun, was enclosed at last in darkness.

CHRISTINA'S FUNERAL HAD BEEN magnificent, but her memorial was not. "She was poorly Interr'd," wrote one contemporary, "in a Cave of the Sacrasty of Saint Peter."[13] This was partly true: she had in fact been buried in the crypt of the basilica, alongside the cardinal archbishops of that great church, but for some time there was barely a stone to record her last place of rest. In her last will and testament, she herself had stated that she wanted nothing but the simplest epitaph: *For Almighty God,* it was to read, *Christina Lived Sixty-Three Years.* For a long time, she did not have even this. The pope eventually commissioned a marble relief, but the money was lacking to complete it, a wryly fitting memorial in itself to an incorrigibly spendthrift queen. Finally, in the early years of the new century, a new pope, one of Christina's last surviving friends, decided that the sculpture would be paid for from the basilica's own building fund.[14]

She might have had a memorial sooner had Azzolino lived to oversee it, but the cardinal did not long outlive the queen. He had time to attend to some of the matters pertaining to her will, and to make his own, and from time to time he was seen walking slowly through the streets, "in deepest Mourning, with the greatest Lowliness and Dejection."[15] Christina's papers were transferred from the Riario to his own residence on the Borgo Nuovo, and there, with the help of her secretary and his own, the cardinal sifted through them, setting aside the official documents, and, as the queen had wished, burning all her private correspondence. In so doing, he revealed his own temperament, both systematic and sentimental, for it seems that, before burning her letters, he read them all chronologically, through them living once again the excitement of Christina's early days in Rome, the thrill of their political intrigues, and the first fine, careless rapture of their love. The letters included many that he himself had written to her; despite his frequent injunctions that she should burn them, Christina had never found it in her heart to do so, and they remained to the cardinal now, fragile, futile remnants of a friendship that had been, as he wrote, "my greatest glory."[16] But what Christina's love had forbidden, so death would now prevent. Azzolino's worsening illness interrupted the quiet work of destruction. A small, yellowing treasure would survive him.[17]

He was by now very unwell, and he seems to have known that his own end was approaching. For his own peace of mind, and perhaps on Christina's belated behalf as well, he attempted to bring about a reconciliation with the French ambassador, who had long detested them both, but this took the last of his strength, and his efforts were in any case unsuccessful. The French court repaid them with the rumor that he had killed the queen in order to come the more quickly into his inheritance.

If Azzolino heard of it, it made no difference to him. Christina's death had left him oblivious to smaller wounds, no matter how vicious. "I am inconsolable," he wrote, "and I shall always be inconsolable."[18] His loss, he said, had been "appalling," and he could recover neither health nor spirits. Now, with the lovely Roman spring blossoming around him, Azzolino began to fade. He died on the eighth of June, just fifty days after Christina, having received a last blessing from the pope.

His funeral was held, as hers had been, in the Chiesa Nuova, and there he was buried, next to the chapel of Saint Filippo Neri, to whom he had been so long devoted. Romolo Spezioli—Azzolino's physician, and Christina's—provided a marble tablet to mark his grave.

THE QUEEN'S LAST WILL and testament was read only hours after her death, at ten o'clock in the morning, by the notary who served her household. Azzolino had not waited to hear it, nor had he needed to; he knew already that he had been appointed her heir, and the pope, Innocent XI, her executor. There were a few particular bequests: the three chaplaincies at St. Peter's, Bernini's marble bust to the pope, jewels to the kings of France and Spain, and jewels, too, to the Habsburg emperor, grandson of her father's old enemy, and to Friedrich, the elector of Brandenburg, son of "the Great Elector" who might once have become her husband. To the young king of Sweden, the son of Karl Gustav, who had loved Christina so long and so well, she left nothing at all. But there were various sums for the servants, and linen for the ladies-in-waiting, and laboratory equipment for her alchemist, and for one little boy of her household, of whom she had been especially fond, a substantial pension of two hundred scudi per annum.

Everything else that Christina had owned, or claimed, now passed to Azzolino's estate. True to form, she had been far more generous than she could afford to be; there was in fact no money to pay for the chaplaincies, or the servants' legacies, or the little boy's pension, nor indeed, to redeem the jewels, long pawned. The cardinal had not been able to sort out the various claims and counterclaims that now came to rest at the door of his own heir, Pompeo Azzolino, a young cousin, a minor noble, generally known as the cardinal's nephew, and whispered to be his son. "Who would ever have believ'd," wrote one disdainful commentator, "that a little Gentleman of Marca d'Ancona should become Heir to the Daughter of the Great Gustavus Adolphus the Terror of Germany."[19] The "Great Gustavus Adolphus" would not have been flattered by the reflection, for the eventual heir had inherited mostly debts, including one of seventy thousand scudi, which Christina had borrowed from one of the papal library funds; it would have been enough to pay the rent on Azzolino's palazzo for 150 years. Pompeo had no choice but to sell what he could to discharge the debts and meet the obligations of the will, ignoring the Swedish king's demand for the return of the paintings and other valuables that Christina had taken with her from Stockholm. The pope bought her fine collection of books and manuscripts "for a piece of Bread"—in fact, for eight hundred scudi, an absurdly small sum, given that the manuscripts alone numbered almost two thousand.[20] A papal nephew bought her coins and medallions, "of which there was so fine Setts in all sorts of Metals."[21] The paintings and sculptures found their way to different great houses across the Continent; the furniture and plate were similarly dispersed. Nothing was returned to Christina's distant homeland, but the Swedes at least reclaimed the lands ceded on her abdication to their erstwhile queen. At the end of it all, there was "no great matter" left for Pompeo Azzolino, which might at least have pleased Christina. She had never liked him much.

EPILOGUE

YBIL HAD FORESEEN Azzolino's death, as she had foreseen Christina's, but when she announced that the pope himself was soon to die, she was seized and thrown into the Castel Sant'Angelo. Yet she proved to be right: the "severe and angry Pope" breathed his last in the fierce dog days of a Roman August, but it did not help Sybil. A later pope, taking pity on her, transferred her to a convent, where she "pin'd away for Grief."

Christina's other servants were left to shift as best they could. There was no money for them, so they did as they had always done; they took whatever light fingers or strong arms could take, and went on their way. The queen's nobler friends, missing her provocative soul, revived her old academy and made her its spiritual patron, and without her they actually made something of it. Freed of her long insistence on the time-worn subjects of love, they began to talk of literary reform, and their meetings at the Riario became important cultural events in the new, analytical century.

Sweden's greatness did not last. Christina's life had coincided with her country's time as a major continental power. In 1709, a terrible defeat at the Battle of Poltava drove the Swedes back into the northern periphery of Europe, and left the way clear for the rise of another great northern power—Peter the Great's Russian empire. As Gustav Adolf had feared, Russia had at last "learned her strength," and his own land had paid the price of it. Its blaze of glory over, Sweden retired to fulfill a quieter destiny; its king, Karl XII, Christina's own great-nephew, took a final ironic refuge with a friendly Ottoman sultan.

In Stockholm, memories of the late queen began to fade. Those who had known her had grown old and gone; half a century had passed since the drama of her abdication. One spring afternoon in 1697, fire broke out at the

Tre Kronor Castle, the home of Sweden's kings where she had spent her childhood and her youth, where she had fallen in love, and begun her library, and taken her first awkward steps into political intrigue. The flames swept through the buildings, destroying the library, the great hall, the private apartments. What Christina had not smuggled away had now been taken more surely, leaving only ashes, and regrets. Karl Gustav's own son lay dead in the chapel. Christina's world was passing away.

She had lived, inescapably, beneath the vast shadow of her great father, the most romantic figure in Swedish history, and, in a sense, she had spent her life doing battle with him. But a stronger legacy had driven her, the legacy of her passionate, chaotic mother, stronger within herself than she could ever dare to recognize. She could not look unblinking upon her own nature, for in so doing she would have destroyed her very sense of self.

It is a sad tale, in a way, a tale of promise unfulfilled, and of strength thwarted by weakness. Yet, endearing and exasperating, so much less than she might have been, Christina at the end somehow came into her own, sitting in the sun in her garden, eating her boiled chestnuts, clouting a lazy servant now and then, a "plump, shabby, indomitable figure dressed in rusty black."

"I don't know that I ever really tried to overcome my faults," she wrote toward the end of her life. They were the faults of a queen, and the faults of an unloved child, the faults of a fearful woman, and those of a gifted mind with yet no kernel of genius, and of a desperate, hopeless seeking to be great. Christina was a child of her heroic time, a misshapen pearl of the Baroque, lustrous and precious despite its imperfections. From grand and lovely portraits, her blue eyes gaze out across the centuries, still challenging, still hoping, while the fireworks of her story, at first sight dazzling, at times even lurid, settle at last to a more human glow.

NOTES

BIRTH OF a prince

1. Johan Hand, quoted in Berner (1982), *Gustav Adolf: Der Löwe aus Mitternacht,* p. 184.
2. Maria Eleonora's maternal grandfather was the Herzog (Duke) of Prussia, and her great-uncle the Herzog of Cleves. Gustav Adolf was the nephew of Erik XIV of Sweden and his sister Maria Elisabeth. All four were held to be insane.
3. As well as his native Swedish, Gustav Adolf spoke and wrote German fluently from childhood, and later spoke French, Italian, and Dutch. He also understood Spanish and English, and had some knowledge of Russian and Polish. In keeping with the times, he knew Latin and Greek, as well.
4. Quoted in Berner (1982), *Gustav Adolf: Der Löwe aus Mitternacht,* p. 179.
5. The quotation is from Roberts (1973), *Gustavus Adolphus and the Rise of Sweden,* pp. 52–53.
6. The union was the 1397 Union of Kalmar. Margareta was queen of Denmark by inheritance, of Norway by marriage, and of Sweden by election. The Union was shattered in 1523, when Gustav Eriksson Vasa, Christina's great-grandfather, led the Swedes to revolt against Danish dominance within it.
7. In 1567, at Kalmar Castle, Sweden's King Erik XIV, Gustav Adolf's machiavellian and pathologically suspicious uncle, had murdered one of Sweden's nobles with his own hands. Erik was deposed the following year and died in 1577, probably poisoned on the orders of his brother, King Johan III, with the agreement of the Senate.
8. Livonia: part of present-day Latvia.
9. Letter from Gustav Adolf to Johann Kasimir, August 29, 1621, quoted in Berner (1982), *Gustav Adolf: Der Löwe aus Mitternacht,* p. 201.
10. Letter from Gustav Adolf to Johann Kasimir, May 21, 1625, paraphrased from Stolpe (1966), *Christina of Sweden,* p. 34.
11. Letter from Maria Eleonora to Gustav Adolf's sister, the Princess Katarina, quoted in Berner (1982), *Gustav Adolf: Der Löwe aus Mitternacht,* p. 201.
12. Maria Eleonora appears to have suffered from temporary or periodic phonemic paraphasias in her speech. An acquired aphasia of this kind would be consistent with a stroke, presumably suffered during childbirth, or possibly an aneurism or a benign brain tumor. (In 1629, she had a serious illness of the eyes, which may also have been a stroke symptom.) A condition of this kind could have improved

with time; Charles Ogier, for instance, who spoke with Maria Eleonora in French, makes no mention of a language difficulty during his visit to Sweden in the later 1630s. I am most grateful to Anne Buckley for her analysis of the queen's abnormal use of language, which, at a distance of centuries, must of course remain conjectural.

13. Margareta is more commonly known by her previous married name of Slots (possibly a contraction of Slotsdotter). She was the daughter of a Dutch *Kaufherr* (merchant) named Abraham Cabiljau, and was in fact married three times. Her first husband was an officer of engineers named Andries Sersanders, to whom she was married during her affair with Gustav Adolf. Her second marriage was to a papermaker named Arendt Slots, and her third to a fireworks maker, Jakob Trello. All three of Margareta's husbands were Dutchmen. She is not known to have had any other children. The year of her birth is unknown, but she died in 1669, and was buried in the little church of Vada, north of Stockholm, next to her third husband. Gustav Gustavsson, later the Count of Vasaborg, was to have five children of his own.

14. The eighth of December, 1626, is the day of Christina's birth according to the Julian calendar, then still in use in Sweden, as elsewhere in Protestant Europe. By the Gregorian calendar, already in use throughout Catholic Europe, and later to be adopted in Protestant lands, it was the eighteenth of December, 1626.

15. Quoted in Raymond (1994), *Christine de Suède: Apologies*, p. 92.

16. Quoted in *ibid*.

17. The Latinate spelling of her first name came into use only after her conversion to Catholicism.

18. Gustav Adolf's mother was Christine von Holstein Gottorf (or Gottorp), 1573–1625. She had married Karl IX in 1592, following the death of his first wife, Marie Christine von der Pfalz, in 1589. Gustav Adolf's grandmother Christine was the daughter of the Landgrave of Hesse, Philipp der Grossmütig (the Magnanimous). Christina was also the name of a daughter of the noble family of Flemingh, in Finland, with whom, before his marriage, the young Gustav Adolf had been in love. Two love poems survive that he appears to have dedicated to her. One of these, written in German, is an acrostic, using the first letter of each line to spell out her name. Very little is known about the Finnish Christina, but although the needs of the Swedish state required him to take a different bride, Gustav Adolf's feeling for her may have survived in the name he gave to his daughters. It may be significant that both infants were named Christina, rather than Christine, which was their grandmother's name, but this is more likely to have been simply the use of the usual Swedish version.

19. The most common such malformation is congenital adrenal hyperplasia (CAH), which accounts for about sixty percent of all cases of ambiguous genitalia in the newborn. CAH is the result of a biochemical defect that prevents certain steroids from being produced in sufficient quantities. As the body pushes the adrenal gland harder to increase the steroid level, more and more testosterone is made, encouraging masculinizing traits in females. A female with CAH is technically a pseudohermaphrodite (organically female but in appearance quite masculine).

However, if Christina did suffer from CAH, hers would necessarily have been a mild case, since she would otherwise almost certainly have died in infancy through salt wasting or insufficient cortisol. Other intersex conditions can also cause masculinizing traits in female sufferers, particularly after puberty. I am most grateful to Melissa Cull of the U.K. Adrenal Hyperplasia Network for her help with information on intersex disorders.

20. The conjecture of some kind of intersexuality is not discounted by the results of the autopsy following Christina's death in 1689 (which makes no mention of the external genitals), nor by the examinations of Christina's remains that were carried out at the opening of her grave in 1966 (see Hjortsjö [1966], *The Opening of Queen Christina's Sarcophagus in Rome*). The soft tissues had, of course, disintegrated, and the undoubtedly female skeleton is consistent with a diagnosis of intersexuality.

DEATH OF A KING

1. Evelyn, *The Diary of John Evelyn* (1908 ed.), p. 3 (entry from the year 1624).
2. The precedent for political change by defenestration had been set in Prague two hundred years before, on July 30, 1419, when Hussites (followers of the executed religious reformer Jan Hus) threw the German mayor out of the town hall window, so sparking the "Hussite War" against German papal armies. The later defenestration of May 23, 1618 was a deliberate imitation of the first. The classic history of the Thirty Years' War in English, first published in 1938, is C. V. Wedgwood's *The Thirty Years War*. For a Czech-oriented history of the war, see Polišenský (1971), *The Thirty Years War*.
3. Habsburgs had held the throne of the Holy Roman Empire continuously since 1438, and also before this, though not successively. They were to continue to hold the throne until 1806, when the Holy Roman Empire was dissolved by a new emperor, Napoleon I of France. The Habsburg archdukes, still holding most of their eastern territories, thenceforth styled themselves Emperors of Austria until 1867, the first year of the Dual Monarchy of Austria-Hungary. The Habsburg Austro-Hungarian Empire continued until the end of the First World War.
4. The Protestant Union, also known as the Evangelical League, had been formed in 1608. A Catholic League was formed shortly afterwards in opposition to this.
5. Gustav Adolf's letter of November 1627 to Axel Oxenstierna, quoted in Roberts (1973), *Gustavus Adolphus and the Rise of Sweden*, p. 68.
6. The Truce of Altmark, September 16, 1629. With the corollary Treaty of Tiegenhoff (February 18, 1630), Gustav Adolf made a separate peace with the Hanseatic city of Danzig (now Gdansk).
7. In her classic account of the Thirty Years' War, C. V. Wedgwood discounts Tilly's less flattering epithet of "butcher of Magdeburg." See Wedgwood (1992), *The Thirty Years War*, pp. 286–91.
8. The logistical systems that Wallenstein had developed for military supply were arguably not to be surpassed until the twentieth century. For a colorful summary of his character and early career, see Wedgwood (1992), *The Thirty Years War*,

pp. 170–73. For an example of his celebrated military supply system, see *ibid.*, p. 316.

9. For a brief discussion of Gustav Adolf's innovations in military tactics, see Roberts (1967), *Essays in Swedish History*, chapter 3. For a more detailed discussion of general military advances during the period, see Anderson (1988), *War and Society in Europe of the Old Regime 1618–1789*.

10. The Battle of Breitenfeld took place on September 8/18, 1631. For an engaging account of the battle, see Wedgwood (1992), *The Thirty Years War*.

11. Letter from Christina to Gustav Adolf, originally in German, quoted in Arckenholtz (1751–60), *Mémoires concernant Christine, reine de Suède*, vol. IV, p. 190.

12. The Battle of Lützen took place on November 6/16, 1632. See Wedgwood (1992), *The Thirty Years War*.

13. In 1600, in the infamous "Bloodbath of Linköping," five prominent Swedes, including four members of Sweden's noblest families, had been executed on the orders of Gustav Adolf's father, King Karl IX, to ensure his own hold on power. The Danish king Kristian II had set the precedent in Swedish history by his own "Stockholm Bloodbath" of 1520, in which no fewer than eighty-two Swedish nobles, rebelling against Danish rule, were beheaded in the market square.

14. It now appears that this is not the precise place where Gustav Adolf died but that it is within fifty yards or so of that place.

15. Maria Eleonora's letter quoted in Stolpe (1966), *Christina of Sweden*, p. 35.

16. Nyköping Castle had been the scene of the "Banquet of Nyköping," a particularly gruesome tale in Sweden's history. In 1317, King Birger Ladulås invited his brothers, Erik and Valdemar, to a banquet in the castle, but on their arrival, the two were thrown into the castle dungeon, where they were left to starve to death.

17. Quoted in Raymond (1994), *Christine de Suède: Apologies*, p. 130.

18. This document from a 1634 convocation of the *riksdag*, quoted in Garstein (1992), *Rome and the Counter-Reformation in Scandinavia*, p. 41, note 6.

THE LITTLE QUEEN

1. Raymond (1994), *Christine de Suède: Apologies*, p. 101.

2. Since 1604, it had been accepted that female heirs, in the absence of male heirs, could inherit the Swedish throne, but in 1627, at the second meeting of the Swedish *riksdag*, Gustav Adolf persuaded the parliamentarians to state expressly that, in default of male heirs, his daughter would succeed to the throne after him. Christina herself was to rescind this provision. It would not be until 1980 that female heirs could once again inherit the Swedish throne.

3. Raymond (1994), *Christine de Suède: Apologies*, p. 110.

4. *Ibid.*, p. 111.

5. The peace that Gustav Adolf had made with Muscovy was the Peace of Stolbova of February 27, 1617, by which the Swedes gave up their claim to the Russian throne but gained the provinces of Keksholm and Ingria (Ingermanland). The Russians arrived at Nyköping in the summer of 1633, at about the same time that Maria Eleonora arrived there with the body of the late king.

6. Raymond (1994), *Christine de Suède: Apologies*, p. 130.
7. Baron Gabriel Gustavsson Oxenstierna died in 1641, and was replaced as High Steward by Count Per Brahe.
8. Raymond (1994), *Christine de Suède: Apologies*, p. 114.
9. Christina writes mistakenly that Karl Karlsson Gyllenhjelm had been imprisoned in Poland for *eighteen* years.
10. For details of the long-drawn-out negotiations for the betrothal of Christina and Friedrich Wilhelm, see Schulze (1898), *Die Project der Vermählung Friedrich Wilhelms von Brandenburg mit Christina von Schweden*, and Arnheim (1903–5), *Gustav Adolfs Gemahlin Maria Eleonora von Brandenburg*.
11. Gustav Adolf's visit to Berlin to promote the betrothal between Christina and Friedrich Wilhelm took place in June 1631.
12. Arnheim (1903–5), *Gustav Adolfs Gemahlin Maria Eleonora von Brandenburg*, Jahrbuch 7, p. 177.
13. The Senate met on April 5, 1633. This report is drawn from Arnheim (1903–5), *Gustav Adolfs Gemahlin Maria Eleonora von Brandenburg*, Jahrbuch 7, pp. 176ff.
14. Raymond (1994), *Christine de Suède: Apologies*, p. 94.
15. *Ibid.*, p. 133.
16. *Ibid.*, pp. 130–33.
17. *Ibid.*, p. 132.
18. For details of Maria Eleonora's memorial plans, see Axel-Nilsson, "Pompa memoriae Gustavi Adolphi magni," in von Platen (1966), *Queen Christina of Sweden: Documents and Studies*. The French envoy was Charles Ogier.
19. Raymond (1994), *Christine de Suède: Apologies*, p. 132.
20. These quotations from the directive of the *riksdag* are taken from Clarke (1978), "The Making of a Queen," p. 230.
21. Raymond (1994), *Christine de Suède: Apologies*, p. 131.

love and learning

1. The quotation is from Roberts (1973), *Gustavus Adolphus and the Rise of Sweden*, p. 93, and see *passim* for an overview of Gustav Adolf's educational reforms.
2. Christina's letter of October 28, 1636 to Johan Matthiae, quoted in Arckenholtz (1751–60), *Mémoires concernant Christine*, vol. IV, p. 191.
3. These included parts of Curtius's *History of Alexander the Great*, Christina's favorite text. With her tutor over the next few years, she read more Curtius, Livy, Cicero's speeches against Catiline, Sallust's *Catiline* and *Jugurtha*, Caesar's commentaries, and the plays of Terence, a standard and not unduly demanding program. See Clarke (1978), "The Making of a Queen," p. 230. Clarke suggests that Christina must have read the "major Latin poets" and "the more difficult prose writers" after Matthiae's departure in 1643. He compares her progress unfavorably with that of the "really precocious" Edward VI of England in the previous century (see p. 232).
4. Clarke (1978), "The Making of a Queen," p. 124.
5. Quoted in Masson (1968), *Queen Christina*, p. 58.

6. See Camden (1972 ed.), *History of the Most Renowned and Victorious Princess Elizabeth*. Elizabeth (1533–1603) was the daughter of Henry VIII and his second wife, Anne Boleyn. She became queen on the death of her Catholic half-sister, Mary Tudor, in 1558.

7. Raymond (1994), *Christine de Suède: Apologies*, pp. 121–22.

8. *Ibid.*, p. 121, note 25. Raymond notes that this passage is omitted from Arckenholz.

9. *Ibid.*, p. 123.

10. *Ibid.*, pp. 122–23.

11. Paraphrased from Masson (1968), *Queen Christina*, p. 60.

12. From the *Esquisse de la Reine Christine Auguste*, in Arckenholtz (1751–60), *Mémoires concernant Christine*, vol. IV, pp. 287–90.

13. Raymond (1994), *Christine de Suède: Apologies*, p. 136.

14. The Frenchman, Charles Ogier, is quoted in Stolpe (1966), *Christina of Sweden*, p. 32.

ACORN BENEATH AN OAK

1. The Treaty of Brömsebro was signed between Sweden and Denmark in August 1645. Sweden gained the Baltic islands of Gotland and Ösel and the western provinces of Jämtland and Härjedalen, and also (for thirty years only) the southern province of Halland, on the Sound.

WARRING AND PEACE

1. Juncker Päär, quoted in Roberts (1967), *Essays in Swedish History*, p. 137, note 77.

2. Paraphrased from Johan Ekeblad, quoted in Roberts (1967), *Essays in Swedish History*, p. 137, note 77.

3. Sweden became involved in the slave trade in 1647 through the efforts of Dutch entrepreneurs working for the Swedish Africa Company.

4. The most prominent of these Dutch entrepreneurs was Louis de Geer, who purchased vast tracts of land in Sweden and set up mining, manufacturing, banking, retail, and other operations within the country. For an overview, see Rich and Wilson, eds. (1967), *Cambridge Economic History of Europe*, vol. IV, pp. 565–66. One notable Swedish exception was Christina's own mistress of the wardrobe, Madame Beata Oxenstierna (née De la Gardie), who had built up a substantial clothing business, and even owned a fleet of ships.

5. Johan Matthiae's book, published in 1645, was the *Idea boni ordinis in Ecclesia Christi* (Plan for Good Order in the Church of Christ).

6. From the text of the Treaty of Münster. British Foreign Office translation.

7. The Dutch poet was Jan Zoet. The Turks finally took Crete on September 6, 1669, after twenty-four years of fighting.

8. By the Treaty of Westphalia, Sweden also gained the Pomeranian town of Stettin, the island of Rügen, the two Mecklenburg towns of Wismar and Warnemünde,

and the secularized bishoprics of Verden and Bremen (though not the town of Bremen itself).

9. This warning appears in Gustav Adolf's last letter to Axel Oxenstierna (November 9, 1632); see Wedgwood (1938), *The Thirty Years War*, p. 337. By the Treaty of Westphalia, France gained "sovereignty over Metz, Toul, and Verdun; Pinerolo; the Sundgau in southern Alsace; Breisach, garrison rights in Philippsburg; the *Landvogtei* or 'Advocacy' of ten further Alsatian cities." See Davies (1997), *Europe*, p. 567.

10. During the 1640s, revolts against the Spanish Habsburgs had broken out in Catalonia, Portugal, and Sicily. See Elliott (1963), *Imperial Spain*.

11. The Czech was the educational reformer Comenius (1592–1670). Silesia's Protestants were guaranteed the right to practice their own religion. See Cauly (1995), *Comenius*, p. 262.

12. The bull *Zelus Domus Dei* (Jealousy of the House of God) was issued by Pope Innocent X on November 26, 1648, following his receipt of the terms of the Treaty of Münster. The quotation is taken from Davies (1997), *Europe*, p. 568.

13. Following the Peace of Westphalia, the Dutch held the Rhine River, the Danes the Elbe, and the Swedes the Oder.

14. As late as October 1648, with the preliminary treaties already signed, leaving Prague in Catholic hands, Comenius was writing to urge Axel Oxenstierna to save the city for the evangelical cause. See Cauly (1995), *Comenius*, pp. 262–63.

15. See Cavalli-Björkman (1999), *La Collection de la reine Christine à Stockholm*.

16. The paper was the *Wöchentliche Zeitung*. See Reinken (1966), *Deutsche Zeitungen über Königin Christine: 1626–1689*, pp. 44–47.

17. The "ell" was about 45 inches, or 115 centimeters.

PALLAS OF THE NORTH

1. In 1632, Gustav Adolf looted a major collection from his doubtful ally, Maximilian of Bavaria, who responded by taking Generals Torstensson and Horn hostage. See Cavalli-Björkman (1999), "*La Collection de la reine Christine à Stockholm*," p. 298.

2. See Blom (2002), *To Have and to Hold*, pp. 27–49. The phrase "encyclopedia of the visible world" is from Evans (1997), *Rudolf II and His World*, p. 177.

3. Christina's letter of May 22, 1652, to the Duca di Bracciano is quoted in Cavalli-Björkman (1999), *La Collection de la reine Christine à Stockholm*, p. 302.

4. The period of civil war in France that lasted from 1648 to 1653 is generally known as *La Fronde* (the word *fronde* means a slingshot). There were in fact two rebellions, though the causes overlapped and many people changed allegiance over the years. During the First Fronde, or the *Fronde du Parlement* (1648–51), the Parliament rebelled against the principle of absolute monarchy, as exercised by the child-king's regents, his mother, Anne of Austria, and Cardinal Mazarin. During the Second Fronde, or the *Fronde des Princes* (1651–53), the nobility, led by the Prince de Condé (*le Grand Condé*), rebelled against the power of Cardinal Mazarin himself, and sought to drive him permanently out of France. For an

overview of the Fronde in English, see Dunlop (1999), *Louis XIV*, chapter two. See also Pujo (1995), *Le Grand Condé*. For a contemporary account, see de Wicquefort, ed. (1978), *Chronique Discontinue de la Fronde*.

5. The unnamed Swedish scholar is quoted in Arckenholtz (1751–60), *Mémoires concernant Christine*, IV, p. 232.
6. Dorpat: now Tartu, in eastern Estonia.
7. Letter from Chanut, quoted in Kermina (1995), *Christine de Suède*, p. 39.
8. Chanut's brother-in-law was Claude Clerselier (1614–84), Descartes's translator and posthumous editor.
9. Paraphrased from a letter from Celsius to Georges de Scudéry, quoted in Arckenholtz (1751–60), *Mémoires concernant Christine*, IV, p. 233.
10. See Åkerman (1991), *Queen Christina of Sweden and Her Circle*, p. 39, note 83.
11. The epithet likened Christina to Pallas Athene, the Greek goddess of wisdom.

Tragedy and comedy

1. Descartes's letter to Chanut, November 1, 1646, quoted in Descartes (1989 ed.), *Correspondance avec Elisabeth*, p. 246.
2. This was a common riposte at the time to the implications of Copernicus and Galileo that the universe must be infinite. Descartes responded in a letter to Chanut of June 6, 1647. See *ibid.*, pp. 263–64.
3. Descartes's letter to Chanut, June 6, 1647, quoted in *ibid.*, pp. 263–64.
4. Descartes's letter to Chanut, June 6, 1647, quoted in *ibid.*, p. 268.
5. Descartes's letter to Christina of November 20, 1647, quoted in *ibid.*, pp. 270–72.
6. Descartes's letter to Chanut, November 1, 1646, quoted in *ibid.*, p. 247.
7. Descartes's letter to Chanut, February 21, 1648, quoted in *ibid.*, p. 276.
8. Descartes's letter to Chanut, May 1648, quoted in *ibid.*, p. 279.
9. Descartes's letter to Christina, February 26, 1649, quoted in *ibid.*, p. 284.
10. Descartes's letter to Chanut, February 26, 1649, quoted in *ibid.*, p. 281.
11. Descartes's first letter to Chanut of March 31, 1649, quoted in *ibid.*, p. 285.
12. Descartes's second letter to Chanut of March 31, 1649, quoted in *ibid.*, p. 287.
13. Descartes's letter to Chanut of February 26, 1649, and second letter to Chanut of March 31, 1649, quoted in *ibid.*, pp. 283 and 288–89.
14. Descartes's letter to Chanut, November 1, 1646, quoted in *ibid.*, p. 246. Princess Elisabeth had been the dedicatee of his *Principles of Philosophy*, published in Amsterdam in 1644.
15. Descartes's letter to Chanut, March 6, 1646, quoted in *ibid.*, p. 241.

Hollow crown

1. Poland was already stricken by the Chmielnicki cossack uprising from which so many refugees would make their way to Sweden. In the years 1648–54, Bogdan Chmielnicki (Khmelnytsky), led an unsuccessful rebellion of Dnieper cossacks and peasants against King Wladyslaw IV Vasa and his successor, Jan II Kazimierz Vasa.

2. Christina's declaration of January 1649 is quoted in Kermina (1995), *Christine de Suède*, p. 67.
3. Charles I was executed on January 30, 1649. See Fraser (1973), *Cromwell*, pp. 290–91. The quotation is from Andrew Marvell's "An Horatian Ode upon Cromwel's Return from Ireland," lines 63–64.
4. Christina's coronation took place on October 20, 1650 by the old Julian calendar then still in use in most of Protestant Europe. By the Gregorian calendar, later adopted generally, the date was ten days later. The coronation had initially been planned for August 1647.
5. Both quotations are from Kermina (1995), *Christine de Suède*, pp. 132–33.
6. For a detailed description of the robe and other aspects of the coronation, see Gudrun Ekstrand, "A Robe of Purple Velvet for Queen Christina," in von Platen (1966), *Queen Christina of Sweden: Documents and Studies*. The crowns were removed for reuse in 1774.
7. In the event, Jordaens's paintings, which depicted Ovid's tale of Psyche, were hung in the royal library in the Tre Kronor Castle in Stockholm. They were destroyed in the castle fire of 1697.
8. Christina's second coronation arch was painted by Nicolas Vallari.
9. Father Antonio Macedo's eyewitness report is quoted by Sten Karling in "L'Arc de triomphe de la Reine Christine à Stockholm," in von Platen, ed. (1966), *Queen Christina of Sweden: Documents and Studies*.
10. The De la Gardies' country house, at the bay of Edsviken near Stockholm, is still standing. It is now known as Ulriksdal.
11. For details of Christina's silver throne, see Carl Hernmarck, "Der Silberthron Christinas," in von Platen (1966), *Queen Christina of Sweden: Documents and Studies*.
12. Christina informed the Senate (the *råd*) on August 7, 1651.
13. They included the mayor of Stockholm, Nils Nilsson; Senator Bengt Skytte; and Bishop Abo Tersenius. The two condemned men refused to reveal the names of some of the most powerful men involved.

THE ROAD TO ROME

1. Christina's letter of August 1651 to the General of the Jesuit Order, quoted in Garstein (1992), *Rome and the Counter-Reformation in Scandinavia*, p. 635.
2. Diary of John Evelyn (1908), p. 330, entry for September 23, 1680. Evelyn had been speaking with "Signor Pietro," one of Christina's court musicians who had lived in Stockholm before her abdication.
3. Descartes's letter to Christina, November 20, 1647, quoted in *Descartes, Correspondance avec Elisabeth*, p. 272.
4. Quoted in Kermina (1995), *Christine de Suède*, p. 302.
5. Garstein (1992), *Rome and the Counter-Reformation in Scandinavia*, p. 703.
6. Quoted in Garstein (1992), *ibid.*, p. 705.

7. The French had been at war with the Spanish Habsburg Empire since 1643, and had signed no mutual treaties at the Peace of Westphalia. The war would continue until the Peace of the Pyrenees in 1659.

8. Christina's letter of December 5, 1653, to Magnus De la Gardie, quoted in Arckenholtz (1751–60), *Mémoires concernant Christine*, vol. I, p. 359. The letter was translated into Latin for Christina by Nicolaas Heinsius, a celebrated Latinist at her court.

ABDICATION

1. Christina's letter of December 19, 1653, to Madame Saumaise, quoted in Arckenholtz (1751–60), *Mémoires concernant Christine*, vol. I, pp. 233–34.

2. Whitelocke (1855), *A Journal of the Swedish Embassy*, Vol. 2, p. 282.

3. *Ibid.*, pp. 220–21.

4. *Ibid.*, p. 223.

5. *Ibid.*, p. 231.

6. *Ibid.*, p. 314.

7. *Ibid.*, p. 316.

8. Raymond (1994), *Christine de Suède: Apologies*, p. 102.

CROSSING THE RUBICON

1. Garstein (1992), *Rome and the Counter-Reformation in Scandinavia*, p. 725.

2. Arckenholtz (1751–60), *Mémoires concernant Christine*, vol. III, p. 174.

3. From a report to John Thurloe, quoted in Masson (1968), *Queen Christina*, p. 220.

4. The phrase is Antonia Fraser's. See Fraser (1973), *Cromwell*, p. 428.

5. Andrew Marvell, "The Character of Holland," in *The Complete Poems*, p. 88, lines 1–8.

6. For a description of Antwerp at about this time, see Schama (1999), *Rembrandt's Eyes*, chapter 4. The quotation is from p. 159.

7. I am indebted to Dr. Mary Frandsen of the University of Notre Dame in Indiana for bringing this letter to my attention. See Sächsische Hauptstaatsarchiv (Dresden), Loc. 8563/1, fol. 858r-v.

8. Letter from Charles Longland in Livorno to John Thurloe, July 3, 1654, quoted in Garstein (1992), *Rome and the Counter-Reformation in Scandinavia*, p. 726.

9. For details regarding the plays Christina attended, see Lanoye (2001), *Christina van Zweden*.

10. Quoted in Garstein (1992), *Rome and the Counter-Reformation in Scandinavia*, p. 698, note 8.

11. Pallavicino (1838), *Descrizione del primo viaggio fatto a Roma*, p. 40.

12. *Ibid.*, p. 1.

13. Christina's letter of November 5, 1655, to Pope Alexander VII, quoted in *ibid.*, p. 42.

ROME AT LAST

1. Pope Alexander VII's address to the cardinals, quoted in Garstein (1992), *Rome and the Counter-Reformation in Scandinavia,* pp. 749–50.
2. Quoted in Rodén (2000), *Church Politics in Seventeenth-Century Rome,* p. 114.
3. The city of Castro, in the region of Bolsena, had been held as a papal fief by the Farnese family since 1537. Pope Urban VIII had tried to take it from the family for the benefit of his own nephews. In 1649, the Duca Ranuccio II Farnese was implicated in the murder of the proposed new bishop of Castro, and the city was destroyed on the orders of Pope Innocent X.
4. The Palazzo Farnese was commissioned by Cardinal Alessandro Farnese, later Pope Paul III, and was built during the years 1514–89. It had four successive architects: Sangallo, Michelangelo, Vignola, and Della Porta. Since 1875 it has housed the French Embassy in Rome.
5. See Dempsey (1995), *Annibale Carracci: The Farnese Gallery, Rome.*
6. Christina's letter to Belle of January 6, 1656, paraphrased from Masson (1968), *Queen Christina,* pp. 263–64.
7. It was first performed on January 31, 1656. Revivals of Marazzoli's operas *Dal male il bene* and *Le armi e gli amori* were also given in Christina's honor at the Palazzo Barberini. Tenaglia's *Il giudizio di Paride,* now lost, was staged for her at the Palazzo Pamphili, while at the Jesuits' Collegio Germanico, she was honored with Carissimi's *Il sacrificio d'Isaaco* and *Giuditta,* also both lost.
8. Magnuson (1982), *Rome in the Age of Bernini,* p. 190. The Jews' races were finally banned by Christina's friend, Pope Clement IX, in 1668.
9. A painting by Filippo Gagliardi and Filippo Lauri of this Barberini *Giostra delle Caroselle* is held in Rome's Museo di Roma.
10. A modest family of the time could live on about 150 livres per year.
11. Barbara Strozzi (1619–77) and Leonora Baroni (1611–70). Both were singers and composers. See Bowers (1986), *The Emergence of Women Composers in Italy.*

LOVE AGAIN

1. Extract from Gregorio Leti's *Il Livello Politico* of 1678, quoted in Rodén (2000), *Church Politics in Seventeenth-Century Rome,* p. 81.
2. For a detailed description of a cardinal's required clothing and other appurtenances, see *ibid.,* pp. 78–79.
3. The code they used was not deciphered until the very end of the nineteenth century, by the Swedish diplomat Baron de Bildt, an important editor of Christina's letters. For details of the codes used, see the appendices in Bildt (1899), *Christine de Suède et le Cardinal Azzolino: Lettres inédites.*
4. See Rodén (2000), *Church Politics in Seventeenth-Century Rome,* pp. 93–94 for a list and brief biography of each of the eleven original members of the *Squadrone Volante.* The description is Pallavicino's.

5. Raymond (1994), *Christine de Suède: Apologies*, p. 126.
6. Andrew Marvell, "A Letter to Dr. Ingelo," in *The Complete Poems*, pp. 240–41, line 60. The English translation of the Latin original is by A. B. Grosart.
7. The phrase belongs to Baron de Bildt, editor of Christina's letters to Azzolino.

Fair Wind For France

1. Letter of April 14, 1656 from Charles Longland in Livorno to John Thurloe in London, quoted in Garstein (1992), *Rome and the Counter-Reformation in Scandinavia*, p. 767.
2. Quoted in Castelnau (1944), *Christine Reine de Suède*, p. 152.
3. Christina's meeting with Malaval may in fact have taken place on her second visit to Marseilles, in the summer of 1657.
4. Quoted in Bernard (1970), *The Emerging City: Paris in the Age of Louis XIV*, p. 69.
5. Quoted in Gobry (2001), *La Reine Christine*, pp. 203–4.
6. Paraphrased from Gobry (2001), *La Reine Christine*, pp. 205–6.
7. Menestrier had probably been trained in the "memory palace" technique, a Renaissance practice that survived in Jesuit schools. Reciting lists of unconnected words was a performance feature of this technique. See Spence (1985), *The Memory Palace of Matteo Ricci*.
8. The horseshoe staircase was built by Jean Androuet du Cerceau in 1634, in the reign of Louis XIII. The gardens had been laid out by his father, Henri IV, the first Bourbon king of France. See Dunlop (1985), *Royal Palaces of France*. The "marvellous fountain" quotation is from Dunlop (1999), *Louis XIV*, p. 68.
9. The quotation is from Count Orlov (1824), *Voyage dans une partie de la France*, Letter XXII.
10. Madame de Montpensier, *Mémoires* (1985 ed.), Vol. I, p. 399.

The Rising Sun

1. Molière, *Les Précieuses ridicules*, act I, scene ix.
2. Loret's verse is quoted in Gobry (2001), *La Reine Christine*, pp. 210–11.
3. Patru, *Harangue de Monsieur Patru faite en 1656*, p. 78. Olivier Patru was elected to the Académie in 1640, and at his first attendance gave such an eloquent speech of thanks that a *harangue de réception* was thenceforth required of all new *académiciens*. Despite more than forty years' membership of the Académie, Patru never did win the *prix d'éloquence*.
4. Quoted in Bernard (1970), *The Emerging City: Paris in the Age of Louis XIV*, p. 88.
5. Madame de Montpensier (1985 ed.), *Mémoires*, vol. I, p. 408.
6. See La Rochefoucauld, "Portrait de M.R.D. fait par lui-même," in La Rochefoucauld (1999 ed.), *Maximes*, pp. 253–54.
7. The phrase is Émile Zola's.
8. Molière, *Les Précieuses ridicules*, act I, scene ix.

9. Madame de Motteville's *Mémoires*, quoted in Gobry (2001), *La Reine Christine*, pp. 219–20.
10. Marie Mancini's ungenerous contemporary was the Comte de Bussy-Raboutin, quoted in Dunlop (1999), *Louis XIV*, p. 48.
11. Madame de Montpensier (1985 ed.), *Mémoires*, vol. I, p. 409.
12. Christina's letter to Azzolino, quoted in Neumann (1936), *The Life of Christina of Sweden*, pp. 194–95.
13. Madame de Montpensier (1985 ed.), *Mémoires*, vol. I, pp. 409–10.

FONTAINEBLEAU

1. Quoted in Kermina (1995), *Christine de Suède*, p. 199.
2. Lascaris's letter to Azzolino of December 28, 1656, quoted in Neumann (1936), *The Life of Christina of Sweden*, p. 198. The original pun is a play on Lascaris's name (*lasca*, a kind of fish), and *cefalo* (another kind of fish, but also meaning an erection).

AFTERMATH

1. From Madame de Motteville's *Mémoires*, quoted in Gobry (2001), *La Reine Christine*, p. 242.
2. Evelyn (1908 ed.), *The Diary of John Evelyn*, pp. 329–30; entry for September 23, 1680.
3. Cardinal Mazarin's letter to Christina, quoted in *ibid.*, p. 243.
4. Christina's letter to Cardinal Mazarin, quoted in Castelnau (1944), *La Reine Christine*, p. 206.
5. Paraphrased from Stolpe (1966), *Christina of Sweden*, pp. 241–42.
6. Christina's letter to Chanut, quoted in Neumann (1936), *The Life of Christina of Sweden*, p. 207.
7. Christina's letter of November 15, 1657, to Francesco Maria Santinelli, paraphrased from Neumann (1936), *The Life of Christina of Sweden*, pp. 207–8.
8. Mazarin's letter to Karl X Gustav, quoted in Kermina (1995), *Christine de Suède*, p. 206.
9. Christina's letter to Mazarin, quoted in *ibid.*, pp. 209–10.
10. See Madame de Montpensier (1985 ed.), *Mémoires*, vol. II, pp. 23–24.
11. See *ibid.*, pp. 32–33.
12. From Madame de Motteville's *Mémoires*, paraphrased from Masson (1968), *Queen Christina*, p. 299.
13. University debates throughout France found uniformly that Christina had acted legally. The German philosopher and jurist Leibniz later also found in her favor. He attempted to meet her during his visit to Rome in 1689, but his arrival in the city coincided with her last illness.
14. Quoted in Masson (1968), *Queen Christina*, p. 211.
15. Mazarin's letter to Ambassador Terlon, quoted in *ibid.*, p. 212.

OLD HAUNTS, NEW HAUNTS

1. Now the Palazzo Corsini, on the Via della Lungara, toward the district of Traste-vere. The palazzo was extensively rebuilt during the eighteenth century.
2. The most notable differences were those over the 1659 Treaty of the Pyrenees be-tween France and Spain, in which the pope had refused to allow Louis a free hand in the granting of bishoprics and abbeys.
3. Quoted in Kermina (1995), *Christine de Suède*, p. 223.
4. Quoted in *ibid.*, pp. 223–24.
5. Paraphrased from Masson (1968), *Queen Christina*, p. 317.
6. From a report from the Venetian ambassador, quoted in Neumann (1936), *The Life of Christina of Sweden*, p. 224.
7. The basis of Christina's manuscript collection was the loot brought back to Sweden during the Thirty Years' War, particularly from Prague. The queen added to this by buying complete collections of manuscripts, notably those belonging to Hugo Grotius, Gerard Vossius, Pierre Bourdelot, and, above all, Alexandre Petau. It was a diverse collection, including classical, religious, medical, and philosophical works, a large number of Old French literary texts, and astronomical treatises, including some autographed by the Danish astronomer Tycho Brahe. See *Christina, Queen of Sweden: A Personality of European Civilization* (1966), pp. 529–30.
8. Edward Browne writing of Christina in January 1665, quoted in Evelyn (1908), *The Diary of John Evelyn*, p. 329, note 1.
9. Hollingworth (1927), *The History of the Intrigues and Gallantries of Christina, Queen of Sweden*, p. 216.
10. *Ibid.*, p. 215.
11. *Ibid.*, p. 146.

DEBACLE

1. Letter from Christina in Hamburg to Hugues de Lionne, December 11, 1666, quoted in Bildt (1899), *Christine de Suède et le Cardinal Azzolino*, p. 284.
2. Letter from Christina in Hamburg to Azzolino, November 17, 1666, quoted in *ibid.*, p. 265.
3. Letter from Christina in Hamburg to Azzolino, September 29, 1666, quoted in *ibid.*, pp. 232–33.
4. Letter from Christina in Hamburg to Azzolino, September 29, 1666, quoted in *ibid.*, p. 235. The Great Fire of London burned from September 12 to 16, 1666. Peace was not made until the following May.
5. Letter from Christina in Hamburg to Azzolino, June 22, 1667, quoted in Bildt (1899), *Christine de Suède et le Cardinal Azzolino*, pp. 369–70.
6. Karl X Gustav had taken Hälsingborg from the Danes in 1658. It is now once again part of Denmark, and is known as Helsingør.
7. Christina's letter to Macchiati of June 24, 1667, quoted in Bildt (1899), *Christine de Suède et le Cardinal Azzolino*, p. 368.

8. Letter from Christina in Jönköping to King Karl XI, May 24, 1667, quoted in *ibid.*, p. 354.
9. Letter from Pontus De la Gardie in Norrköping to King Karl XI, May 21/31, 1667, quoted in *ibid.*, p. 355.
10. Letter from Count Pontus De la Gardie at Skarhult to King Karl XI, May 27/June 6, 1667, quoted in *ibid.*, p. 357.
11. Letter from Christina in Hälsingborg to the French ambassador, Terlon, at Copenhagen, June 6, 1667, quoted in *ibid.*, p. 356.
12. Letter from Christina in Hamburg to Azzolino, June 15, 1667, quoted in Arckenholtz, (1751–60), *Mémoires concernant Christine*, vol. II, pp. 113–16.
13. Hollingworth (1927), *The History of the Intrigues and Gallantries of Christina, Queen of Sweden*, pp. 49–50.
14. Extract quoted in Bildt (1899), *Christine de Suède et le Cardinal Azzolino*, pp. 378–79.
15. Hollingworth (1927), *The History of the Intrigues and Gallantries of Christina, Queen of Sweden*, pp. 50–52.
16. *Ibid.*, pp. 52–53.
17. Letter from Christina in Hamburg to Azzolino of August 3, 1667, quoted in Bildt (1899), *Christine de Suède et le Cardinal Azzolino*, pp. 381–82.

MIRAGES

1. Christina's letter to Azzolino of November 17, 1666, quoted in Bildt (1899), *Christine de Suède et le Cardinal Azzolino*, p. 265.
2. Christina's letter to Azzolino of January 5, 1667, quoted in *ibid.*, p. 292.
3. Letter from Christina in Hamburg to Azzolino of August 4, 1666, quoted in *ibid.*, pp. 189–90.
4. Letter from Christina in Hamburg to Azzolino, April 20, 1667, quoted in *ibid.*, pp. 338–39.
5. Letter from Christina in Hamburg to Azzolino, July 28, 1666, quoted in *ibid.*, pp. 183–84.
6. Christina's letter to Azzolino of June 23, 1666, quoted in *ibid.*, p. 163.
7. Christina's letter to Azzolino of November 3, 1666, quoted in *ibid.*, p. 257.
8. Christina and Azzolino used two codes, the *cifra grande* and the *cifra piccola*— the "big code" and the "little code." The "big code" was used for official correspondence, and the "little code" reserved for personal phrases. Both were deciphered at the end of the nineteenth century by the Swedish diplomat Baron Carl de Bildt, who was then serving in Rome. Both codes are given in full in the appendices to Bildt (1899), *Christine de Suède et le Cardinal Azzolino*. In the preface to his book (pp. xxvff.), Baron de Bildt explains how he discovered the key to them. The coded passages from Christina's letters are quoted in italics.
9. Christina's letter to Azzolino of January 12, 1667, quoted in Bildt (1899), *Christine de Suède et le Cardinal Azzolino*, p. 295. The words in italics were coded.
10. Letter from Christina in Hamburg to Azzolino of January 26, 1667, quoted in *ibid.*, p. 305. The words in italics were coded.

11. Letter from Christina in Hamburg to Azzolino of March 9, 1667, quoted in *ibid.*, p. 321.
12. Christina's letter to Azzolino of October 5, 1667, quoted in *ibid.*, p. 392.
13. Christina's first letter of July 18, 1668, to Azzolino, quoted in *ibid.*, p. 456.
14. Quoted in Raymond (1994), *Christine de Suède: Apologies*, p. 73.
15. Quoted in *ibid.*, p. 80.
16. The existing manuscripts for *La Vie de la reine Christine, faite par elle-même, dédiée à Dieu* date from June 11, 1681. They are held in the Royal Archives in Stockholm (Azzolinosamlingen K430) and in the manuscript collection of the Bibliothèque Interuniversitaire de Montpellier.
17. Azzolino's letter to Marescotti is quoted in Bildt (1899), *Christine de Suède et le Cardinal Azzolino*, p. 449.
18. The quotations from Christina's letter to Marescotti are from *ibid.*, p. 447.
19. Christina's letter of August 1, 1668, to Azzolino, quoted in *ibid.*, pp. 460–61.
20. Quoted in Raymond (1994), *Christine de Suède: Apologies*, pp. 135–36.
21. Hollingworth (1927), *The History of the Intrigues and Gallantries of Christina, Queen of Sweden*, p. 214.
22. Christina's second letter of July 18, 1668, quoted in Bildt (1899), *Christine de Suède et le Cardinal Azzolino*, p. 457.
23. Christina's letter of August 20, 1668, to Azzolino, quoted in *ibid.*, pp. 467–68.

GLORY DAYS

1. The libretto, which Christina herself wrote, together with the Abbé Alessandro Guidi, was *Clore e Damone*. It does not appear to have survived.
2. Hollingworth (1927), *The History of the Intrigues and Gallantries of Christina, Queen of Sweden*, p. 190.
3. Christina's letter to her French agent, paraphrased from Masson (1968), *Queen Christina*, pp. 360–61.
4. From John Evelyn's Diary, quoted in Hibbert (1985), *Rome*, p. 179.
5. The description of Bernini is quoted in *ibid.*, p. 197.
6. Most of Christina's sculptures are now housed in Madrid's Prado Museum. They were sold by Azzolino's nephew to the Odescalchi family, and bought from them in the eighteenth century by King Felipe V of Spain, grandson of Louis XIV.
7. Christina's letter to Azzolino of August 1, 1668, quoted in Bildt (1899), *Christine de Suède et le Cardinal Azzolino*, p. 464. Clement IX (Giulio Rospigliosi) was elected pope in June 1667, and died in December 1669.
8. It is now known as the Palazzo Torlonia.
9. Paraphrased from Rodén (2000), *Church Politics in Seventeenth-Century Rome*, pp. 217–18.
10. Paraphrased from *ibid.*, p. 226.
11. Bowers (1986), "The Emergence of Women Composers in Italy," p. 141.
12. Quoted in Masson (1968), *Queen Christina*, p. 376.
13. The declaration of August 15, 1686, is quoted in Neumann (1936), *The Life of Christina of Sweden*, pp. 221–22.

14. See Hollingworth (1927), *The History of the Intrigues and Gallantries of Christina, Queen of Sweden*, pp. 155–56.
15. Christina's letter to Pope Innocent XI is quoted in *ibid.*, pp. 161–62.
16. *Ibid.*, p. 217.
17. Azzolino's letter to Christina of December 1679, quoted in Rodén (1986), "Drottning Christina och Kardinal Decio Azzolino: Kärleksbrev från det sista decenniet" (K 415, no. 29), p. 71.
18. Letter from Azzolino to Christina, early 1680, quoted in *ibid.*, (K 415, no. 8), p. 72.
19. The cardinal's letter of "infinite thanks" to Christina is quoted in *ibid.*, (K 415, no. 15), p. 72.
20. Quoted in Raymond (1994), *Christine de Suède: Apologies*, p. 128.
21. *Les Sentiments*, no. 386, quoted in Bildt (1906), *Pensées de Christine*, p. 239. The reference to Christina's father is quoted in Raymond (1994), *Christine de Suède: Apologies*, p. 89.
22. See La Rochefoucauld (1999 ed.), *Maximes: suivies des Réflexions diverses*.
23. *Les Sentiments*, nos. 53, 203, 377, 288, and 402, quoted in Bildt (1906), *Pensées de Christine*, pp. 179, 49, 237, 60, and 245.
24. *Les Sentiments*, no. 428, quoted in *ibid.*, p. 250.
25. *Les Sentiments*, no. 429, quoted in *ibid.*, pp. 250–51.
26. For a discussion of Azzolino's *Relatione de' Pontefici*, see Rodén (2000), *Church Politics in Seventeenth-Century Rome*, chapter 9.
27. Marie-Louise Rodén refers to fragmentary sources from 1679, now held in Riksarkivet (the National Archive) in Stockholm. See *ibid.*, p. 154, note 30.
28. *Les Sentiments*, nos. 202 and 203, quoted in Bildt (1906), *Pensées de Christine*, pp. 202–3.

JOURNEY'S END

1. Paraphrased from Francois-Maximilian Misson's description of April 1688, quoted in Masson (1968), *Queen Christina*, p. 384.
2. See Baldinucci's *Vita di Gian Lorenzo Bernini*.
3. *Les Sentiments*, nos. 85 and 380, quoted in Bildt (1906), *Pensées de Christine*, pp. 184 and 237–38.
4. Hollingworth (1927), *The History of the Intrigues and Gallantries of Christina, Queen of Sweden*, p. 184.
5. *Ouvrage du Loisir*, no. 1084, quoted in Bildt (1906), *Pensées de Christine*, p. 160.
6. Hollingworth (1927), *The History of the Intrigues and Gallantries of Christina, Queen of Sweden*, p. 198.
7. *Ibid.*, p. 199.
8. *Ibid.*, pp. 207–8.
9. Modern medical reviewers suggest that Christina may have suffered from adult-onset diabetes.
10. Christina's letter to Johan Olivecrantz, quoted in Kermina (1995), *Christine de Suède*, p. 304.

11. See Hollingworth (1927), *The History of the Intrigues and Gallantries of Christina, Queen of Sweden*, pp. 200–1.

12. Azzolino's letter to Cardinal Cibo at the Vatican Secretariat of State, April 19, 1689, quoted in Rodén (1987), "The Burial of Queen Christina of Sweden in St. Peter's Church," Appendix. Azzolino dated the letter April 28, 1689.

13. Hollingworth (1927), *The History of the Intrigues and Gallantries of Christina, Queen of Sweden*, p. 211.

14. Christina's memorial was commissioned in 1696 by Pope Innocent XII Pignatelli and was sculpted by Carlo Fontana. The funds to complete it were provided in 1701 by Pope Clement XI Albani.

15. Hollingworth (1927), *The History of the Intrigues and Gallantries of Christina, Queen of Sweden*, p. 212.

16. From Azzolino's letter to his cousin Francesco, quoted in Rodén (2000), *Church Politics in Seventeenth-Century Rome*, p. 299.

17. Christina's official correspondence remained in the possession of the Azzolino family, together with the cardinal's own papers, until 1925, when they were purchased by the Swedish state; they are now held in Riksarkivet (the National Archive) in Stockholm under the title of *Azzolino-samlingen* (the Azzolino Collection). In 1985, Azzolino's private papers (*L'Archivio Azzolino*) were donated to the Biblioteca Comunale in the Italian city of Jesi.

18. Azzolino's letter of April 30, 1689, to Giovanni Mattia del Monte, quoted in Rodén (2000), *Church Politics in Seventeenth-Century Rome*, p. 308, note 56.

19. Hollingworth (1927), *The History of the Intrigues and Gallantries of Christina, Queen of Sweden*, p. 222.

20. Christina's books and manuscripts were bought by Pope Alexander VIII (Pietro Ottoboni), and in 1690 the manuscripts were donated to the Vatican Library, where they are now classified together as *Codices Reginenses Graeci et Latini*. Christina's collection of printed books remained in the possession of the Ottoboni family until the mid-eighteenth century, and was then dispersed to individual collectors.

21. Hollingworth (1927), *The History of the Intrigues and Gallantries of Christina, Queen of Sweden*, p. 222. Christina's collection of medallions was bought by Don Livie Odescalchi, nephew of Pope Innocent XI Odescalchi. In 1797, Pope Pius VI Braschi ceded the collection to France as part of the Treaty of Tolentino. See Bildt (1908), *Les Médailles romaines de Christine de Suède*.

BIBLIOGraPHY

Accademia nazionale dei Lincei, Convegno internazionale: *Cristina di Svezia e la musica*: Actes d'un congrès international organisé à l'Accademia nazionale dei Lincei, Rome, December 5–6, 1996 (Atti dei convegni lincei 138, 1998).

Åkerman, Susanna, *Queen Christina of Sweden and Her Circle: The Transformation of a Seventeenth-Century Philosophical Libertine* (Leiden: E.J. Brill, 1991).

Anderson, M.S., *War and Society in Europe of the Old Regime 1618–1789* (Guernsey, Channel Islands: Sutton Publishing, 1998).

Arckenholtz, Johann Wilhelm, *Mémoires concernant Christine, reine de Suède pour servir d'éclaircissement à l'histoire de son regne et principalement de sa vie privée, et aux événements de son tems civile et litéraire*, 4 vols. (Leipzig and Amsterdam: 1751–60).

Arnheim, Fritz, "Gustav Adolfs Gemahlin Maria Eleonora von Brandenburg: Eine biographische Skizze," in *Hohenzollern Jahrbuch* 7, 8, and 9, 1903–05.

Baldinucci, Filippo, *Vita di Gian Lorenzo Bernini* (Milan: Edizioni del Milione). First published in 1682 by Vincenzo Vangelisti, Florence.

Bernard, Leon, *The Emerging City: Paris in the Age of Louis XIV* (Durham, North Carolina: Duke University Press, 1970).

Berner, Felix, *Gustav Adolf: Der Löwe aus Mitternacht* (Stuttgart: Deutsche Verlags-Anstalt, 1982).

Bildt, C.D.N., Le Baron de, *Christine de Suède et le Cardinal Azzolino: Lettres inédites, 1666–68, avec une introduction et des notes* (Paris: E. Plon, Nourrit et Cie, 1899).

——, *Pensées de Christine, reine de Suède* (Stockholm: P. A. Norstedt & Söner, 1906).

——, *Christine de Suède et le conclave de Clement X, 1669–1670* (Paris: E. Plon, Nourrit et Cie, 1906).

——, *Les Medailles romaines de Christine de Suède* (Rome: 1908).

Bjurström Per, *Feast and Theatre in Queen Christina's Rome* (Stockholm: Nationalmuseum, 1966).

Blok, F.F., *Nicolaas Heinsius in dienst van Christina van Zweden* (Delft: Ursulapers, 1949).

Blom, Philipp, *To Have and to Hold: An Intimate History of Collectors and Collecting* (London: Allen Lane, 2002).

Bowers, Jane, "The Emergence of Women Composers in Italy, 1566–1700," in J. Bowers and J. Tick (eds.), *Women Making Music* (Urbana, Illinois: University Press, 1986).

Boyer, F., "Les Antiques de Christine de Suède," in *La revue archéologique*, Paris, 1932.

Callmer, Christian, *Drottning Kristinas samlingar av antik konst*, Svenska Humanistiska Förbundet 63 (Stockholm: P. A. Norstedt & Söner, 1954).

——, *Königin Christina, ihre Bibliothekare und ihre Handschriften: Beiträge zur europäischen Bibliotheksgeschichte*: Acta Bibliothecae Regiae Stockholmiensis, 30 (Stockholm: Kungliga Biblioteket, 1977).

Camden, William, *History of the Most Renowned and Victorious Princess Elizabeth, Late Queen of England*, ed. Wallace T. MacCaffery (Chicago: Chicago University Press, 1972). Originally published in 1615 and 1625, with the annotations of Sir Francis Bacon, as *Annales Rerum Gestarum Angliae et Hiberniae Regnante Elizabetha*.

Cametti, A. "Cristina di Svezia, l'arte musicale e gli spettacoli teatrali in Roma," in *Nuova antologia* 155 (1911).

Cassirer, Ernst, *Descartes, Corneille, Christine de Suède* (Paris: Librairie Philosophique J. Vrin, 1997). Trans., Madeleine Francès and Paul Schrecker.

Castelnau, Jacques, *La Reine Christine, 1626–1689* (Paris: Payot, 1981). First published as *Christine, reine de Suède* (Paris: Hachette, 1944).

Catteau-Calleville, Jean Pierre Guillaume, *Histoire de Christine Reine de Suède avec un précis historique sur la Suède, depuis les anciens temps jusqu'à la mort de Gustave-Adolphe* (Paris, 1815).

Cauly, Olivier, *Comenius* (Paris: du Félin, 1995).

Cavalli-Björkman, Görel, "La Collection de la reine Christine à Stockholm," trans. Louise Hadorph Holmberg, in *1648: Paix de Westphalie—l'art entre la guerre et la paix*; Proceedings of the Colloquium (Münster and Paris: Westfälisches Landesmuseum and Klincksieck-Musée du Louvre, November 1999).

Chantelou, Paul Fréart de, *Journal du voyage du cavalier Bernin en France*: unedited manuscript published and annotated by Ludovic Lalanne (Paris: Gazette des beaux-arts, 1885).

Chanut, Pierre-Hector, *Correspondence diplomatique suède, 1646–1651* (Paris, Archives du Ministère des Affaires Étrangères).

——*Mémoires et négotiations* (Paris, 1675).

Christina, Queen of Sweden, *Brev fran sex decennier*, translated by Sven Stolpe (Stockholm: Natur och kultur, 1960).

Christina, Queen of Sweden: A Personality of European Civilization: Catalogue of the exhibition of June 29–October 16, 1966, arranged in collaboration with The Royal Library, The Royal Collections, The Royal Armoury, The Royal Cabinet of Coins and Medals, and The National Record Office, Stockholm (Stockholm: Nationalmuseum, 1966).

Christine de Suède, *Tesmoignage de la Reyne Christine de Suede, en faveur de Monsieur Des-Cartes*, in Lettre circulaire de Claude Du Molinet aux génovéfains, datée de Sainte-Geneviève de Paris, le 1ᵉʳ février 1669.

Christine, reine de Suède, *Analecta reginensia: Extraits des manuscrits latins de la reine Christine conservés au Vatican*, ed. Wilmart, André Dom Henri-Marie-André, O.S.B. (Città del Vaticano: 1933, Studi e testi, 59).

Christout, Marie-Françoise, *Le Ballet de Cour au XVII siècle* (Geneva: Minkoff, 1987).

Claretta, G., *La Regina Christina di Svezia in Italia 1655–1689* (Turin, 1892).

Clarke, M.L. "The Making of a Queen: The Education of Christina of Sweden," *History Today* 28 (April 1978).

Conversazioni, Enrica, ed., *L'Archivio Azzolino Conservato dal Comune di Jesi* (Jesi: Biblioteca e Archivio Storico Comunale, 1988).

Cristina di Svezia: Mostra di documenti vaticani (Vatican City, 1989).

Davies, Norman, *Europe: A History* (Oxford: Oxford University Press, 1997).

Demetz, Peter, *Prague in Black and Gold: Scenes from the Life of a European City* (New York: Hill and Wang, 1997).

Dempsey, Charles, *Annibale Carracci: The Farnese Gallery, Rome* (New York: George Braziller, 1995).

Descartes, René, *Correspondance avec Elisabeth et autres lettres* (Paris: Flammarion, 1989).

Dunlop, Ian, *Royal Palaces of France* (London: Hamish Hamilton, 1985).

——, *Louis XIV* (London: Chatto and Windus, 1999).

Elliott, J.H., *Imperial Spain, 1469–1716* (London, 1963).

Evans, R.J.W., *Rudolf II and His World* (London: Thames and Hudson, 1997). First published by Oxford University Press, 1973.

Evelyn, John, *The Diary of John Evelyn* (London: Macmillan, 1908). First published in 1818.

Franklin, Alfred, *Christine de Suède et l'assassinat de Monaldeschi au château de Fontainebleau d'après trois relations contemporaines* (Paris: Émile Paul, 1912).

Fraser, Antonia, *Cromwell: Our Chief of Men* (London: Weidenfeld and Nicolson, 1973).

Garstein, Oskar, *Rome and the Counter-Reformation in Scandinavia: The Age of Gustavus Adolphus and Queen Christina of Sweden 1622–1656* (Leiden: E.J. Brill, 1992).

Gobry, Ivan, *La Reine Christine* (Paris: Pygmalion, Gérard Watelet, 2001).

Hibbert, Christopher, *Rome: The Biography of a City* (London: Penguin, 1985).

Hjortsjö, Carl-Herman, *Queen Christina of Sweden: A Medical-Anthropological Investigation of Her Remains in Rome* (Lund: Gleerup, 1966).

——, *Drottning Christina: Gravöppningen i Rom 1965* (Lund: Corona, 1967).

Hollingworth, Phillip, trans., *The History of the Intrigues and Gallantries of Christina, Queen of Sweden. And of Her Court, Whilst She Was at Rome. Faithfully Render'd into English, from the French Original* (London: The Cayme Press, 1927). First published in 1697.

Jansson, Arne, ed., *Johan Rosenhanes dagbok 1652–1661* (Stockholm: Kungl. samfundet för utgivande av handskrifter rörande Skandinaviens historia, 1995).

Kermina, Françoise, *Christine de Suède* (Paris: Perrin, 1995).

Kniff, Henrik, *Kristina Alexandra och operan i barockens Rom* (Stockholm: ARTES 3, 1997).

La Rochefoucauld, Le duc de, *Maximes: suivies des Réflexions diverses, du Portrait de La Rochefoucauld par lui-même et des Remarques de Christine de Suède sur les Maximes* (Paris: Garnier, 1999).

Lanoye, Diederik, *Christina van Zweden: Koningin op het schaakbord Europa,* *1626–1689* (Leuven: Davidsfonds, 2001).

Lawrenson, T.E., *The French Stage in the XVIIth Century* (Manchester: Manchester University Press, 1957).

Le Bel, Père, "Relation de la mort du marquis Monaldeschi," in Orlov, Count Grigorii Vladimirovitch, *Voyage dans une partie de la France, ou Lettres descriptives et historiques adressées à Mme la comtesse Sophie de Strogonoff,* 3 vols. (Paris: Bossange, 1824).

Lekeby, Kjell, *Kung Kristina—drottningen som ville byta kön* (Stockholm: Vertigo, 2000).

Linderen, J., "The Swedish 'Military' State, 1560–1720," *Scandinavian Journal of History,* 10, 1985.

Magnuson, Torgil, *Rome in the Age of Bernini,* 2 vols. (Stockholm: Almqvist and Wiksell, 1982).

Marvell, Andrew, *The Complete Poems* (London: Everyman, 1993).

Masson, Georgina, *Queen Christina* (London: Secker and Warburg, 1968).

Michel, Patrick, *Mazarin, prince des collectionneurs: Les collections et l'ameublement du Cardinal Mazarin (1602–1661)—histoire et analyse* (Paris: Broché, 1999).

Montanari, Tomaso, "Cristina di Svezia, il cardinale Azzolino e il mercato veronese," *Ricerche di Storia dell'arte* 54 (1994).

——, "Il Cardinale Decio Azzolino e le collezioni d'arte di Cristina di Svezia," *Studi Secenteschi* XXXVIII (1997).

——, "La Dispersione delle collezioni di Cristina di Svezia. Gli Azzolino, gli Ottoboni, e gli Odescalchi," *Storia dell'arte* 90 (1997).

——, "Bernini e Cristina di Svezia: Alle origini della storiografia berniniana," in A. Angelini, *Gian Lorenzo Bernini e i Chigi tra Roma e Siena* (Milan: Silvana, 1999).

——, "Cristina di Svezia, il cardinale Azzolino e le mostre di quadri a San Salvatore in Lauro," in Vera Nigrisoli Wärnhjelm, ed., *La Regina Cristina di Svezia, il Cardinale Decio Azzolino jr. e Fermo nella seconda metà del Seicento* (Fermo: Città di Fermo/Fondazione Cassa di Risparmio di Fermo, 1997).

Montecuccoli, Count Raimondo, *I Viaggi: Opera inedita pubblicata a cura di Adriano Gimorri* (Modena: Società Tipograficà Modenese, 1924).

Montfaucon, Bernard de, *Les Manuscrits de la Reine de Suède au Vatican* (Vatican City: 1964).

Montpensier, Anne Marie Louise Henriette d'Orléans, *Mémoires,* 2 vols. (Paris: Fontaine, 1985).

Motteville, Madame de (1806), *Mémoires pour servir à l'histoire d'Anne d'Autriche* (Paris, 1855).

Neumann, Alfred, *The Life of Christina of Sweden* (London: Hutchinson and Co., 1936).

Ogiers, Charles, *Från Sveriges Storhetstid: Franske legationsseckreteraren Charles Ogiers dagbok under ambassaden i Sverige 1634–1635,* trans. from Latin by Sigurd Hallberg (Stockholm: P.A. Norstedt & Söner, 1914).

Orlov, Count Grigorii Vladimirovitch, *Voyage dans une partie de la France, ou Lettres descriptives et historiques adressées à Mme la comtesse Sophie de Stroganoff*, 3 vols. (Paris: Bossange, 1824).

Oxenstierna, Axel, *Rikskansleren Axel Oxenstiernas skrifter och brefvexling* (Stockholm: P.A. Norstedt & Söner, 1888–89 and 1888–1930); series 1, vols. 1–8; series 2, vols. 1–12.

Pallavicino, Sforza, *Drottning Kristinas väg till Rom*, trans. Karin Stolpe and Monica Stolpe (Stockholm and Rome: Italica, 1966).

——, *Descrizione del primo viaggio fatto a Roma dalla regina di Svezia Cristina Maria* (Rome: Salviucci, 1838).

Patru, Olivier, "Harangue de Monsieur Patru faite en 1656, à la Reine Christine de Suède, au nom de l'Académie Française," in *Recueil des harangues prononcées par Messieurs de l'Académie française dans leurs réceptions & en d'autres occasions différentes: depuis l'establissement de l'Académie jusqu'à présent*, vol. 1 (Amsterdam, 1709).

Polišenský, J.V., *The Thirty Years War* (London: B.T. Batsford Ltd., 1971).

Pujo, B., *Le Grand Condé* (Paris: Albin Michel, 1995).

Raymond, Jean-François de, *La Reine et le philosophe: Descartes et Christine de Suède* (Paris: Lettres Modernes, 1993).

——, *Christine de Suède: Apologies* (Paris: Les Editions du Cerf, 1994). This edition includes *La Vie de la reine Christine, faite par elle-même, dédiée à Dieu, L'ouvrage du loisir*, and *Les Sentiments*.

——, *Pierre Chanut, ami de Descartes: un diplomate philosophe* (Paris: Beauchesne, 1999).

Regteren Altena, J.Q. van, *Les Desseins italiens de la reine Christine de Suède*, Analecta reginensia 2, Nationalmusei Skriftserie 13 (Stockholm, 1966).

Reinken, Liselotte von: *Deutsche Zeitungen über Königin Christine: 1626–1689* (Münster: C.J. Fahle, 1966).

Retz, Paul de Gondi, Cardinal de, *Mémoires* (Paris: Garnier, 1987).

Rich, E.E. and C.H. Wilson, eds., *The Cambridge Economic History of Europe, vol. 4: The Economy of Expanding Europe in the Sixteenth and Seventeenth Centuries* (Cambridge: Cambridge University Press, 1967).

Riesman, D., "Bourdelot, a Physician of Queen Christina of Sweden," *Annals of Medical History*, new series, 9 (1937), 191.

Roberts, Michael, *Essays in Swedish History* (London: Weidenfeld and Nicolson, 1967).

——, *Sweden as a Great Power 1611–1697: Government, Society, Foreign Policy* (London: Edward Arnold, 1968).

——, ed., *Sweden's Age of Greatness, 1632–1718* (London: Macmillan, 1973).

——, *The Swedish Imperial Experience, 1560–1718* (Cambridge: Cambridge University Press, 1979).

——, *From Oxenstierna to Charles Twelfth: Four Studies in Swedish History* (Cambridge: Cambridge University Press, 1991).

——, *Gustavus Adolphus: Profiles in Power* (London: Longman, 1992). Originally published as *Gustavus Adolphus and the Rise of Sweden* (London: English Universities Press, 1973).

Rodén, Marie-Louise, "Drottning Christina och Kardinal Decio Azzolino: Kärleks-brev från det sista decenniet," *Personhistorisk Tidskrift* 82 (1986).

——, "The Burial of Queen Christina of Sweden in St. Peter's Church," *Scandinavian Journal of History*, 1987, vol. 12, no. I.

——, ed., *Politics and Culture in the Age of Christina* (Rome and Stockholm: Suecoromana, 1997).

——, *Church Politics in Seventeenth-Century Rome: Cardinal Decio Azzolino, Queen Christina of Sweden, and the Squadrone Volante* (Stockholm: Almqvist & Wiksell International, 2000).

Sardi, Cesare, *Cristina Regina di Svezia in Lucca nel 1658: Ricordi storici* (Lucca: Giuseppe Giusti, 1873).

Schama, Simon, *The Embarrassment of Riches: An Interpretation of Dutch Culture in the Golden Age* (London: Fontana, 1991).

——, *Rembrandt's Eyes* (London: Allen Lane, 1999).

Schulze, R., *Das Projekt der Vermählung Friedrich Wilhelms von Brandenburg mit Christina von Schweden* (Halle, 1898).

Spence, Jonathan D., *The Memory Palace of Matteo Ricci* (London: Penguin, 1984).

Stolpe, Sven, *Från stoicism till mystik: Studier i drottning Kristinas maximer* (Stockholm: Bonniers, 1959).

——, *Drottning Kristina: vol. I, Den svenska tiden* (Stockholm: Bonniers, 1960).

——, *Drottning Kristina. vol. II, Efter tronavsägelsen* (Stockholm: Bonniers, 1961).

——, *Christina of Sweden*, ed., Sir Alec Randall (London: Burns and Oates, 1966).

Sundström, Einar, "Notiser om drottning Kristinas italienska musiker," *Svensk Tidskrift för Musikforskning* 43 (1961).

Thurloe, John, *A Collection of the State Papers of John Thurloe, Esq., Secretary to the Council of State and the Two Protectors Oliver and Richard Cromwell* (London, 1742).

Tydén-Jordan, Astrid, *Drottning Kristinas kröningskaross 1650* (Stockholm: Livrustkammaren, 1990).

Von Platen, Magnus, ed., *Queen Christina of Sweden: Documents and Studies*, Analecta Reginensa I, Nationalmuseum Skriftserie Nr. 12 (Stockholm: P.A. Norstedt & Söner, 1966).

Webster, Charles, *From Paracelsus to Newton: Magic and the Making of Modern Science* (Cambridge: Cambridge University Press, 1982).

Wedgwood, C.V., *The Thirty Years War* (London: Pimlico, 1992). First published in London by Jonathan Cape, 1938.

Weibull, Curt, *Monaldescos död: Aktstycken och berättelser* (Göteborg: Erlander, 1937).

——, *Drottning Kristina: Studier och Forskningar* (Stockholm: Natur och Kultur, 1946).

Weibull, Martin, *Drottning Kristina och Klas Tott: några historiska beriktiganden* (Lund: Berlingska boktryckeri, 1892–93).

Whitelocke, Bulstrode, *A Journal of the Swedish Embassy in the Years 1653 and 1654*, 2 vols. (London: Longman, Brown, Green, and Longmans, 1855).

Wicqueford, A. de, *Chronique discontinue de la Fronde*, ed. Robert Mandrou (Paris: Fayard, 1978).

Wilmart, Andreas, *Codices Reginenses Latini* (Vatican City: Biblioteca Apostolica Vaticana, 1937).

Wittrock, Georg, *Regering och allmoge under Kristinas förmyndare: Studier rörande allmogens besvår* (Uppsala: Almqvist & Wiksell, 1948).

——, *Regering och allmoge under Kristinas egen styrelse: Riksdagen 1650* (Uppsala: Almqvist & Wiksell, 1953).

Wrangel, Fredrik Ulrich, *Première visite de Christine de Suède a la Cour de France, 1656* (Paris: Firmin-Didot, 1930).

ILLUSTRATION CREDITS

Count Magnus Gabriel De la Gardie. Oil painting by Matthias Merrian, 1649. © Statens Konstmuseer, Stockholm.

Queen Christina, age almost 24. Oil painting by David Beck. © Statens Konstmuseer, Stockholm.

Countess Ebba Sparre, Christina's "Belle." Oil painting ca. 1653 by Sébastien Bourdon. © National Gallery of Art, Washington, D.C. (Samuel H. Kress Collection).

Queen Christina and Belle visit Claude Saumaise, *Drottning Kristina på besök hos professor Saumaise.* Gouache by Niclas Lafrensen, 1794. © Statens Konstmuseer, Stockholm.

Queen Christina at the age of 26, a hunting portrait. Oil painting by Sébastien Bourdon. © Museo Nacional del Prado, Madrid.

René Descartes. Oil painting by David Beck. © Institut Tessin, Paris.

Queen Christina and her "academy," *Drottning Kristina omgiven av lärda, däribland Descartes.* Oil painting by Louis-Michel Dumesnil. © Statens Konstmuseer, Stockholm.

Queen Christina's abdication, *Drottning Kristinas tronavsägelse på Uppsala slott den 6 juni 1654.* Engraving by Willem Swidde, after an original by Erik Dahlberg. © Statens Konstmuseer, Stockholm.

Queen Christina drives through the market square of Antwerp. Oil painting by Erasmus de Bie. © Statens Konstmuseer, Stockholm.

Queen Christina. Oil painting by Sébastien Bourdon, 1653. © Statens Konstmuseer, Stockholm.

Queen Christina. Oil painting by Abraham Wuchters, 1660. © Statens Konstmuseer, Stockholm.

Pope Alexander VII (Fabio Chigi). Bust by Gian Lorenzo Bernini. © Statens Konstmuseer, Stockholm.

Cardinal Decio Azzolino, junior. Bust by Pietro Balestra, ca. 1670. © Statens Konstmuseer, Stockholm.

Queen Christina and Cardinal Azzolino together in the cardinal's library. Engraving by an anonymous artist. From a Dutch book of 1697, *Het Leven en Bedryt,* III.21.833, at Skoklosters slott. © LSH foto Samuel Uhrdin.

Cardinal Jules Mazarin. Oil painting by Pierre Mignard, "le Romain." © Photo Réunion des musées nationaux: Harry Bréjart.

La Duchesse de Montpensier, *"la Grande Mademoiselle."* Oil painting from the school of Pierre Mignard, "le Romain." © Photo Réunion des musées nationaux: D. Arnaudet/G. Blot.

Louis XIV of France. Engraving by Robert Nanteuil. © Trustees of the British Museum.

Bernini's oval piazza in front of St. Peter's Basilica. Engraving by Giovanni Battista Falda. © Trustees of the British Museum.

Bernini's Piazza Navona in Rome. Engraving by Giovanni Battista Falda. © Trustees of the British Museum.

The queen's arrival at Fontainebleau. Engraving by Israel Silvestre the Younger. ©
Trustees of the British Museum.

Pope Clement IX (Giulio Rospigliosi). Oil painting by Carlo Maratti. © Hermitage Museum, St. Petersburg.

Pope Innocent XI (Benedetto Odescalchi). Engraving by A. Clouet. © Biblioteca Apostolica Vaticana (Vatican).

The Palazzo Riario, now the Palazzo Corsini, in Rome. Engraving by Giovanni Battista Falda. © Statens Konstmuseer, Stockholm.

Queen Christina in 1662. Etching by an unknown artist. © Statens Konstmuseer, Stockholm.

Queen Christina in 1667. Oil painting attributed to Wolfgang Heimbach. © Statens Konstmuseer, Stockholm.

Queen Christina in 1687. Oil painting by Michael Dahl. © Grimsthorpe and Drummond Castle Trust, Ltd., Lincolnshire.

Queen Christina toward the end of her life. Bust attributed to an unknown Italian artist. © Statens Konstmuseer, Stockholm.

Queen Christina's funeral procession in April 1689. Engraving by Robert van Audenaerd. © Statens Konstmuseer, Stockholm.

INDEX

About the author

About the book

Read on

Insights,
Interviews,
& More . . .

Meet Veronica Buckley

VERONICA BUCKLEY was born in New Zealand in 1956. After her studies in language, music, and philosophy, she worked as an orchestral cellist before taking up academic scholarships at the Universities of London and Oxford. Following her doctoral research in history, she worked in the management of technical information in finance, computing, and the oil industry. An abiding interest in literature eventually led her away from corporate life and into writing. She now lives in Paris with her husband, writer Philipp Blom. ❧

You've Got to Have a Dream

I FIRST ENCOUNTERED Queen Christina more than twenty-five years ago, when I was a student preparing an essay on the moral philosophy of Descartes. My tutor introduced me to the correspondence between the two—I believe it took me years to return the books to him.

It was pretty dry reading, looking back, but I was young and earnest, and keen to explore all the big ideas of the great and the good, especially ideas put forward by a woman—this was the 1970s, after all; a woman with ideas was as daring then as a public smoker is today. I took it into my head to write a biography of her, and set about learning Swedish with a patient immigrant—a tall, blonde, stately woman, quiet of voice and calm of manner—in short, the very opposite of the Queen Christina I was gradually getting to know.

Being young and earnest is a great thing, on the whole, but it's to no real advantage when you're trying to persuade other people that you can write a biography. I'm glad I didn't know then that it would take twenty-five years to bring my dream to fruition. The usual events intervened, the business of life took its course, and it wasn't until the year 2000 that Queen Christina reappeared, determined not to be overlooked a second time, in my unsuspecting life. ▶

> 66 Being young and earnest is a great thing, on the whole, but it's to no real advantage when you're trying to persuade other people that you can write a biography. I'm glad I didn't know then that it would take twenty-five years to bring my dream to fruition. 99

You've Got to Have a Dream *(continued)*

My husband was writing a book about why people collect things, and one evening, in the darkness of a London winter, he began to talk of a fabulous cabinet that had found its way to the court of Queen Christina. And so, in the light of a lovely, flaming fire (yes, a real fire, which somehow makes everything seem possible), the spark was rekindled, and I found that a still-enthusiastic middle age can have advantages of its own, at least as far as publishers are concerned.

The cabinet came to rest in the Chancellor's room at Uppsala University, and I settled down to write my first book. I have loved spending time in Christina's engaging, maddening, always unpredictable company, and I will miss her in my daily life. But others will get to know her now, and besides, one of my longest-held dreams has at last come true.

Christina
Myth and Reality

ONE'S FIRST THOUGHTS of Queen Christina today, at least outside of Sweden, are generally thoughts of Greta Garbo, feisty and fabulous in the title role of Rouben Mamoulian's 1933 film, *Queen Christina*. A flirtatious cross-dresser in trousers and high boots, Garbo's Christina is tough, amusing, clever, brave, and always, always gorgeous. Battling the odds, defying the rules, calling the shots at every turn, she is everything to all unconventional men, and to unconventional women too. No wonder Garbo's Christina became an instant icon for those who dared—and for those who didn't dare—to be different.

Was it true? Was Queen Christina really the kind of woman that Garbo so unforgettably portrayed? Did she drink and joke and fight with men, and then sleep with them?

We know at least that she wasn't a drinker of anything harder than the occasional fruit juice—Christina's most outrageous tipple was a vaseful of rosewater stolen, at the age of six, from her mother's dressing table.

Did she like fighting? She certainly liked to talk about it, and even adopted a special martial pose when she did so, with her chin up and one foot planted determinedly in front of the other. She was a definite clouter of servants, in the time-honored seventeenth-century way, and wasn't averse to ordering the occasional execution; still, she insisted, "I'm not cruel. I have never killed an animal without feeling real sympathy for it." She was a marvelous horsewoman and a very good ▶

> " A flirtatious cross-dresser in trousers and high boots, Garbo's Christina is tough, amusing, clever, brave, and always, always gorgeous. "

5

Christina (*continued*)

shot, but her ultimate dream of leading an army into battle was never realized. Despite her trousers and the short sword she liked to wear, she was a woman; there was never going to be any real fighting.

Was she amusing, like Garbo's Queen? She was very amusing, though not everyone laughed at her jokes. Her wit was too sharp and her information too accurate. She always kept up with the latest gossip and always left a trail of embarrassed aristocrats behind her wherever she went. Did she sleep with men? That needs a paragraph of its own.

Christina liked men. She liked their company, their manners, the things they talked about—grand politics, soldiering, sex. She was a great swearer, not only of dramatic oaths, though she was certainly fond of these, but also of crude language, and sometimes really foul language. Her attitude to sex in the abstract was broad enough for the roughest seventeenth-century soldier. She staged lewd plays and told ribald jokes, and was known to wink an eye at promiscuity and even rape within her own household.

But when it came to herself, her attitude was very different. She regarded the act of sex as an act of submission of woman to man, a power contest and nothing more, with the woman always the loser. There is some truth in Garbo's seductive, even predatory Queen. "I am passionate by nature," Christina said. But she couldn't imagine surrender without submission. "I could never bear to be *used* by a man," she wrote, "the way a peasant uses his fields."

And her dashing lover in Mamoulian's film,

> ❝ Her attitude to sex in the abstract was broad enough for the roughest seventeenth-century soldier. ❞

the Spanish diplomat Don Antonio Pimentel
de Prado? Garbo pulled rank, in true Christina
style, to have Lawrence Olivier thrown out of
the role, replacing him with her real-life lover,
silent movie idol John Gilbert. Like Gilbert,
the real Pimentel was tall and handsome. He
was a soldier—a general, in fact—a man of
naturally commanding disposition, though,
unlike Gilbert, the real Pimentel was apparently
completely bald. He did spend many hours
closeted in private with the Queen, and he was
generally believed to be her lover. But
it was all a misunderstanding: the courtiers
misunderstood, Pimentel misunderstood,
Christina herself misunderstood. She thought
he had been sent to Sweden to further her
secret conversion to Catholicism, and kept
dropping him hints, and locking him away
with her, until he must have begun to wonder
himself whether her intentions toward him
were entirely honorable. Certainly he made
several attempts to get away from her court,
without waiting for permission from Spain.
Pimentel was not Christina's lover, despite the
charm of their candlelit tête-à-tête in
Mamoulian's film.

   The Swedes rumored that she had other
lovers, the Danes swore she did, the French
and Spanish made political capital out of it, and
the Italians took it all for granted. But it
probably wasn't true. "God knows I am
innocent of all the things they've said to
blacken my name," Christina said, and I think
she was innocent—protected, or prevented,
by her own native distaste for what she
believed the act of sex to be. She did fall in
love with her courtier Magnus De la Gardie
and, more scandalously, with a prince of
the Church, Cardinal Azzolino. The gossip- ▶

Christina *(continued)*

mongers said she had several children by both men, but it doesn't seem likely as the children never came to light.

Did she sleep with women then? There was always a great deal of talk about Christina's supposed lesbianism. She is said to have seduced young girls at court, diplomats' wives, opera singers, and even nuns. She did have a special fondness for one gentle young beauty in Sweden, her "Belle," Mamoulian's Elizabeth Young. Belle and Christina commonly shared a bed, no unusual matter in the seventeenth century for two unmarried women, but the Queen enjoyed the provocative possibilities of the situation, and often drew deliberate attention to it. She liked to write suggestive letters, beyond the limits of *précieuse* effusiveness, to the pretty women she met, and she liked to be alone with them, to share confidences, and to exert influence over them. The rooms in her Roman palazzo had a good many paintings and sculptures of women, all determinedly naked despite repeated protestations from the Pope.

The rumors about Christina, and her own insinuations, took no time at all to ossify into supposed fact. She was widely believed to be lesbian or bisexual, or possibly, in mitigating afterthought, a hermaphrodite. Her declared reluctance to marry added weight to the charge, and there was plenty of circumstantial evidence to be brought to bear: her mannish way of walking, her love of hunting, her gruff voice, and her flat shoes—to a continent full of courtiers eager for scandal, it all showed clear sexual aberration. Christina herself did nothing to quench the little flames, so that three hundred years later her reputed bisexuality (and Garbo's too) ensured an anxious time for the censors of Mamoulian's film script. But nothing crept in. Hollywood in the 1930s simply wasn't ready for it.

Unlike Garbo's Queen, the real Christina probably didn't have much of a sex life at all. Was she at least charismatic like Garbo, assured, striking? Garbo's Christina is supremely sure of herself. Never faltering, she maintains an absolute self-confidence until the end of the film, standing on the deck of her ship, staring into the far distance, "thinking," as Mamoulian insisted, "of nothing." But had she been able to see a little further still, a shadow of self-doubt must surely have crossed that luminously beautiful face. Mamoulian's film, ending as it does with the Queen's voyage away from Sweden, evades the tragedy inherent in her abdication. Life in Sweden had meant a crown, a kingdom, a court, subjects, wealth, power, and respect. The real Christina mistook the accidents of her birth for personal qualities that she could retain, no matter what she did, to the end of her life. Until this point, standing on the deck, sailing away on a summer wind, she was supremely confident too. Just twenty-seven years of age (Garbo was also twenty-seven), she had no notion—and because of that, no fear— of the compromises, the tradeoffs, the backing into corners, that a life without power or money would bring. Garbo's Christina is exhilarated and exhilarating, and we look on, admiring and envious. The real Christina we watch with exasperation—how reckless! how foolish!—or with a vague concern; young people, so headstrong, never see what they're doing until it's too late....

Christina was certainly headstrong, willful, and impulsive. She wanted to determine things for herself, but she was born to a life ▶

Christina (*continued*)

in which the smallest details—down to how she held her knife and fork—were already determined for her, as a matter of state policy. Should her table manners be Swedish or French? The one showed national pride, the other international savoir faire. Christina didn't care either way. She was just as happy to forgo knives and forks altogether, and pick up her food with her hands. And as with table manners, so with the rest of her life. She had her own way of doing things, and on the whole she held to it despite opposition and against all odds, ducking and diving where necessary, noisily insistent by preference.

Her absolute determination to run things for herself, to reject the straightjacket role of a dynasty-bearing princess, has turned her, in our day, into something of a feminist icon. Her rejection of the role prepared for her was genuine: appalled by the idea of conceding authority to a husband, and especially by the prospect of childbearing, she outright refused to marry.

But it's too much to claim that Christina was a feminist, or that she held any kind of protofeminist ideas. If anything, she was an antifeminist. She despised most women, and was bitterly resentful of being a woman herself—"my greatest defect," she called it, and all the many failures of her life she laid to its account. "Women are weak in soul and body and mind," she wrote. "Women should never be rulers, and I am so convinced of this that I would have barred my daughters from the succession, if I had married. It would have been a betrayal of my kingdom to leave it to girls. I know I am speaking against my own interest—but then, I have always made a

> 66 Should her table manners be Swedish or French? The one showed national pride, the other international savoir faire. Christina didn't care either way. She was just as happy to forgo knives and forks altogether, and pick up her food with her hands. And as with table manners, so with the rest of her life. 99

point of speaking the truth, whatever it has cost me."

Christina's determination and her unconventionality, though they might be feminist models of a kind, were not the result of any political or social considerations on her own part. She was by nature different from most of her contemporaries, and she was naturally headstrong, and the result of both was her loudly unconventional life.

I have always made a point of speaking the truth, whatever it has cost me. This is true, as far as it goes, though Christina seldom admitted uncomfortable truths about herself. She said what she thought to anyone, about anyone. She was clever and perceptive, and sufficiently analytically minded to keep dozens of Jesuits amused in her company for more than three decades. But where she herself was concerned, her insight failed her—in fact, she rarely dared to think critically about herself. It would have been too expensive for her, psychologically; she would have had to examine the myth she had created of herself, and for herself. The myth was a paradox: it sustained her, yet it destroyed her ability to accomplish anything real.

Christina was the daughter of a king, and it was her good fortune, and her misfortune, that Gustav Adolf the Great was one of the finest men of the century, a dazzling legend even while he lived, "the lion of the north." To match him, to surpass him, was Christina's driving need. It was impossible and she took refuge, quickly and enduringly, in an extravagant myth of herself as brilliant, powerful, invincible—even, flying in the face of the clearest evidence, tall. Christina's myth of her own personal greatness was the most ▶

Christina *(continued)*

destructive element of all in her impulsively destructive life.

What remains of the true Queen Christina? Where do our myths of her, and her myth of herself coincide and, perhaps, merge into the truth? In her irrepressibility, perhaps, and her unconventionality, striding about in her trousers and high boots, defiantly turning her back on duty and disaster, while keeping her blues eyes fixed on the ever-bright fantasy future ahead. Endearing and alarming, she demands our attention, whether we like it or not. In Christina we see, as she herself saw, a superstar quality which kept us all hooked for more than three hundred years. Whether intrigued or appalled, we find ourselves continually fascinated. ∾

Modern-Day
Christinas

Queen Rania Al-Abdullah of Jordan for her
undoubted star quality

Crown Princess Masako of Japan for being
at the crossroads between cultures

*Crown Princess Maha Chakri Sirindhorn
of Thailand* for her potential to benefit
her country

Elizabeth Sprague Coolidge for being a
patroness of the arts

Margaret Thatcher for her love of
unchallenged power

And, alas, too many royals of the last hundred
years who have put pleasure before duty.

Favorite **Royals**

The Goodies

Cleopatra of Egypt, 69–30 B.C., for her
 absolute chutzpah

Elizabeth I of England, 1533–1603, for her
 intelligence (and her survival skills)

Gustav II Adolf of Sweden, 1594–1632, for his
 double gift of vision and pragmatism

The Baddies

Roman Emperor Caligula, 12–41, for his
 grotesque and willful abuse of power

Vlad the Impaler, 1431–76, for being Vlad the
 Impaler

Tsar Nicholas I "the Stultifying" (my epithet)
 of Russia, 1796–1855, for sending
 Dostoevsky to Siberia

Don't miss the next book by your favorite author. Sign up now for AuthorTracker by visiting www.AuthorTracker.com.